Mercury

Mercury
A HISTORY OF QUICKSILVER

LEONARD J. GOLDWATER

YORK PRESS

Baltimore, Maryland

Copyright © 1972 by York Press, Inc.
All rights reserved
Manufactured in the United States of America
Library of Congress Catalog Card Number 73-189648
ISBN 0-912752-01-7

Contents

Foreword

This is an unusual book by an unusual man, and it is unlikely that it could have been written by anyone else. It is primarily a history, but the historian's attitude is tempered on every page with long experience in occupational medicine and the problems of the use and abuse of mercury. It is not the first historical study written by a man who has been involved with the practical aspects of its subject. It may possibly be the first in which a single therapeutic agent is traced vertically through the layers of recorded medical history.

As long ago as 1936 the author was studying the effects of mercury on employees in the hat industry. In 1956, ten years before the general alarm over mercury, he was involved in WHO studies of mercury, lead, and arsenic pollution and was working on analytical procedures for large scale human studies. Just as he has questioned alarmist reports based on inadequate analysis in science, so he is skeptical of received dogmas in history. It is his particular delight to examine some point which has established itself in the literature by repetition and to find a flaw in the premise on which it was first based. He could have made a first-class career as a historian and it is gratifying that in "retirement" he has been able to devote two years of intensive library research, travel, and inquiry to a complex and timely subject. Speaking personally, it was a lasting pleasure to have him do much of his work on this book in the Trent Collection of Duke University Medical Center Library.

G. S. T. Cavanagh
Curator, Trent Collection,
Duke University Medical Center Library

Preface

Somerset Maugham, having built a comfortable fortune on his writings, advised aspiring authors not to be preoccupied with having their works published, sold, or read. One should write, he said, because of an inner urge to write and for the sheer joy of the experience, without thought of fame or fortune.

I have written this book to satisfy an inner urge and have found joy in the experience. In part, the urge came from a desire to see if I could write a publishable book, a sort of self-testing. Thanks to a generous grant-in-aid from the Commonwealth Fund, it was not necessary for me to be preoccupied with any financial return. As an extra dividend, the two years of support provided an opportunity for me to find a new, congenial "home" in Duke University. Now the book has been published, thus satisfying all of Maugham's requisites.

Frequently I am asked why I became interested in mercury. Impressed by the accomplishments of Dr. Robert A. Kehoe in the toxicology of lead, I decided in 1956 to attempt to become the "Kehoe" of mercury. I predicted that some day it would be useful to have some knowledge about the behavior of this fascinating metal. Happily the day did not come too soon, not before I had had more then ten years of intensive research at Columbia University into the absorption and excretion of mercury in man. Shortly after 1956, I began to collect material for use in writing a monograph.

What I have tried to do is to trace the story of mercury from earliest time up to the present, describing its most important uses in and contributions to medicine, science, technology and, to some extent, economics. Much of this is relevant to an understanding of the explosive impact mercury has had on man's environment in the period starting in the 1960's.

Since prehistoric times, mercury and its compounds have had an aura of mystery and mysticism, figuring prominently in the occult arts and in folklore.

Unfortunately, a number of time-honored beliefs about the metal are more closely related to folklore than to science or history. Even today, new folklore is in the making and has begun to masquerade as science. Perhaps it is a bit churlish that I have attempted to separate fact from fancy and have cast doubt on such old favorites as the Mad Hatter being a victim of mercury poisoning.

For the most part I have consulted primary sources of information or have drawn on my own observations. I have done most of the translating from foreign languages with occasional assistance in the ancient languages from Professor William H. Willis, Dr. John L. Sharpe, III and Dr. Paul Meyvaert, all of Duke University. Help in the interpretation of Egyptian terminology was generously provided by Professor John R. Harris, formerly of Oxford but now at the University of Copenhagen. Samples of ancient materials for chemical analysis were kindly supplied by Mr. P. R. S. Morey of the Ashmolean Museum, Dr. Robert Wauchope of the Middle American Research Institute at Tulane University and Dr. Frederick J. Dockstader, Director of the Museum of the American Indian in New York.

Among the dozens of individuals who assisted in the preparation of the manuscript, there are several who deserve special mention. High on the list stands Mr. G. S. T. Cavanagh, Curator of the Trent Collection at Duke University, who placed in my hands countless gems from which I selected a large number for inclusion in the text, who read some of the early drafts and made many useful suggestions. The entire Duke University Medical Center Library staff deserves a vote of thanks for procuring materials from other libraries and for help in finding obscure reference works. Miss Myrl Ebert, Librarian at the Health Sciences Library of the University of North Carolina, and the staff of that library, rendered similar assistance. I am especially grateful to Mr. Roger A. Crane, editorial consultant to the Commonwealth Fund, for seeing me through some periods of discouragement and for much wise advice. Mr. S. H. Williston* provided me with valuable technical information, particularly that dealing with mining methods. Dr. Francisco Guerra of the Wellcome Historical Medical Library, Mr. George D. Painter of the British Museum, Dr. David M. Rogers of the Bodleian Library, Dr. John B. Blake of the National Library of Medicine (USA) and their respective staffs gave liberally of their time. Professor David E. Eichholz of Bristol University, Professor A. Leo Oppenheim of the Oriental Institute (University of Chicago) and Professor William H. Stahl of Brooklyn College helped in unravelling some knotty problems. The important task of checking the references was diligently and patiently performed by Miss Margaret F. Jones. My wife Charlotte showed admirable forbearance throughout and assisted significantly in preparing the index. My brother Walter helped in more ways than I can mention.

---

*Mr. Williston died in December 1971.

Introduction

Mercury judiciously manag'd, seems to me, to be the only true Panacea, and universal Antidote, sought by wise, and boasted of by pyrotechnical Enthusiasts. Mercury seems pointed out and impress'd by the Signature of the God of Nature, for the Cure, at least for the Relief, of intelligent Creatures, made miserable by hereditary Diseases by natural appetites irregularly indulg'd, by Ignorance, bad Example and Frailty in the human Kind especially made so by high Food, and spiritous Liquors mostly.*

It is "The hottest, the coldest, a true healer, a wicked murderer, a precious medicine, and a deadly poison, a friend that can flatter and lie."†

So it has been with quicksilver throughout its recorded history, rarely described in temperate terms. That it should have held a unique fascination for man is not surprising. It has caused untold numbers of deaths and contributed to endless destruction; it has also participated in countless ways in some of the most important advances in medicine, science, and technology. The prominence of its contributions is in contrast to its relative rarity among known elements. New uses are constantly being found for this versatile metal; its full potential has not yet been realized. As its common name, quicksilver, implies, it is still very much alive as a force in man's everyday life. As is true of all active chemicals, it must be understood if it is to be a healer and not a murderer, a precious medicine and not a deadly poison, and a friend that neither flatters nor lies.

---

*Cheyne, G. 1742. The Method of Cure of Diseases of the Body and Mind. London. p. 119.
†Woodall, J. 1639. The Surgeon's Mate or Military & Domestic Surgery. London. p. 256.

Part I

1

Occurrence of Mercury

The Origin of Mercury

Mercury, according to recent estimates, originated on Earth about four and a half or five billion years ago, at the time the Earth itself came into being (Ahrens 1965; Day 1963; Bowen 1966). There is no certainty as to the duration of Earth's formative period but it is generally believed that all metals did not originate at the same time; mercury's "creation" probably came relatively late. Igneous rocks solidified before sedimentary rocks were formed; therefore metals incorporated into the former are older than those which are found in the latter. According to this hypothesis, mercury is younger than lead, zinc, tin, copper and nickel.

Various theories have been advanced to explain the origin of the individual elements. One such theory holds that mercury was formed during what is termed the second parting of the elements. Prior to that period, many rocks had already become solidified, but the ions of mercury were too large to permit them to fit into the crystal lattices of these rocks during solidification. As a result, mercury was laid down only in cracks and fissures in the older formations, rather than being incorporated into the rock structure, and is relatively abundant in sedimentary deposits. Geological distribution similar to that of mercury applies to a number of other rare elements, including gold.

Occurrence

Whether or not an element can be said to be present anywhere depends on the sensitivity of our methods of detection. It is possible to measure as little as one molecule, but the procedure is too complicated for analyzing the thousands of specimens which would have to be tested before an incontrovertible statement on distribution could be made. Strictly speaking, a value of zero should be applied only to specimens which do not contain a single molecule of the element being sought. In actual practice it is much more common to assign a value of "zero" to all amounts below the limit of sensitivity of whatever analytical procedure is used. This is particularly true in testing for metals in biological materials, where applicable methods require that considerably more than one molecule per cell be present in order to give a positive result. In analytical chemistry, the term "zero" is meaningless unless the sensitivity of the method is stated.

For mercury, methods are available which have sufficient sensitivity as well as practicability to permit chemists to obtain an accurate picture of the distribution of this element in the environment. Mercury, at least in trace amounts, seems to be ubiquitous on our planet, a fact that was pointed out by Stock as early as 1934 (Stock and Cucuel 1934). It would be more accurate, perhaps, to say that the metal has been found wherever it has been sought in the lithosphere, hydrosphere, biosphere, and atmosphere of the Earth. Mercury has not been found in the atmosphere of the sun (Lockyer 1878), but mercury lines have been identified in the spectra of some of the stars (Watteville 1906). The probability of a mercury atmosphere on the warmer planets of our solar system is high (Williston 1968), and the element has been found in samples of moon-rock (Reed, Goleb, and Jovanovic 1971). Modern analytical techniques have rendered obsolete Mellor's statement that "Mercury is neither abundantly nor widely distributed . . . ."

The Mercury Cycle

Mercury circulates and is re-distributed in nature. (See Fig. I-1.) Man's activities can affect the mercury cycle. Man-made modifications, particularly unnatural concentrations, have been of increasing concern in relation to possible effects on the health of living organisms.

Not only does man play a role in redistribution, but he also creates countless new combinations of mercury with other elements. Many of these combinations constitute a greater threat to health than does elemental mercury. There are a few mercury compounds, discussed in later chapters, which have been used for the benefit of man's health.

Lithosphere

Most of the rock formations which make up the earth's crust contain no more than traces of mercury. Sandstones and limestones contain the element in concentrations of about 0.03 parts per million (ppm). (Older reports express the values as grams per ton, which is roughly speaking about the same thing as ppm.) Granites contain about 0.08 ppm, but shales are much richer in mercury. Average figures for shales are about 0.4 ppm of mercury, but

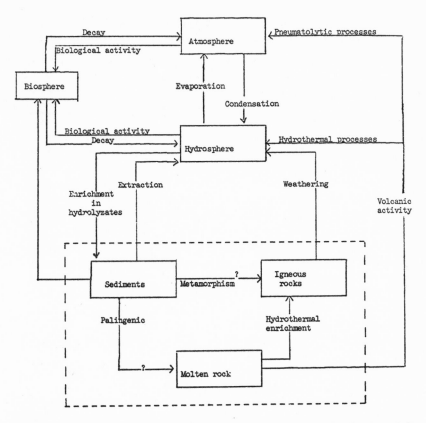

Figure I-1 The Mercury Cycle (From *Geochemistry*, by K. Rankama and G. Sahama. Chicago: University of Chicago Press. 1950 [Copyright 1950. Copyright under International Copyright Union by The University of Chicago. All rights reserved.)

samples have been found which contain as much as one hundred times this amount. As noted above, the greater concentration of mercury in shales than in granites, sandstones, and limestones is related to the younger geological age of sedimentary formations, of which shales constitute one type (Ahrens 1965; Day 1963; Aller 1961; Mason 1958; U.S. Department of the Interior 1970).

Mercury is found in greater than trace amounts in at least 30 naturally occurring ores of which cinnabar (HgS) is practically the only one which has a sufficiently high concentration to justify commercial exploitation for the recovery of the metal. Some mercury-containing minerals are listed in Table I-1.

A traditional practice among mineralogists has been to name newly-discovered compounds after the place of discovery or in honor of the discoverer or other prominent personage (Dana 1944-62; Desautels 1968). As can be seen in Table I-1, this has resulted in minerals of similar composition bearing more than one name (Arquerite and Moschellandsbergite, for example),

TABLE I-1

Mercury-Bearing Minerals*

Name	Formula	Where Found
Arquerite	$AgHg_3$, Ag_5Hg_3, Ag_6Hg, Ag_2Hg_5	Chile
Barcenite	Antimonate of Mercury	Mexico
Bordosite	$AgHgI$, $AgCl \cdot 2HgCl$	Chile
Calomel	$HgCl$	Texas, U.S.A.
		Yugoslavia
		Germany, Italy
Cinnabar	HgS	All continents except Antarctica
Coccinite	Hg_2OCl (?)	Mexico
Coccinite	HgI_2	Australia
Coloradoite	$HgTe$	Colorado, U.S.A.
Eglestonite	Hg_4Cl_2O	Texas, U.S.A.
Gold Amalgum	Au_2Hg_3, Au_2Hg_5	California and Oregon, U.S.A.
		Borneo
Guadalcazarite	$Hg_5 \cdot Zn \cdot Se$	Mexico
Hermesite	Tetrahedrite + Hg	Bavaria, Germany
Idrialite	$HgS + C_3H_2$	Yugoslavia
Iodargyrite	$AgHgI$ (?)	Germany, Spain, France,
		Congo, Chile, U.S.A., U.S.S.R.,
		Australia
Kalgoorlite	$Ag_2Au_2HgTe_6$	Australia, Colorado
Kleinite	$Hg \cdot NH_4Cl \cdot SO_4$ (?)	Texas, U.S.A.
Kongsbergite	$AgHg$	Norway
Lehrbachite	$HgSe + PbSe$	Germany
Leviglianite	$HgS + Zn$	Italy
Livingstonite	$HgSb_4S_7$	Mexico
Magnolite	Hg_2TeO_4	Colorado, U.S.A.
Metacinnabarite	HgS (dimorphous)	Mexico
Montroydite	HgO	California, Texas, U.S.A.
Moschellandsbergite	Ag_2Hg_3	Sweden, Bavaria, Germany,
		France
Mosesite	$Hg_6(NH_3)_2Cl_2(SO_4)(OH)_4$	Texas, Nevada, U.S.A.
Onofrite	$ZnS \cdot 6HgS$	Oregon, U.S.A.
Potarite	Pd_3Hg_2	British Guiana
Schwatzite	Tetrahedrite + Hg	Tyrol
Terlinguaite	Hg_2OCl	Texas, U.S.A.
Tiemannite	$HgSe$	Harz Mts., Germany,
		Utah, U.S.A.
Tocornalite	$(AgHg)I$	Chile

*Adapted from Dana's The System of Mineralogy (1944–62), with additions.

and in the case of coccinite, in the same name's being applied to different compounds.

Seven stable isotopes of mercury occur in nature and 18 additional radio-isotopes have been prepared. The former, with their mass numbers (A) and the latter with their mass numbers, half lives and modes of decay are listed in Table I-2.

Exact values for the abundance of the various elements in the lithosphere are difficult to establish because of the complicated structure of the earth's crust and because of the varied concentrations of the elements in different

TABLE I-2

Isotopes of Mercury

| *Naturally-Occurring Stable* | | | | |
No. *Mass (A)*	*Natural Abundance (%)*	*Isotope*	*Radioactive* *Half Life*	*Mode of Decay*
196	0.146	<195	0.7 min.	a
198	10.02	<191	~3 hrs.	
199	16.84	<191	1.5 hrs.	EC
200	23.13	189	23 min.	EC
201	13.22	190	20 min.	EC
202	29.80	191	57 min.	EC
204	6.85	192	5.7 hrs.	EC, β+
		193m	11 hrs.	EC + IT
		193	~6 hrs.	EC
		194m	0.4 sec.	IT
		194	130 days	EC
		195m	40 hrs.	EC, IT
		195	9.5 hrs.	EC
β– = negative beta emission		197m	24 min.	EC, IT
β+ = positron		197	65 min.	EC
a = alpha		199m	42 min.	IT
EC = Orbital electron capture		203	47 days	β–
IT = Isomeric transition		205	5.2 min.	β–
m = isomer		206	7.5 min.	β–

Adopted from various United States Government reports.

geological formations. A fairly good approximation can be made, however, of the abundance of mercury in relation to other elements.

Geochemists have estimated that about 97 or 98 percent of the lithosphere is made of eight elements: oxygen, silicon, aluminum, iron, calcium, magnesium, sodium and potassium. When titanium, phosphorus, hydrogen, and manganese are added, there remains less than 0.6 percent for the combined total of all of the other 78 naturally occurring elements. Mercury is sixteenth from the bottom of the list, the elements which are less abundant than quicksilver being: neon, ruthenium, rhodium, palladium, tellurium, rhenium, osmium, iridium, platinum, gold, polonium, radon, radium, actinium and protactinium (Day 1963; Aller 1961). Some estimates place selenium below mercury in abundance. According to one estimate mercury comprises about 2.7×10^{-6} percent of the lithosphere (Aller 1961). Discrepancies in the values for abundance of the elements given by various geochemists may be due, at least in part, to some using figures for the entire earth and others just for the crust (lithosphere).

Petrologists in the Soviet Union have reported on an analytical method with which they are able to accomplish quantitative separation in about five minutes and complete analysis in about a half hour. The sensitivity permits measurement of amounts of mercury in the range of $n \times 10^{-8}$ grams. In studies embracing more than 800 samples of rock, the Soviet scientists have found a concentration of mercury ranging from about 10 ppm to 0.0003 ppm. The higher values were present in a few samples of pegmatites, but a large majority of the samples contained less than 0.1 ppm (Studies on Mercury Deposits, 1964 transl.). Results of the same order of magnitude have been reported for soil samples from British Columbia (Warren, Delavault, and Barasko 1966).

An illustration of the extent to which man-made redistribution of mercury can change the concentrations in the lithosphere is found in a report from Japan, published in 1964 (Fujimura). Dust from a heavily travelled street in Tokyo contained 0.018 to 0.02 ppm and dirt from fields and rice paddies showed 0.36 to 0.38 ppm of mercury, the high value for the latter being explained as being due to the use of mercurial fungicides. Stock and Cucuel in 1934 reported finding mercury in concentrations of about one part per million in street dust of Berlin and Karlsruhe and one-third that amount in dust from a foot-path in the mountains. A summary of early published values for mercury in the lithosphere is given in Table I-3. More recent data are recorded in a 1970 publication of the U.S. Geological Survey.

Hydrosphere

Interest in the presence of mercury in the earth's waters goes back at least as far as the latter part of the 18th century and possibly to 1600 (Partington 1965). In 1799, J. L. Proust (1754–1826), best known for his recognition of the law of constant proportions, published the first of several papers on mercury in sea water. He lists a number of earlier chemists, including Robert Boyle and G. F. Rouelle, who knew about the presence of mercury in sea salt. Proust's analytical procedure was based on the amalgamation of mercury with gold and tin, and his findings were purely qualitative, but he did demonstrate that mercury was present in ocean water. It is one of about 60 elements so far demonstrated as being present in the sea.

TABLE I-3

Mercury in the Lithosphere

Stock & Cucuel, Germany, 1934	0.058–0.079 g/ton
Preuss, Germany, 1941	0.01–0.1 g/ton
Saukov, U.S.S.R., 1946[a]	0.064–0.09 g/ton
Aydin'yan et al., U.S.S.R., 1962	0.003–10.0 ppm
	most <0.1 ppm
Warren et al., Canada, 1966	0.01–10.0 ppm
	most <0.25 ppm
Anderson, Sweden, 1966–7[b]	0.02–0.92 ppm

Entire lithosphere 2.7×10^{-6} per cent Hg. Less abundant elements than mercury are neon, ruthenium, rhodium, palladium, tellurium, rhenium, osmium, iridium, platinum, gold, polonium, radon, radium and actinium.

NOTE: g/ton = 1/907,000 or approximately 1 ppm

[a]Cited in Studies on Mercury Deposits, 1964 transl.
[b]Cited in Westöö 1967.

Between the time of Proust and the last quarter of the 19th century there was only limited interest in mercury in the hydrosphere. One exception is found in the competition among balneologists and hydrotherapists in France to establish the supremacy of this or that watering place; the competition resulted in efforts to determine which constituents were responsible for the peculiar virtues of the different waters. The springs at Rocher were credited with having worked an almost miraculous cure in the case of a lady who was suffering from syphilis. When the chemist who analyzed a sample of the water found mercury (along with a number of other metals) he was convinced that he had the explanation for the potency of the Rocher water. He gives a figure for mercury of 0.0080 per liter, but does not state what unit he is talking about! It might have been grams, milligrams, or something else (Garrigou 1877). His analyses were later repeated by chemists representing competing spas who failed to find mercury (Willm 1879).

Waters from thirty-four mineral springs were analyzed by Bardet in 1913, using a spectographic technique applied to the dried residue of spring waters. The published results do not give quantitative data but Bardet does state that he found lead in all samples, silver and tin in most, and mercury rarely. That he should have found any mercury at all is surprising in view of the analytical method employed.

An explanation of the migration of mercury from the lithosphere to the hydrosphere can be found in the work of Becker, published in 1887 and 1888. He demonstrated the mechanisms by which cinnabar in nature becomes soluble and thus is able to pass into the hydrosphere. His reports, however, do not contain any quantitative data on mercury in waters.

Some idea of the state of the art of quantitative analysis as applied to mercury in waters is given in an article by Raaschou published in 1910, "A Microchemical Method for the Determination of Mercury." Raaschou was interested in water solutions produced by the human body, namely, urine, and his microdetermination was suitable for samples containing less than two milligrams per liter. The method, based on distillation and condensation of mercury, is complicated and time-consuming, but can measure as little as 30 μg in 100 ml. of urine. He reviews in his article earlier analytical methods, including one employing the colorimetric procedure with diphenylcarbazide.

Many of the analytical methods of the 19th and early part of the 20th centuries would hardly measure up to modern standards. Little or no consideration was given to possible impurities in reagents, to other possible sources of contamination, or to possible losses of mercury during the preparation of samples for analysis.

For more than twenty years following its publication in 1934 (Stock and Cucuel), the work of Stock stood out as the only major contribution to knowledge of mercury in the hydrosphere. His findings are still the basis for figures given in standard textbooks, although the debt to Stock is not always acknowledged. His values for mercury, in micrograms per liter are:

Sea Water (North Sea)	0.03
Rain Water	0.05–0.47 (mean, 0.20)
Rhine River	0.10
Spring Water	0.010–0.05
Tap Water (Berlin)	0.010
Tap Water (Karlsruhe)	0.01–0.05

The analytical procedure perfected and utilized by Stock (1938) depended on the amalgamation of mercury with copper, liberation of the mercury and measuring the size of the droplet. This method is probably more sensitive and possibly more accurate than any used by early investigators, but it is too time-consuming to be practicable for large-scale surveys.

Extensive studies of mercury in the hydrosphere have been carried out by Soviet scientists, following the development of a rapid, sensitive analytical method (Aidin'yan 1962). Their investigations have covered some of the major rivers in Eastern Europe as well as the Black Sea, Sea of Azov, various parts of the Mediterranean Sea, and the Atlantic and Indian Oceans. Values for mercury ranged from 0.9×10^{-7} to 2.8×10^{-6} grams per liter, with little difference between the rivers and the seas (1×10^{-6} grams per liter is 0.0001 ppm, or 0.1 μg per liter, or approximately 0.0001 grams per ton). The Soviet figures are, therefore, of the same order of magnitude as those reported by Stock.

Man-made alterations in distribution of mercury in the hydrosphere have been of increasing concern in several parts of the world, particularly in Japan and Sweden, and more recently in North America. Some of the disturbances are a result of the manufacture and use of organo-mercurial pesticides, great quantities of which have been employed by Japanese rice growers to control the fungus disease known as "rice blast." A report published in 1964 (Fujimura) gives mercury levels for rivers and ditches in the Tokyo area of 0.004 to 0.1 ppm. These values are stated to be approximately ten times as high as in rural areas and are explained as being due to contamination from industrial wastes. No figures are given for surface waters in rice-growing districts and no data are available on mercury levels for the time preceding the start of the pesticide industry. Rain water and snow collected in an urban area in Japan showed mercury concentrations of 0.0006 ppm.

Contamination of streams and ditches is not the only cause for Japanese concern over the pollution of the hydrosphere with mercurial compounds. An episode centering around Minamata Bay, on the southwest coast of the island of Kyushu, has attracted attention not only in Japan but also in the entire world of toxicology.

A mysterious illness, subsequently named Minamata disease, first began to appear in the latter part of 1953, the highest incidence being among fishermen and their families (Kurland, Faro, and Siedler 1960; Tokuomi et al. 1961; Takeuchi 1961). Later, when it was recognized that sea birds and household cats were being affected, attention was focused on fish and shell fish as etiological factors. This in turn led to a study of the water in Minamata Bay and to the identification of mercury in a factory effluent as the cause of the disease. Although the mercury discharged in the factory wastes was believed to be in an inorganic form, the manifestations of Minamata disease were those of poisoning with alkyl (organic) mercurials, causing degenerative changes in the central nervous system, often with fatal outcome. At first it was thought that there had been a chemical biotransformation of the inorganic into an alkyl (methyl) mercurial, either in fish or in the mud at the bottom of the bay. A change of this type has been shown to be possible (Wood, Kennedy, and Rosen 1968; Ui 1968). Ultimately, however, methyl mercury compounds were found along with the inorganic forms in the factory effluent (Ui 1968). This apparently closed the ring of evidence as to the etiology of Minamata disease.

Scarcely had the Minamata problem been solved and the outbreak brought under control when in 1965 a second, similar episode occurred in the Agano River delta near Niigata on the northwest coast of the Island of Honshu. Once again methyl mercury was incriminated, with shellfish or fish being the mode of transmission, but there has been difficulty in establishing which of two or three possible sources was responsible for contaminating the water (Ui 1968; Tsuchiya 1969 [personal communication]).

Late in 1967, the Swedish Medical Board found it necessary to ban the sale of fish from about forty Swedish rivers and lakes, due to the finding of high concentrations of methyl mercury in fish caught in these waters. A general survey of surface waters revealed widespread pollution with organomercurials, but in most instances the levels were not sufficiently high to call for restrictive action. The source of the mercury is believed to have been paper pulp mills which use mercurial compounds as slimicides (Barnes 1967). Serious poisoning of wild life had been noted as early as 1960 and this led to extensive studies of the environment. The results were published in 1967 (Westöö; Tejning) but no new data on mercury in the hydrosphere are included in these reports.

In the United States, the Sport Fishing Institute in 1969 sounded an alarm about mercury in rivers, suggesting that it may be more of a threat than DDT residues (Water Newsletter 20 August 1969). Within a few months the problem had attracted nationwide attention.

Values for mercury in various waters are summarized in Tables I-4 and I-5. A comprehensive review of the subject was published in 1970 by the U.S. Geological Survey.

TABLE I-4
Mercury in the Hydrosphere

Proust, France, 1799	Sea water	Hg present
Stock and Cucuel, Germany, 1934	North Sea	0.03 μ/l (ppb)
	Fresh waters	0.01-0.1 μg/l
	Rain water	0.05-0.48 μg/l
Aydin'yan, U.S.S.R., 1962	Rivers, lakes, seas	0.09-0.28 μg/l
Fujimura, Japan, 1964	Rain water	0.6 μg/l
	Ditches	4-100 μg/l

TABLE I-5

Mercury in Selected Streams, 1970. Analyses by U.S. Geological Survey.
(Data are provisional and subject to revision.)

Stream and location	Date	Mercury (Hg) (micrograms per liter)
Pemigewasset River at Woodstock, N.H.	June 8	3.1
Merrimack River above Lowell, Mass.	June 8	1.2
Hudson River downstream from Poughkeepsie, N.Y. . . .	Apr. 7	.1
Delaware River at Port Jervis, N.Y.	Apr. 23	<.1
Susquehanna River at Johnson City, N.Y.	July 6	.1
Chemung River near Wellsburg, N.Y.	July 6	.2
North Branch Potomac River near Barnum, W.Va.	June 3	1.2
Pascagoula River at Merrill, Miss.	June 9	3.0
Ohio River near Grand Chain, Ill.	June 26	.1
Maumee River at Antwerp, Ohio	June 10	6.0
Oswegatchie River at Gouverneur, N.Y.	June 16	1.2
Rainy River at International Falls, Minn.	May 14	<0.1
Wisconsin River near Nekoosa, Wis.	June 10	2.4
Bighorn River at Kane, Wyo.	June 30	<.1
Floyd River at Sioux City, Iowa	June 9	.2
North Platte River near Casper, Wyo.	June 23	.1
Missouri River at Hermann, Mo.	June 24	.2
Mississippi River near Hickman, Ky.	June 25	<.1
St. Francis River at Marked Tree, Ark.	June 19	.1
San Antonio River near Elmendorf, Tex.	June 11	<.1
Colorado River near Yuma, Ariz.	June 18	<.1

From U.S. Geological Survey, *Water Resources Review,* July 1970, p. 7.

Atmosphere

On purely theoretical grounds, it is reasonable to assume that mercury is
present everywhere in the earth's atmosphere. The relatively high vapor pressure of the metal at ordinary temperatures and the operation of the mercury
cycle (Fig. 1) support this assumption.

Mercury may be present in the atmosphere as a vapor or as an aerosol, and
its presence may be due to natural or man-made causes. All forms of mercury
are capable of being converted into aerosols and thus remain suspended in the
atmosphere for long periods of time.

Concentrations of mercury in the atmosphere under natural conditions are
very low and consequently require the most sensitive methods for analysis.
(Measurements are made in nanograms, or 1×10^{-9} grams.) Studies on
naturally occurring atmospheric mercury have been limited in number and extent. Man-made accumulations of mercury in the air of factories and mines
have been subjected to extensive quantitative study since the early decades of
the 20th century but the same is not true for the general environment.

More than 400 years ago the metallurgist Vannoccio Biringuccio (1480–
1539) made some practical observations on the use of mercury vapors in the
atmosphere for detecting deposits of ore:

In the month of April or May before sunrise when the weather is tranquil and quiet, the places where this quicksilver is found are recognized by certain thick and dense vapors that rise above these places but do not go very high because they are heavy. Some who are experienced in this follow the sign almost as one does water. . . . (Biringuccio, p. 82)

Despite this clear, early description of the detection of mercury vapor as an aid in prospecting for valuable ores, modern Soviet geochemists have claimed, and are generally given, credit for being the first to use this procedure. The basic work is that of Saukov in 1946 (cited in *Studies on Mercury Deposits*, 1964 transl.). While it is true that the Russian geochemists, and those in other countries who have used the detection of mercury vapor in the atmosphere as an aid in prospecting, have had the advantage of more refined instruments than those used by Biringuccio, there is no available evidence to show that latter-day results have been any better than those of four or five hundred years ago.

Refinements in what might be called the Biringuccio-Saukov technique have been developed in the United States by Williston (1964; 1968). According to him the Russians are able to detect mercury vapor at a lower limit of 1×10^{-6} percent, whereas an instrument he has designed can measure 1/10,000th of that amount. In absolute terms this means that Williston is able to detect as little as 1×10^{-11} grams of mercury in a sample of air. Atmospheric studies he has made in the neighborhood of Palo Alto, California, have shown readings mainly within a range of one to ten nanograms (1 to 10×10^{-9} grams) per cubic meter of air, with occasional higher peaks up to 50 micrograms per cubic meter. Hourly and daily fluctuations were common, with readings on clear, sunny days generally higher than those during times when the skies were overcast, but smog was always accompanied by high levels of mercury (Williston 1968). Earlier observations had led Williston to believe that the background value for mercury vapor in the atmosphere is in the range of 1 to 4×10^{-8} grams per cubic meter. His measurements have not included mercury in particulate form, that is, dust or aerosol.

A study embracing twenty air samples from randomly selected homes and simultaneously taken samples from adjacent outdoor air in New York City showed a range of 0.001–0.041 micrograms of mercury per cubic meter of air for the indoor and of 0.001–0.014 micrograms per cubic meter for the outdoor samples. These determinations were part of an evaluation of suspended particulate matter and did not include mercury present as vapor.* Surveys conducted in Japan have shown mercury concentrations up to 10 micrograms per cubic meter of air along a busy highway and from 1.2 to 2.4 micrograms per cubic meter in less polluted areas (Fujimura 1964). Values similar to these have been reported from the Soviet Union (Leites 1952), but both Japanese and Russian data are scanty.

---

*L. J. Goldwater. Unpublished data.

Preoccupation with such easily measurable gross pollutants of the atmosphere as carbon monoxide, sulphur dioxide, and lead seems to have diverted attention away from mercury and other less obvious pollutants whose action, nevertheless, may be of equal or greater significance than that of the "favorites." Combinations of pollutants in the air may have effects on health which are more than simply additive. Mercury's unique potency as a catalyst in chemical reactions suggests at least a potential role for this ubiquitous metal in some of the poorly understood phenomena in air pollution. The role might be either beneficial or harmful. First steps toward evaluation of mercury in the atmosphere are now (1971) being taken. Subsequent correlation of the mercury findings with other observations might reveal whether or not mercury is having any significant effect on health.

Some published values for mercury in the atmosphere are summarized in Table I-6 and additional data can be found in U.S. Geological Survey Professional Paper 713 (U.S. Department of the Interior 1970).

Biosphere

Prior to the middle of the 20th century few studies had been made on the amount of mercury in plants and animals. The textbook *Geochemistry* by Rankama and Sahama, published in 1950, devotes a single paragraph to the subject. The authors state that both plants and animals are able to concentrate mercury, a fact that has been amply confirmed (Klein and Goldberg 1970), and that some marine algae build up a mercury concentration more than a hundred times greater than that of the sea water in which they grow. They state that "Droplets of metallic mercury have been found in the seed capsules of *Holosteum umbellatum* growing in some mercury-rich soils." This story is related by Needham (Needham and Wang Ling, Vol. 3, pp. 378-379) in a slightly different version, but neither he nor Rankama gives a substantiating reference. Needham cites a work by Henckel which is said to describe a plant found in China ". . . which contained so much mercury that it would give rise to an amalgam as soft as butter when ground with copper filings." A third al-

TABLE I-6

Mercury in the Atmosphere

Leites, U.S.S.R., 1952	0.05-1.0 (approx.)
Cholak, 1952	
Cincinnati, U.S.A.	0.03-0.21 (2 samples—dust)
Charleston, W. Va., U.S.A.	0.17 (one sample—dust)
Goldwater, N.Y.C., 1962*	0.001-0.041 (dust)
Fujimura, Japan, 1964	1.2-10.0 (total)
Williston, California, U.S.A., 1966 (1968)	0.01-0.04 (vapor)
Williston, California, U.S.A., 1968	
Winter	0.0005-0.025 (vapor)
Summer	0.001-0.050 (vapor)

NOTE: All values in $\mu g/m^3$ of air
*Unpublished data.

lusion to mercury in plants made by Needham has to do with the extraction of mercury from purslane (*Portulaca oleracea*). The author does not give his source of this statement but a detailed account of the process can be found in Postlethwayt's *Universal Dictionary of Trade and Commerce*—of all places. Marvels such as these sound as though they came straight out of Pliny. Postlethwayt, writing in 1774, ascribes the purslane story to "a famous Chinese author" whose name is not given.

Studies of the mercury content of vegetation were made by H. V. Warren and his associates in British Columbia in 1962 and 1964 and reported in 1966. One set of 12 samples was taken from trees growing in soil which had been contaminated by the operations of a mercury recovery plant. The samples showed concentrations of mercury from 1.5 to 30 parts per million of ash. Other sets of samples taken from trees in areas of little or no contamination with mercury showed values ranging only from 0.03 to 4.7 ppm. The entire study embraced 141 samples, principally from trees. Stock in 1938 had reported the mercury content of unashed wood to be $0.15-0.24$ $\mu g/100$ g or $0.0015-0.0024$ ppm.

As part of an investigation of the phytotoxicity of several forms of mercury, Hitchcock and Zimmerman of the Boyce Thompson Institute for Plant Research studied the mercury content of a number of plants, particularly roses. In reviewing similar work done by others they cite the finding by Dimond and Stoddard of 0.2 ppm of mercury in the petals of normal roses and 0.07 ppm in the leaves. The analytical method (Association of Official Agricultural Chemists 1930) used by Hitchcock and Zimmerman was "not suitable for determining small amounts of mercury of the order of 1 ppm or less on a dry weight basis" so they do not add any new data on "normal" mercury in plants. Zimmerman and Crocker state, however, that "it is probable that many plants, possibly all, contain mercury, and that different parts of the plant would contain different amounts of mercury." The Boyce Thompson investigators clearly demonstrated the uptake of mercury by plants grown in soils and atmosphere which had been artificially enriched with the metal or its compounds. Confirmation of these early observations is found in the work of Shacklette.

Mercury in Foodstuffs

Among the early published studies of mercury in foods are those of Stock and Cucuel in 1934 and of Gibbs and his co-workers in 1940 and 1941. Their results are summarized in Tables I-7 and I-8. Without giving details on his analytic method or its sensitivity, Gibbs states that ". . . mercury was found in tissues of dog, cat, rabbit, rat, frog, cow, hog, fish, in adults, newborn and fetuses" (Gibbs, Pond, and Hanssmann 1941).

In comparing the work of Stock with that of Gibbs, the point of greatest interest is not the similarity or dissimilarity of their findings but their antithetical interpretations. Stock is noted for his overemphasis on the dangers in-

TABLE I-7

Concentration of Mercury in Foods
Based on Analyses by Stock and Cucuel

Vegetables	Mercury in µg/100 g fresh weight	Meats	Mercury in µg/100 g fresh weight
Sprouts	0.7	Hog	
Lettuce	0.9	Flesh	0.6–1.3
Dried Peas	1.2	Blood	0.1
Dried Beans	4.6	Lung	0.9
Carrots	0.8	Brain	0.14
Red Beets	2.8	Heart	0.28
Turnips	0.5	Liver	0.7
Radishes (Radieschen)	3.4	Kidney	2.5
Radishes (Retticle)	1.4	Veal Liver	2.0
Horseradish	4.4	Veal Kidney	6.7
Potatoes	0.2–0.4	Beef Liver	1.3
		Beef Kidney	6.7
Fruits			
Strawberries	0.7	*Dairy Products*	
Cherries	0.5	Milk (per liter)	0.6–4.0
Apples	0.4	Cheese	0.9–1.0
Pears	1.2	Butter	0.2
Peaches	1.2		
		Miscellaneous	
Grains		Fish (salt water)	2.4–11.0
Rye Flour	3.6	Fish (fresh water)	2.8–18.0
White Flour	2.0	Egg	0.22
Mixed Bread Flour	2.6	Algae	1.8–3.7
"Edelweiss blütenmehl"	2.4	White wine (per liter)	0.4–1.0
		Wine vinegar (per liter)	1.4
Vegetable fats		Beer (per liter)	0.07–1.4
Olive oil	1.2		
Cocoanut fat	6.2		
Palm fat	8.6		
Palmin	11.5		
Peanut oil	2.2		

Adapted from Stock and Cucuel 1934
NOTE: Samples were from a student mess; average portion was 700 g; Hg intake,
0.25–2.6 µg daily total; estimated average daily Hg intake, 5 µg.

herent in the smallest quantities of absorbed mercury, while Gibbs minimizes
any such risk. Paradoxically, however, Stock and Cucuel in 1934 suggested
that mercury probably plays a biological role as either a positive or negative
catalyst; they were among the first proponents of this idea.

An investigation of the mercury content of foodstuffs, conducted at
Columbia University in 1964, embraced 75 different items of diet (Cerkez).
No attempt was made to secure samples from outside the New York area, and
the analyses were confined to a single sample of each food item. The analyses
were performed in duplicate by a photometric method sensitive in the nanogram range (Jacobs 1967, pp. 355–357). The results, shown in Table I-9, represent the average of the two duplicate analyses. Composite results for the

TABLE I-8

Concentration of Mercury in Foods (Based on Work of Gibbs)

Description	Mercury μg/100 g Fresh Weight	Description	Mercury μg/100 g Fresh Weight
Milk (fresh)*	0.325–0.708	Green peas (pureed)*	0.498–2.462
Milk (evaporated)	0.208–0.358	Spinach (canned)	2.034–2.462
Bread (white)	0.498–0.600	Pork roast	3.864–4.368
Liver (hog)*	0.156–2.910	Beef roast	0.598–0.894
Salmon*	0.854–0.498	Corn (canned)	0.854
Tomato juice	0.0–0.009	Pork and Beans (canned)*	0.498–1.610
Hamburger	0.084–0.498	Peaches (canned)	0.854
Green beans (canned)	1.362–1.610	Rolled oats cereal	0.156–0.498
Haddock fish	0.156–0.498	Dog food	1.994–2.604
Ralston wheat cereal	0.498–0.598	Small animal feed	1.360–2.460

No mercury was found in:

Fresh turnips	Fresh cabbage	
Fresh beets	White potatoes	
Fresh carrots	Salted soda crackers	
	Fresh Eggs*	

After Gibbs, Shank, and Pond 1940.
NOTE: For all samples, tests were performed on three aliquots. Ranges in table show variability in the tests. Samples were obtained in the area of Memphis, Tennessee. To convert μg/100 g to ppm move decimal point two places to the left. Sensitivity of method not stated.
*Two different samples analysed.

various categories of foodstuffs are summarized at the end of Table I-10. All results apply to fresh weight.

Values for mercury in foods found in several studies are summarized in Table I-10. This table invites unwarranted comparisons and conclusions. The strongest possible caveat is in order. The data from the various investigations should not and cannot be compared for these reasons:

1. The analyses were done in different laboratories using different analytical methods.
2. Details of the procedure for collection and analysis are not given in sufficient detail to permit critical evaluation.
3. In the case of Gibbs' results, the wide spread between the figures of the three aliquots of the same sample leads to questions about the reliability of the findings.
4. The number of samples analyzed is too small to justify any kind of generalization.
5. While the values reported from Japan are generally much higher than the others, they reflect a situation in which mercurial pesticides had been in widespread use for several years. No information is available on the mercury content of foods before the use of these pesticides was started. It may *not* be concluded, therefore, that the mercury necessarily came from pesticides and not from native soils, water or air.
6. Except for one or two food categories there is no uniform consistency among the four sets of data. Where agreement is found it may be purely accidental.

TABLE 1-9

Concentration of Mercury in Foods

Description	Mercury μg/100g	Description	Mercury μg/100g
Meat Products		**Dairy Products**	
Fresh Veal	0.94	Fresh Milk	0.81
Fresh Beef	0.46	Cheese	8.20
Fresh Lamb	0.12	Butter	14.10
Fresh Liver	0.52		
Cooked Chicken	4.00	**Grain Products**	
Cooked Beef	1.10		
Canned Chicken	15.00	Bread	0.81
Cooked Liver	10.00	Flour	0.24
Fresh Pork	0.36	Maize Flour	0.22
Bacon	8.00	Macaroni	2.50
Kosher Salami	2.20	Cornflakes	1.25
Pork Sausage	1.00	Crackers	0.60
Bologna Sausage	1.10	Rice	0.23
Broiled Turkey	1.80		
Broiled Beef	3.10	**Fruits**	
Liver Sausage	5.70		
		Apple	2.10
Fish		Banana	3.10
		Pear	1.40
Fresh Halibut	1.63	Grapes	0.97
Fresh Smelt	0.00	Orange	0.76
Fresh Shrimp	0.77	Orange Juice	0.40
Fresh Hake	1.13		
Fresh Carp	1.94	**Nuts**	
Fresh Salmon	0.99		
Cooked Salmon	5.90	Brazil nut	5.69
Smoked Salmon	4.90	Almond	1.70
Canned Tuna	3.76	Filbert	3.20
Canned Herring	0.89	Walnut	5.60
		Cashew	5.10
Vegetables			
		Miscellaneous Food Products	
Fresh Lettuce	0.15		
Fresh Tomato	0.66	Tea	0.0
Fresh Tomato Skin	0.495	Coffee	4.0
Raw Potato	0.00	Chocolate	1.30
Raw Carrots	2.04	Halvah	0.86
Canned Carrots	0.18	Mustard	3.80
Raw Cabbage	0.70	Margarine	6.70
Celery Leaves	0.19	Egg White	1.10
Green Onion	0.71	Egg Yolk	6.20
Onion	0.33		
Leek	1.90	**Beverages**	
Canned Peas	0.25		
Radish	0.18	Beer	0.44
Raw Green Pepper	0.32	Grapefruit Juice	0.13
Parsnip	0.12	Coca Cola	0.18
Tomato Puree	0.72	Drinking Water	0.03

Summary of Table I-9

The Range and Mean Values of Mercury in Foods and Beverages

No. of Samples	Description	Hg in µg/100 g Range	Mean
16	Meats	0.12–15.0	3.455
10	Fish	0.00–5.90	2.191
16	Vegetables	0.00–2.04	0.559
3	Dairy products	0.81–14.10	7.773
7	Grain products	0.23–2.50	0.837
6	Fruits	0.40–3.10	1.455
5	Nuts	1.70–5.69	4.258
8	Miscellaneous foods	0.86–6.70	3.120
4	Beverages	0.03–0.44	0.195
Total 75			

From Cerkez and Goldwater [unpublished data].

Prior to 1970 there was great paucity of data on mercury in foods. Such meagre information as was available came from a mere half dozen countries; we have no information on the mercury content of foodstuffs for several entire continents. Before the late 1960's virtually nothing had been published on the form in which mercury is present in foods, not to mention its solubility, the amount absorbed from diets and its metabolic fate. In spite of the great voids in knowledge, governmental agencies have established levels of permissible mercury concentrations in foods. The Food and Drug Administra-

TABLE I-10

Comparison of Mercury Concentrations in Foods as reported in five studies

Foodstuff	Cerkez and Goldwater[a] (1964)	Stock (1934)	Stock (1938)	Gibbs (1940)	Fujimura (1964)
Meats	0.12–15.0	0.10–6.7	0.50–2.0	0.084–4.36	31–36
Fish	0.0–5.9	2.4–18.0	2.50–18.0	0.156–1.36	3.5–54
Vegetables (Fresh)	0.0–2.0	0.20–4.4	0.50–3.5	0	3.0–6.2
Vegetables (Canned)	0.18–0.72	–	–	0.498–2.46	–
Milk (Fresh)	0.81	0.06–0.40	0.06–0.40	0.325–0.708	0.32–0.70
Butter	14.1	0.2	7.0–28.0	–	–
Cheese	8.2	0.9–1.0	–	–	–
Grains	0.23–2.50	2.0–3.6	2.5–3.5	0.156–0.60	1.2–4.8
Fruits (Fresh)	0.40–3.10	0.40–1.2	0.50–3.5	–	1.8
Egg White	1.10	–	–	–	8.1–12.5
Egg Yolk	6.20	–	–	–	33–67
Egg (Whole)	–	0.22	0.20	0	–
Beer	0.44	0.007–0.14	0.10–1.50	–	–

NOTE: All values are µg/100 g fresh weight. To convert to ppm move decimal point two places to left. To convert to ppb move decimal point one place to right. FAO/WHO suggested tolerance–50 ng/g (ppb).

[a]Unpublished data.

tion of the United States Department of Health, Education and Welfare in 1938 set a tolerance of "zero" for residues of mercurial pesticides on foods, without, however, making clear what was meant by "zero." On the international scene, the Joint FAO/WHO Codex Alimentarius Commission of the Food and Agricultural Organization and the World Health Organization in 1963 recommended an upper level of 0.05 milligrams of mercury per kilogram (0.05 ppm or 50 nanograms per gram) for all foods except fish and shellfish. (Omission of these two categories was due simply to failure of the responsible committee to submit a report.)

Promulgation of the FAO/WHO standards may not be an unmixed blessing. Regulatory agencies rely strongly on numbers in setting tolerances and so, *faute de mieux*, will find the Codex Alimentarius Commission's figures useful. Food producers and processors may find them troublesome, while scientists will or should view them with scorn since they have been set without adequate scientific basis.

During 1970, as a result of widespread apprehension and a variety of pressures, a massive effort was begun in the United States and in Canada to study a number of environmental aspects of mercury. Dozens of laboratories started analyzing mercury in foods, particularly fish and shellfish, while others focused on water and to a lesser extent on air. Instant experts began to spring up on all sides with ready answers to any and all questions. Politicians began to vie for the dubious honor of espousing the most restrictive legislation and some segments of the news media tried to outdo each other in sensationalism. All in all the performance had much in common with the waves of mass hysteria which swept Europe in medieval and early modern times.

By no means all of the activities dealing with mercury have been wasteful or non-productive. Along with the opportunists, a number of established scientists and laboratories turned their attention to the question and many talented young people were attracted to the field of environmental studies. Numerous valuable conferences were held during 1970 and 1971, attracting outstanding participants from all parts of the world. Scientific journals representing many disciplines began in 1970 to publish an increasing number of articles on mercury, some good, some not so good, as might be expected. As the midpoint of 1971 passed, it was clear that the amount of information on mercury in the environment that will result from two years of effort will exceed the total production of all previous time.

REFERENCES

Ahrens, L. H. 1965. *Distribution of elements in our planet.* New York: McGraw-Hill Book Co.
Aidin'yan, N. K. 1962. The content of mercury in certain natural waters. *Trans. Inst. Geol. Ore Deposits Petro. Mineral. Geochem. Moscow* 70(3):9–14.
Aller, L. H. 1961. *The abundance of the elements.* New York: Interscience Publishers.
Association of Official Agricultural Chemists. Committee on Editing Methods of Analysis. 1930. *Official and tentative methods of analysis.* 3d. ed. Washington, D.C.

Bardet, J. 1913. Étude spectrographique des eaux mi1érales françaises. *Compt. Rend.*
157:224-26.

Barnes, H. J. 1967. Polluted fish sale banned. *Sci. News* 92:564.

Becker, G. F. 1887. Natural solutions of cinnabar, gold and associated sulphides. *Am. J.
Sci.* 33(3):199-210.

Becker, G. F. 1888. Natural solutions of cinnabar, gold and associated sulphides. *Monogr.
U.S. Geol. Surv.* 13:1.

Biringuccio, V. 1942. *Pirotechniq.* Trans. Cyril Stanley Smith and Martha Teach Gnudi
from the first edition of 1540. Cambridge: M. I. T. Press Paperback Edition.

Bowen, H. J. M. 1966. *Trace elements in biochemistry.* New York: Academic Press.

Cerkez, F. 1964. *The evaluation of mercury exposure in human environment with
emphasis on its content in food.* Unpub. Dr. P. H. dissertation, Columbia University
School of Public Health, New York.

Cholak, J. 1952. The nature of atmospheric pollution in a number of industrial com-
munities. Paper read at the 2d National Air Pollution Symposium, 5-6 May 1952,
Pasadena, California.

Dana, J. W. 1944-62. *The system of mineralogy.* 7th ed. rev. and enl. by Charles
Palache, Harry Berman and Clifford Frondel. 3 vols. New York: John Wiley and Sons.

Day, F. H. 1963. *The chemical elements in nature.* London: George G. Harrap and Co.

Desautels, P. E. 1968. *The mineral kingdom.* New York: Madison Square Press.

Dimond, A. E., and Stoddard, E. M. 1955. Toxicity to greenhouse roses from paints containing mercury fungicides. *Conn. Agr. Exp. Sta. New Haven Bull.* 595:19.

Fujimura, Y. 1964. Studies on the toxicity of mercury: Hg series no. 7, 2d report: On
the present status of mercury contamination in environment and foodstuffs. *Japan. J.
Hyg.* 18:10-19.

Garrigou, F. 1877. Sur la présence du mercure dans la source du Rocher. *Compt. Rend.*
84:963-65.

Gibbs, O. S.; Pond, H.; and Hanssmann, G. A. 1941. Toxicological studies on ammoniated mercury. *J. Pharmacol. Exp. Therap.* 72:16.

Gibbs, O. S.; Shank, R.; and Pond, H. 1940. *Report of studies on ammoniated mercury
ointments and red oxide of mercury ointments.* Memphis, Tennessee: The External
Products Research Institute.

Gibbs, O. S.; Shank, R.; Pond, H.; and Hanssmann, G. A. 1941. Absorption of externally
applied mercury. *Arch. Dermatol. Syphilol.* 44:862-72.

Hitchcock, A. E., and Zimmerman, P. W. 1957. Toxic effects of vapors of mercury and
of compounds of mercury on plants. *Ann. N.Y. Acad. Sci.* 65:474-97.

Jacobs, M. B. 1967. *The analytical toxicology of industrial inorganic poisons.* New York:
Interscience Publishers.

Joint FAO/WHO Codex Alimentarius Commission. 1963. *Report of the first session*, 25
June-3 July 1963, Rome. Ref. No. ALINORM 63/12 July 1963. Rome: Food and
Agriculture Organization of the United Nations.

Klein, D. H., and Goldberg, E. D. 1970. Mercury in the marine environment. *Environ.
Sci. Technol.* 4:765-68.

Kurland, L. T.; Faro, S. N.; and Siedler, H. 1960. Minamata disease. *World Neurol.*
1:370-95.

Leites, R. G. 1952. The limit of allowable concentration of mercury vapors in the air of
inhabited areas. In *Limits of allowable concentrations of atmospheric pollutants*, Bk.
1. Moscow: Medgiz.

Lockyer, J. N. 1878. Researches in spectrum analysis in connection with the spectrum of
the sun. *Proc. Roy. Soc.* 28:157-80.

Mason, B. H. 1958. *Principles of geochemistry.* 2d ed. New York: John Wiley and Sons.

Mellor, J. W. 1922-37. *A comprehensive treatise on inorganic and theoretical chemistry.*
16 vols. London: Longmans, Green and Co.

Needham, J., and Wang Ling. 1954-62. *Science and civilization in China.* 4 vols. Cambridge: Cambridge University Press.

Partington, J. R. 1965. *A short history of chemistry.* 3d ed. New York: Harper and Row,
Harper Torchbooks.

Postlethwayt, M. 1774. *Universal dictionary of trade and commerce.* 4th ed. Vol. 2.
London.

Preuss, E. 1940. Spectrographic methods: II. Determination of Zn, Cd, Hg, In, Tl, Ge, Sn, Pb, Sb, and Bi by fractional distillation. *Z. Angew. Mineral.* 3:8-20.

Proust, J. L. 1799. Sur le mercure contenu dans le sel marin. *J. Phys. Chim.* 49:153.

Raaschou, P. E. 1910. Eine mikrochemische Quecksilberstimmungsmethode. *Z. Anal. Chem.* 49:172-204.

Rankama, K., and Sahama, T. G. 1950. *Geochemistry.* Chicago: Chicago University Press.

Reed, G. W.; Goleb, J. A.; and Jovanovic, S. 1971. Surface-related mercury in lunar samples. *Science* 172:258-61.

Shacklette, H. T. 1970. Mercury content of plants. In *Mercury in the environment*, Geological Survey Professional Paper 713, U.S. Department of the Interior, 1970, pp. 35-36.

Stock, A. 1938. Die mikroanalytische Bestimmung des Quecksilbers und ihre Anwendung auf hygienische und medizinische Fragen. *Svensk. Kem. Tidskr.* 50:242-50.

Stock, A., and Cucuel, F. 1934. Die Verbreitung des Quecksilbers. *Naturwissenschaften* 22/24:390-93.

Studies on mercury deposits. 1964. Translation of nine articles from 1960-63 issues of [Akademiya Nauk SSR] *Trudy Institut Geologii Rudnykh Mestorozhdenii, Petrografii, Mineralogii, i Geokhimii.* Joint Publications Research Service, Office of Technical Services, U.S. Department of Commerce, 30 March 1964.

Takeuchi, T. 1961. A pathological study of Minamata disease in Japan. Paper read at the 7th International Congress of Neurology, 13 September 1961, Rome.

Tejning, S. 1967. Mercury in pheasants (Phasianus colchicus, L.) deriving from seed grain dressed with methyl and ethyl mercury compounds. *Oikos* 18:334-44.

Tokuomi, H.; Okajima, T.; Kanai, J.; Tsunoda, M.; Ichiyasu, Y.; Misumi, H.; Shimomura, K.; and Takaba, M. 1961. Minamata disease: An unusual neurological disorder occurring in Minamata, Japan. *Kumamoto Med. J.* 14:47-64.

Ui, J. 1968. A short history of Minamata disease's research and present situation of mercury pollution in Japan. Paper read at Nordisk Kvicksilversymposium, 10-11 October 1968, Lidingö, Sweden.

U.S. Department of the Interior. 1970. *Mercury in the environment.* Geological Survey Professional Paper 713. Washington: U.S. Government Printing Office.

U.S. Geological Survey. 1970. *Water Resources Review.* July.

Warren, H. V.; Delavault, R. E.; and Barasko, J. 1966. Some observations on the geochemistry of mercury as applied to prospecting. *Econ. Geol.* 61:1010-28.

Water Newsletter. 20 August 1969.

Watteville, C. de. 1906. Sur le spectre de flamme du mercure. *Compt. Rend.* 142: 269-70.

Westöö, G. 1967. Methyl mercury compounds in fish. *Oikos Suppl.* 9:13-14.

Williston, S. H. 1964. The mercury halo method of exploration. *Eng. Mining J.* 165: 98-101.

Williston, S. H. 1968. Mercury in the atmosphere. *J. Geophys. Res.* 73:7051-55.

Willm, E. 1879. Sur la présence du mercure dans les eaux de Saint-Nectaire. *Compt. Rend.* 88:1032-33.

Wood, J. M.; Kennedy, F. S.; and Rosen, C. G. 1968. Synthesis of methyl-mercury compounds by extracts of a methanogenic bacterium. *Nature* 220:173-74.

Zimmerman, P. W., and Crocker, W. 1934. Plant injury caused by vapors of mercury. *Contrib. Boyce Thompson Inst.* 6: 167-87.

Occult Arts

Alchemy

Mercury is inseparably connected with alchemy, the entire structure of alchemy having been built on the "tria prima" of mercury, sulfur, and salt. On the other hand, the significance of mercury in alchemical theory and practice is by no means a simple matter; indeed, the alchemists often went to great lengths to assure that the uninitiated would not understand what they were talking about. When the word "mercury" is used in alchemical writings it may signify the metal quicksilver, but this is by no means always the case. A variety of fanciful terms is employed in referring to "mercury," as, for example, "Mercury is our doorkeeper, our balm, our honey, oil, urine, May-dew, mother, egg, secret furnace, oven, true fire, venomous Dragon, Theriac, ardent mine, Green Lion, Bird of Hermes, two edged sword that guards Tree of Life, etc., etc." (Read 1936). A 16th century author is said to have compiled a list of more than 100 synonyms for quicksilver (Crosland 1962, p. 36). There is general agreement among historians that the name "mercury" was given to quicksilver by alchemists, but it is not clear just when this occurred. Identification of the metal with the planet Mercury has been traced to Stephanos of Alexandria in the 7th century A. D. (Crosland 1962).

Alchemical concepts of mercury were based at least in part on the Aristotelian system which dominated all scientific work and "natural philosophy" throughout medieval times and well into the 17th century. The idealization of inanimate objects, which was a part of this system, is illustrated by the frequent use of the expression "mercury of metals." Inherent properties of a mysterious nature, even including life, growth and reproduction were assumed to exist in material things, and mercury represented these properties in metals. Gold was looked upon as the perfect metal and the final stage in a natural process of growth and maturation. A major objective of alchemical practice was to assist nature by speeding up the progress of base metals toward their ultimate perfection. Success in achieving this was known as the "magnum opus" and this was closely related to finding the philosopher's stone and the elixir of life.

Dividing lines between alchemy, chemistry and medicine have not always been entirely clear; mercury has often acted as a common denominator. The alchemy of Taoism in China and that of Tantrism in India, for example, were built on chemical reactions of mercury leading to a medical panacea. Early Islamic alchemical writings, particularly those attributed to Geber (vide infra), showed close alchemical-chemical-medical interrelationships, usually involving quicksilver. Paracelsus (1493–1541), often looked upon primarily as an alchemist, made important contributions to medicine, especially in the therapeutic use of mercurial compounds.

China

According to legend, alchemy originated in China about 4500 B. C., but this notion is almost certainly more romantic than factual (Hutin 1962 translation). The earliest identifiable Chinese references to the alchemical use of quicksilver and cinnabar are found in records dating from the 2nd century B. C. (Dubs 1947). The initial interest is said to have been in a search for an elixir of life that would confer longevity or immortality, with goldmaking coming later (Li Ch'iao-p'ing 1948).

The Taoism of the 3rd century A. D. and later was strongly influenced by alchemy and particularly by mercury (Spooner and Wang 1947). The Taoist religion was founded by Lao-tse about 600 B. C., a fact that has led some historians to state that the beginnings of Chinese alchemy go back to that time. There is little connection between the original and the later Taoism. Lao-tse taught that wisdom is trust, goodness is acceptance and contentment is simplicity; the later Taoists sought salvation through magical potions, elixirs, and rituals that would cure all ills and lead to a long and carefree existence. Mercury was their panacea.

In the 2nd century B. C. in China alchemy could be practiced openly only with royal approval. Cinnabar appears prominently as one of the substances that could be converted to gold, and the use of pots and dishes made of this alchemical gold could bring longevity and even immortality. The first mention of the Chinese word for mercury is found in a book written about 122 B. C. This was not an alchemical book, alchemy being outlawed at that time, but the wording has a strong alchemical flavor (Dubs 1947).

Reduced to its simplest terms, Chinese alchemy as developed by Taoists in the 3rd century is built on the relationships between the Yin and the Yang. The Yin, representing the female principle, embodies cold, darkness, and passivity and resides in the earth. The Yang represents the male and stands for heat, light, and activity and resides in the sun. Conflict and union between the Yin and the Yang give rise to the five elements: earth, wood, metal, water and fire. Various combinations of the five elements give rise to all animate and inanimate objects. Mercury was identified with the Yin, or passive principle. Some of the alchemical formulas employ obvious sex symbolism in describing interaction between the Yin and the Yang (Needham and Wang Ling 1954–62, Vol. 2, Sect. 13).

Most famous of the early Chinese alchemists was Ko Hung who lived in the 4th century A. D. His doctrine was based on the idea that gold was an ideal material and since a man becomes what he eats, the road to perfection was through eating gold. But a true believer was likely to be a poor man and consequently unable to provide himself with gold. It therefore became necessary to find a substitute for the precious metal, and this could be done by making potable gold from cinnabar and several other raw materials. The objective was not based on greed, since the synthetic gold was not to be used in the market place; rather it was religious and sprang from the highest motives.

The Nei P'ien of Ko Hung describes nine kinds of cinnabar about which he says: "By acquiring any one of these nine you will become a genie; there is no need to prepare all nine. Which one is prepared depends entirely on one's preference. If you wish to mount to heaven after taking any one of them, you will go. If you wish to remain a while among men, you will be able to leave and enter anywhere, no matter what the barriers. Nothing will be able to harm you." Detailed directions are given for the preparation of only the first cinnabar, which is known as Flowers of Cinnabar, but some of the powers of the other eight, in addition to enabling the user to mount to heaven, are set forth. The second cinnabar, for example, if smeared on the soles of the feet enables one to walk on water. Several types will cause the fairies to become one's servants, and one variety, if taken with raspberry juice will enable a man of ninety to beget children. Of type number eight Ko says: "When you hold a piece . . . in your hand, every sort of ghost will flee. If it is used for writing over a doorway, no evil powers will dare approach it; thieves, bandits, tigers and wolves will be put to flight." This is surely one of the earliest uses of mercury to ward off evil spirits.

India

Mercury plays a prominent role in the alchemical writings of India as well as of those of China. Interrelationships between mercury, magic, medicine and superstition are part of the picture in both countries. The traditional mysticism of India creates a fertile soil in which all sorts of strange beliefs about mercury have flourished.

Most striking of the similarities between China and India is that of Taoism and Tantrism. Just as the Taoism of Lao-tse degenerated into a weird and often obscene cultism based on mercury, so did the great teachings of Gautama

Buddha give way to an abased form of Tantrism. Both cults grew up at about
the same time and held sway until the 13th or 14th centuries A. D. In the
Tantric rites, Siva represented the male or phallic element while the female
principle was represented by Siva's consort Parvati. This pair has obvious parallel with the Yin and the Yang. The potency of mercury is well illustrated in
the following Tantric recipe: "Take a gold leaf three niskas in weight and
quicksilver nine niskas and run them with acids for three hours. Make the
amalgam into a phallus . . . to be worshipped in due form. By the mere sight
of the phallus of mercury, the sins accumulated by the killing of 1,000 brah-
mins and 10,000 cows are redeemed" (Ray 1956, p. 186).

The beginnings of Indian alchemy have been dated as early as the 2nd
century A. D. and as late as the 8th century. Partington has noted that the
Hindu word for alchemy is Rasasiddhi, which means "knowledge of mercury."

Alchemy in general in its proper sense probably was never intended to be
anything but symbolic, portraying man's striving for perfection and purification. According to this interpretation, those who took the alchemical doctrines literally and perverted them to the transmutation of base metals into
gold were charlatans.

Islam

Living evidence of Arabic interest in alchemy and of the close connection
between the hermetic art and chemistry is found in the vocabulary of modern chemistry. The word alchemy itself is of Arabic derivation, as are alkali,
alcohol, elixir, and a number of other chemical terms. There is little doubt
that interest in alchemy became manifest in the Moslem world very soon after
the founding of the new religion, possibly as early as the middle of the 7th
century (Hutin 1962 translation). Best known of the Arab alchemists who
contributed significantly to knowledge about mercury were Geber (c. 720-
810) (also known as Jabir ibn Hayyan [Multhauf 1966; Holmyard 1968;
Darmstaedter 1922]), Rhazes (864-925), and Avicenna (980-1037). The latter two are better known as physicians than as alchemists—further evidence of
the close relationship of the two disciplines in the Middle Ages.

Geber is something of a controversial character, particularly in relation to
the writings which have been attributed to him (Multhauf 1966); even his
existence has been questioned. Holmyard has given an account of the life and
works of Geber which he concludes by saying ". . . it would seem entirely unjustifiable to deny the existence of a historical alchemist named Jabir ibn
Hayyan who flourished in the eighth century. It would be equally unjustifiable
to minimize the great accretions made to [his] original works . . . ." Among
the contributions to chemistry and medicine which have been ascribed to
Geber are his two methods for preparing red oxide of mercury, a compound
used by Paracelsus in the treatment of syphilis and by Priestley, Lavoisier,
and Scheele in the discovery of oxygen. (The work of Rhazes and Avicenna is
discussed in the section on Islamic medicine, Chapter 14.)

The West

The origins of Western alchemy are scarcely less obscure than those of the East, although no claim to great antiquity has been put forth other than possible linkage to Chaldean astrology. Plato's philosophy of idealism and Aristotle's concepts of the growth and development of inanimate objects are generally considered to have played a part in Western alchemical theory.

There is good evidence that alchemy was well known in the 3rd century A.D. and that the belief in the possibility of transmutation of base metals into gold was accepted (Gibbon n.d.). The emperor Diocletian (245-313) issued an edict about 290 calling for the destruction of all works dealing with alchemy and related subjects. Supposedly he was motivated by fear that artificially created gold would debase the value of the Roman currency or that the makers of precious metals might be able to amass large fortunes with which they could subvert the loyalty of public officials and even foment revolts against the ruling power of Rome.

Despite Diocletian's order at least two important documents dealing with alchemy and other aspects of chemistry somehow survived. These are the Leiden Papyrus X and the Papyrus Graecus Holmiensis, now generally thought to be the oldest surviving original works dealing with chemical and alchemical subjects. There is considerable similarity between the two papyri and they both contain frequent mention of quicksilver, cinnabar and minium. They form a link between the chemical writings of Greece and Rome on the one hand, and those of the medieval period on the other, but they contain little of medical significance, nor do they contribute anything of importance to the advancement of chemistry.

Identification of the metal quicksilver with the planet Mercury, and the subsequent use of the term mercury to designate the metal, is a result of alchemical thought. The concept of seven "planets" had become established as early as the 8th or 9th centuries B. C. and may have been based on Chaldean influences (Crosland 1962). Just when the names of these seven heavenly bodies were first linked with metals is not definitely known, but Origen in the 2nd century A. D. set up a scheme as shown in the first two columns of Table II-1.

Of particular interest is the early identification of the planet Mercury with iron and then with tin, the present association having its origin around the 7th century. Thus, alchemical writings earlier than the 7th century might have used the name or the symbol of the planet Mercury in referring to iron or tin. The term "hermetic art" obviously is derived from the association between the name of the planet Mercury, the metal mercury and the god Hermes. It also emphasizes the dominant role of "mercury" in all of alchemy, the terms hermetic art and alchemy being synonymous.

During the millennium of the 5th to 15th centuries, the practice of alchemy spread throughout the civilized world. Mercury, either as quicksilver or as the metallic "essence," invariably appeared as a central figure. Alchemical writ-

TABLE II-1

Association of Metals with Heavenly Bodies

Origen 2nd Century A.D.		Olympidorus 6th Century A.D.		Stephanos of Alexandria 7th Century A.D.	
Metal	Planet	Metal	Planet	Metal	Planet
Lead	Saturn	Lead	Saturn	Lead	Saturn
Tin	Venus	Tin	Mercury	Tin	Jupiter
Bronze	Jupiter	Electrum	Jupiter	Mercury	Mercury
Iron	Mercury	Iron	Mars	Iron	Mars
Copper	Mars	Copper	Venus	Copper	Venus
Silver	Moon	Silver	Moon	Silver	Moon
Gold	Sun	Gold	Sun	Gold	Sun

FROM: Crosland 1962.

NOTE: Crosland compiled this table from Origen, *Contra Celsum*, 6, 22; Migne, *Patriologia, Series Graecae*, vol. 11, Paris 1857, p. 1323; and Berthelot, *Introduction à la chimie des anciens*, Paris, 1889, pp. 81, 84.

ings, both scholarly and popular, have been produced in huge quantities, and are still being published (Holmyard 1968; Burland 1968). Berthelot's works stand out as classics and Stillman has compiled an excellent bibliography.

Folklore

Quicksilver has figured prominently in folklore, with applications for both good and evil. One of the most widespread uses of quicksilver has been to afford protection against the Evil Eye, "spells" and related misfortunes (Seligmann 1910 and 1927; Olbrich 1927-1942). A typical example of measures to be taken to avoid being bewitched, and said to have been in practice since the middle ages, is to hollow out the kernel of a hazelnut and replace the contents with quicksilver. This amulet is to be hung around the neck, placed under the bedpillow or under the threshold of the door. While the procedure is generally regarded as preventive, it has also been recommended as a cure. Hazelnuts, incidentally, are frequently used as adjuvants to quicksilver.

Another favorite receptacle for quicksilver in amulets is a quill feather. This combination has been used in Esthonia to protect infants against the Evil Eye until permanent "immunity" can be given by baptism. Similar practices have been noted in Sardinia, Spain, Syria, Algiers, Tunis, Persia, Russia, and elsewhere.

Not only are quicksilver amulets supposedly effective as general preventives, but they also can be used to ward off a number of specific ailments. Seligmann (1927) claimed that he had in his possession a small glass tube filled with quicksilver placed in a leather sheath which was carried in London in 1912 as a preventive against rheumatism. In Spain, mercury in a nut shell car-

ried in the coat lining is often believed to prevent erysipelas and in Turkey it
is thought to ward off sore throat if hung around the neck. Other diseases
said to be prevented by mercury amulets are plague, dysentery, colic, and
infestation by lice and other vermin.

Therapeutic uses of quicksilver in the world of folklore cover the entire
range of diseases, but the form of the remedy and the mode of application or
administration vary. Quicksilver also appears frequently in the folklore of veterinary medicine, probably reflecting the importance of livestock in early
times. Seligmann mentions an 18th century Jewish recipe book which gives
instructions for curing poultry of the Evil Eye by combining mercury, salt and
white stones and tying the mixture on the afflicted hen or rooster. People of
the Black Mountains protect their cattle against the "hex" by hanging an amulet containing quicksilver around the neck of the animal. Horses can be protected from evil spirits if in every stall a hole is bored and then filled with
mercury. In North Carolina, cattle can be protected against lice by putting a
string impregnated with mercurial ointment and lard around the neck (Duke
University Library 1952-1964).

Some of the uses of mercurial fungicides and weed killers in present day
agriculture can be said to have had their antecedents in folklore. In northern
Europe, specifically in Prussia, Esthonia, and Finland, quicksilver was used in
several ways to produce better crops (Seligmann 1927). If mercury was
rubbed on the point of a plow or if some of the metal was put into a quill
which is then fastened to the plow or harrow, the field being worked would
be rid of thistles. To prevent the seed from rotting in the ground and from
being eaten by insects the farmer smeared his plow with a mixture of mercury and fat and placed mercury, either free or in a quill, in the sack which
holds the seed to be sowed. Whether or not quicksilver in these uses actually
had herbicidal or fungicidal activity, or helped by providing a deficient trace
metal, is difficult to say.

Herdsmen, hunters, trappers and fishermen have all found uses for mercury, usually as a means of placating or propitiating the appropriate spirits.
The use of mercurialized bread to locate a drowned body may go back as far
as the 13th century and is found in a number of countries, particularly in
Britain (Thompson 1955-1958; Kittredge 1929; N. Rosenberg [personal
communication]). According to J. N. Friend the method was used successfully in locating the body of a 15-year old girl who had been drowned in the
Coventry Canal at Bedworth near Nuneaton in October 1932. A similar attempt on the Avon near Amesbury in May 1925 had failed, possibly, as suggested by one observer, because the bread was of inferior quality, a theory
which is supported by the opinion of Huckleberry Finn.

After Huck had led the townspeople to believe that he had been murdered
and his body thrown into the river, he describes their efforts to recover the
body: "Well, then I happened to think how they always put quicksilver in

loaves of bread and float them off, because they always go right to the
drownded carcass and stop there." Later he recovered one of the loaves and
". . . took out the plug and shook out the little dab of quicksilver, and set my
teeth in. It was 'baker's bread'—what the quality eat; none of your low-down
corn-pone" (Clemens).

An interesting bit of Indian folklore and its application in an actual situation has been recorded by Taylor (1875, pp. 351–352). "Dr. Brown, of
Lahore, states that metallic mercury (para) is often used by the natives of
India in order to injure, aggrieve, or annoy those who have given them of-
fence. They think that when mercury gets into the body, it can only come
out again through the skin, producing sores and leprous spots. In one case
cited by him the question was raised: Is mercury a poison? The sub-assistant
surgeon, to whom the case was referred, stated that in his opinion the metallic mercury was not a poison. This stopped the trial. Dr. Brown examined the
facts, and very properly dissented from this view. The person to whom the
mercury had been given had redness and swelling of the gums, and they bled
on pressure with the finger. The woman vomited twice after taking the mercury, and some globules of the metal were found in the vomited matter. A
conviction was ultimately obtained by altering the indictment and describing
mercury as an 'unwholesome drug' (Medic-Legal Report of Bengal Presidency,
1869, p. 152)."

Mercury appears unexpectedly in the folklore of modern biochemistry
which includes the story of a research worker who had been attempting to
prepare the enzyme enolase in a purified form and had met with no success.
Late one afternoon he thought he was on the verge of solving the problem
and in his excitement and impatience picked up a mercury thermometer
which was handy to use as a stirring rod. Alas, the thermometer broke, liberating its mercury into the beaker containing the test material. The biochemist
was so discouraged by this accident that he left the ruined mixture on his laboratory bench and went home. When he returned the next day he sadly
looked at the beaker but noticed that some crystals had formed. These turned
out to be pure enolase; his problem had been solved by the accidental addition of quicksilver.

Another version of the story of the broken thermometer relates it to the
accidental discovery in 1895 of an efficient means of producing phthalic an-
hydride, an important intermediate in the synthesis of indigo.

> Among the chemists seeking some catalytic agent to stimulate a more com-
> plete conversion of naphthalene to phthalic anhydride was a careless chap
> who went off to lunch one day leaving a thermometer in his apparatus.
> Somehow that thermometer was broken. How this accident happened no-
> body knew. Soon nobody cared, though such carelessness is not easily ex-
> plained away in a well-managed chemical laboratory. The mercury from
> the thermometer bulb turned out to be the long-sought catalyst. (Haynes
> 1942)

Rosicrucianism

Rosicrucianism was closely linked with alchemy during the 17th century, thus involving mercury. One of the prominent members of the order, Michael Maier (1568–1622) expressed himself in typically alchemical fashion. ". . . Mercury is found to be the miraculous splendour and light of the world. It receives therefore the Royal Crown as monarch of visible things under the command of man" (Waite 1961). Maier also speaks of mercury as an antidote against the plague and notes that when it is mixed with certain salts it becomes poisonous. Once again the alchemy-medicine-chemistry relationships appear.

Kabbala

According to Kabbalism, an occult doctrine which seems to have arisen in the 18th century, and which was related to Hebrew interpretations of the Old Testament, Ophiel, the sixth angel, who could transmute quicksilver into a white stone, dealt with everything which related to mercury.

Each of the planetary bodies, according to the Kabbalists, had its own KAME'A; that for Mercury is shown in Figure II-1. The explanation of this is given by Budge as follows:

8	58	59	5	4	62	63	1
49	15	14	52	53	11	10	56
41	23	22	44	45	19	18	48
32	34	35	29	28	38	39	25
40	26	27	37	36	30	31	33
17	47	46	20	21	43	42	24
9	55	54	12	13	51	50	16
64	2	3	61	60	6	7	57

ח	נח	נט	ה	ד	סב	סג	א
מט	יה	יד	נב	נג	יא	י	נו
מא	כג	כב	מר	מה	יט	יה	מה
לב	לו	לה	כח	כט	לח	לט	כה
מ	כו	כז	לז	לו	ל	לא	לג
יז	מז	מו	כ	כא	מג	מב	כד
ט	נה	נד	יב	יג	יב	נא	יו
סד	ב	ג	אס	ס	ו	ז	נז

Figure II-1 The Kamea of Mercury according to the Kabbalists. (From *Amulets and Superstitions*, by E. A. W. Budge. London: Oxford University Press. 1930.)

The total addition of these sixty-four figures in each of three directions is
260, i.e., the total of the numerical values of the consonants in kokab
kesef hayyim, 'star of living silver' (i.e. quicksilver). The number of
TIRIEL, the spirit of the planet, is 2,080, which is the number of the de-
mon of the planet, TAPHTHARTHARATH. As an amulet the square
should be written upon a sheet of tin or quicksilver (Budge 1930).

There is no explanation of how to overcome the technical difficulty of making and writing upon a sheet of quicksilver, but presumably the meaning of
this is symbolic and should not be taken literally.

Plant Lore

Plant lore has been found to have relationships with both folklore and alchemy, as shown in an entry in Folkard's *Plant lore, legends and lyrics* (p.
444).

MOONWORT. The Fern *Botrychium Lunaria* has derived its name from
the crescent shape of the segments of its frond. Perhaps it is this lunar
form that caused it to be so highly esteemed for its magical properties.
The old alchemists professed to be able, by means of Moonwort, which
they called Lunaria minor, or lesser Lunary, to extract sterling silver from
Mercury.

Folk medicine, still widely practiced in many parts of the world, has had
some of its roots in folklore. Drugs with known therapeutic potency, of
which digitalis is an outstanding example, have come into modern medicine
from folklore. Others, if they had any effectiveness at all, could have acted
only as a result of faith or suggestion. Quicksilver stands in a middle position:
certainly it is not without therapeutic potency, but some of its uses in folklore and folk medicine can hardly be explained on the basis of specific pharmacological activity.

Astrology and Chiromancy

Quicksilver does not occupy a prominent place in the arts of astrology or
chiromancy [palmistry] but it does make an occasional appearance as in the
writings of Federicus Chrysogonus (fl. 1528) who included mercury in his
astrological system (Thorndike 1923).

REFERENCES

Berthelot, M. 1889. *Introduction à l'étude de la chimie des anciens et du moyen-âge.*
    Paris: G. Steinheil.
Berthelot, M. 1893. *La chimie au moyen âge.* 3 vols. Paris.
Berthelot, M. 1895. *Les origines de l'alchemie.* Paris.
Budge, E. A. W. 1930. *Amulets and superstitions.* London: Oxford University Press.

Burland, C. A. 1968. *The arts of the alchemists.* New York: Macmillan.

Clemens, S. L. [Mark Twain]. *The adventures of Huckleberry Finn.*

Crosland, M. P. 1962. *Historical studies in the language of chemistry.* London: Heinemann Educational Books.

Darmstaedter, E. 1922. *Die Alchemie des Geber.* Berlin: J. Springer.

Dubs, H. H. 1947. The beginnings of alchemy. *Isis* 38:62–86.

Duke University Library. 1952–1964. *The Frank C. Brown Collection of North Carolina Folklore.* 7 vols. Durham, N. C.: Duke University Press.

Folkard, R. 1884. *Plant lore, legend and lyrics.* London.

Friend, J. N. 1961. *Man and the chemical elements.* 2d ed. London: Chas. Griffin and Co.

Gibbon, E. n.d. *The decline and fall of the Roman Empire.* 2 vols. New York: Modern Library.

Haynes, W. 1942. *This chemical age.* 2d ed. New York: A. A. Knopf.

Holmyard, E. J. 1968. *Alchemy.* Harmondsworth, Middlesex: Penguin Books.

Hutin, S. 1962. *A history of alchemy.* Trans. of the 1951 French ed. by Tamara Alferoff. New York: Walker and Co.

Kittredge, G. L. 1929. *Witchcraft in old and New England.* Cambridge: Harvard University Press.

Ko, Hung. 1966. *Alchemy, medicine, religion in the China of A. D. 320: The Nei p'ien of Ko Hung (Pao-p'u tzu).* Trans. J. R. Ware. Cambridge: M.I.T. Press.

Li Ch'iao-p'ing. 1948. *The chemical arts of old China.* Easton, Pennsylvania: The Journal of Chemical Education.

Multhauf, R. P. 1966. *The origins of chemistry.* London: Oldbourne Book Co.

Needham, J., and Wang Ling. 1954–1962. *Science and civilization in China.* 4 vols. Cambridge: Cambridge University Press.

Olbrich, K. 1927–1942. In *Handwörterbuch des Deutschen Aberglaubens*, ed. Eduard von Hoffman-Krayer and Hanns Bächtold-Stäubli, 10 vols. Berlin: W. de Gruyter and Co.

Partington, J. R. 1965. *A short history of chemistry.* 3d ed. New York: Harper and Row, Harper Torchbooks.

Ray, P. 1956. *History of chemistry in ancient and medieval India*, incorporating *The history of Hindu chemistry* by A. P. C. Ray. Calcutta: Indian Chemical Society.

Read, J. 1936. *Prelude to chemistry.* London: G. Bell and Sons.

Seligmann, S. 1910. *Der böse Blick und Verwandtes.* 2 vols. Berlin: Hermann Barsdorf.

Seligmann, S. 1927. *Die magischen Heil- und Schutzmittel aus der unbelebten Natur.* Stuttgart: Strecker und Schröder.

Spooner, R. C., and Wang, C. H. 1947. The Divine Nine Turn Tan Sha Method: A Chinese alchemical recipe. *Isis* 38 (Pt. 2):235–242.

Stillman, J. M. 1960. *The story of alchemy and early chemistry.* New York: Dover Publications.

Taylor, A. S. 1875. *On poisons in relation to medical jurisprudence and medicine.* 3d American ed. Philadelphia: Henry C. Lea.

Thompson, S. 1955–1958. *Motif index of folk literature.* 6 vols. Bloomington: Indiana University Press.

Thorndike, L. 1923. *History of magic and experimental science.* New York: Macmillan Co.

Waite, A. E. 1961. *The Brotherhood of the Rosy Cross.* New Hyde Park, N. Y.: University Books.

ADDITIONAL READINGS*

Caron, M., and Hutin, S. 1961. *The alchemists.* New York: Grove Press.

Sherr, R.; Bainbridge, K. T.; and Anderson, H. H. 1941. Transmutation of mercury by fast neutrons. *Phys. Rev.* 60:473–79.

*Items listed under this heading here and in subsequent chapters were consulted by the author, but not cited in the text.

3

Mining and Extraction of Mercury

Early European Literature

Although there are more than 30 known ores which contain mercury in more than trace amounts, those that contain cinnabar (mercuric sulfide— HgS) are the only ones that have been subjected to large-scale exploitation. Livingstonite ($HgSb_4S_7$) has been mined as a source of mercury in the Huitzuco district of Mexico. Cinnabar deposits of commercial value have been found on all of the continents except Antarctica. Many of the quicksilver mines mentioned in the literature are no longer being worked; the location and even the existence of some are in doubt. Countries or regions in which mercury mines are known to be, or to have been, located are listed in Table III–1.

Early listings of cinnabar mines can be found in the writings of Theo-phrastus of Eresus (387–372 B. C.) (1956 and 1965 translations), Pliny the Elder (A. D. 23–79) (Plinius Secundus 1938–1963 translation), Dioscorides (fl. 1st cent. A. D.), Strabo (c. B. C. 63–c. A. D. 21), and Vitruvius (fl. 1st cent. A. D.). Needless to say, the geographic references of these Greek and Latin writers have not always been in consonance with later knowledge.

Nothing of major importance was added to the store of information on the location of quicksilver mines for at least 1500 years after the death of Pliny. Even Agricola (1494–1555) in the 16th century did little more than mention the few European quicksilver mines with which he had some personal familiarity (1955 translation). He states that "recently the metal quicksilver was found in Scottish Britain" (1558, p. 405). This statement is reinforced by Neumann who quotes Agricola and gives a reference to ". . . eine Quecksilbergrube zu Apselby (sic) in England (Westmoreland)," which had been described by Hawkins. In support of this, Neumann incorrectly cites "Born-Tetra, Bergbaukunde 1790. I, 200." Presumably the reference which

TABLE III-1

Location of Mercury Mines—Past and Present

Algeria	Ecuador	New Zealand	
Argentine	England (?)	Norway	Tunisia
Australia			Turkey
Austria	France	Peru	
		Phillipines	U.S.A.[a]
Bohemia	Germany	Poland	Alaska
Borneo		Portugal	Arizona
Bosnia	Hungary		Arkansas
Brazil		Rumania	California
	Indonesia	Russia	Idaho
Canada			Nevada
Chile	Italy	Sardinia	Oregon
China		Serbia	Texas
Colombia	Japan	Siberia	Utah
Corsica	Kurdistan	Slovenia	Washington
Croatia	Mexico	Spain	
		Sweden	
Dalmatia		Syra (Cyclades)	
		Syria	

Adopted from Mellor (1922–1937) and Neumann (1904), with additions.

[a]Deposits of minor significance have been found in Colorado, Montana, New Mexico, South Dakota and Wyoming.

Neumann had in mind was I. M. Born and F. W. H. von Trebra, Bergbaukunde, Vol. II, Leipzig: 1790, pp. 444–445. Here can be found a letter from John Hawkins dated London Junii 28, 1789, which contains a list of newly discovered sources of minerals, including "gediegen Quecksilber von Apselby in Westmoreland . . ." (gediegen means native or virgin). There is a town called Appleby in Westmoreland and this is probably the location to which reference is made. However, an exhaustive search in 1968–1969 for evidence of mercury mining in Westmoreland turned up no record of such a mine in local or national records. A leading authority on British mining history, Mr. Robert Annan, confirmed this lack of any existing record that would substantiate the present or past existence of a quicksilver mine in Britain.*

As an example of what was known about quicksilver mines in the 17th century, Nicholas Lemery (1645–1715) may be cited. In his famous *Cours de Chymie* he says "It [mercury] is to be found in many places in Europe, as Poland, Hungary, and even in France. . . ." (Lemery 1686 translation). Another French writer, Jacques Savary des Bruslons (1657–1716), about seventy-five years later, gives a similar list and adds Carniola (part of modern Yugoslavia) and Peru. There is also a long paragraph in Diderot's *Encyclopédie* on quicksilver at Montpellier, including a comment that unfortunately the deposit is located beneath the site on which the city is built and therefore cannot be exploited. John Locke, during his stay in Montpellier around 1675, was told that ". . . in digging cellars etc. they often find here in Montpellier great

*R. Annan 1969: personal communication.

quantitys of running quicksilver . . . amongst the earth" (Dewhurst 1963, p. 65).

An idea of knowledge in 18th century Britain about sources of mercury can be gotten from the *Universal Dictionary of Trade and Commerce* of Malachy Postlethwayt (1707?-1767). The author, who says that his dictionary incorporates "Every thing essential that is contained in Savary's Dictionary," writes that "The places where native cinnabar is produced in plenty are chiefly in Spain and Hungary, and some parts of the East Indies." The omission of Carniola and Peru suggests that Postlethwayt had not studied Savary very carefully and apparently he concluded that the cinnabar or quicksilver shipped by the East India Company came from the East Indies rather than from China.

A new dimension to the record of quicksilver mines was added by Ignaz von Born with the publication of his *Index Fossilium* in 1772. The significance of this work lies not so much in its comprehensive list of European mines (including those of quicksilver) as in its attempt to name, describe and classify the various minerals. Agricola's earlier crude attempts at classification (1955 translation) show nothing of the systematic approach developed by von Born. Another unusual feature of the *Index Fossilium* is its description of the form in which minerals exist and also of the gangue or matrix in which they are found in each mine. Von Born names Idria (now Yugoslavia) as the only source of Hydrargyrum Nativum but lists many sources of cinnabar including Idria, Almadén (Spain), Schneeberg, Tyrol, and several locations each in Hungary, Transylvania, Carniola, and the Palatinate. An anomalous feature of the book is the inclusion of gold from Peru, but not silver or quicksilver.

Early in the 19th century the French mineralogist, J. A. H. Lucas published a two volume work similar to that of von Born in its taxonomic orientation, but more comprehensive and detailed than the *Index Fossilium.* To some extent this book is based on an earlier work (1801) by the crystallographer René-Just Haüy (1743-1822) but it contains much additional material and represents an early attempt at a complete listing and classification of minerals with all of their characteristics and sources. Sources of cinnabar mentioned by Lucas include some mines in Japan, which is rather surprising in view of the very limited production of quicksilver in that country until well into the 20th century. Perhaps Lucas may have confused Japan with China.

Many of the quicksilver mines mentioned by von Born and by Lucas are today either no longer in production or are of minor importance.

A comprehensive list of sources of mercury and mercury-bearing ores was published by Zippe in 1857. A unique and valuable contribution in Zippe's work is found in his long lists of the forms in which mercury is combined with other substances in naturally occurring ores and the locations in which these ores are found, thus amplifying von Born's contributions. He makes a special point that quicksilver combines readily with a large number of other elements, including chlorine, iodine, sulfur, selenium, zinc, antimony, iron, lead, copper, silver, and gold.

Considerable variation exists in the amount of information which is available on quicksilver mines in different countries. Some, particularly the mines at Almadén, have played a significant role in world history and consequently have been responsible for a rich literature.

China

Discussing China first among the countries in which quicksilver mines are found does not necessarily imply that these mines are the oldest known. Difficulties in dating ancient events in China are familiar to all historians but the fact that the Chinese quicksilver mines of Kwei-Chan are commonly spoken of as having been known since "time immemorial" gives them a kind of priority (Biondi 1934). Other Chinese mines are listed by Hintze (1904, Vol. 1, pp. 690–691). Cinnabar has occupied a place of special importance in China starting at least as far back as the Shang-Yin Dynasty (1751–1112 B. C.) (Te-Kong Tong 1967) and continuing up to modern times.

Current data on the Chinese mines are not readily available in the United States, but the production of mercury in China between 1959 and 1966 has been estimated at about 25,000 flasks annually (Shelton 1965; Parker 1968). (A flask of mercury is the standard commercial unit, and contains 76 pounds avoirdupois. See note, p. 41.)

Asia Minor

The ancient Hittite town of Iconium (modern Konya or Konia) is perhaps best known as one of the places visited by Paul (Acts 13:15), and where, by an odd coincidence, the inhabitants called him Mercurius (Acts 14:12). The town is less well known as the place where the religious order of Whirling Dervishes was founded in the thirteenth century A. D. To the archaeologist it is of interest because "important evidence of the extreme antiquity of mining in Asia Minor, probably dating back to the Neolithic period of culture, is afforded by recent excavations of an American mining engineer in some old workings in the neighborhood of Iconium . . ." (Gowland 1920). It was in 1905 that the American engineer, F. F. Sharpless, made the discovery, and his account of it reads like a fairy tale:

> . . . a goatherd who kept his flocks on the almost barren hills near Konia, corralled them at night in a limestone cave within a short distance of one of the buried cities of the Holy Land. One night a refractory goat refused to enter the portal, and the goatherd, picking up a stone to discipline the offending animal, noticed that it was heavy and dark red in color, different from the other stones around there. Laying it aside, he one day broke it

open between two large stones and saw that it was a beautiful rose color
on the inside . . . .

The story goes on to tell how the stone passed into the hands of a mining
engineer who recognized it to be cinnabar. Subsequent investigation led to
the discovery of a subterranean chamber which contained the "remains of
more than 50 human skeletons. Many of the bones were embedded in the
secondary deposit of lime on the floor. There were great quantities of stone
hammers, several pottery lamps, a fair amount of charcoal, several rubbing
stones and some flint arrowheads" (Sharpless 1908; Turner 1908).

Indirect evidence has been adduced which suggests that the ancient cinnabar mines at Konia were in operation about 1500 B. C. If this estimate is
valid, it would establish this mine as one of the oldest for which there has
been any possibility of fixing a date. Of possible relevance to this early mercury mine was the finding in a neighboring buried city of a tablet dedicated
to Zizima, the Phrygian goddess of the mines. The Phrygians are known to
have inhabited this part of Asia Minor around 1500 B. C. and this may give a
clue to the age of the mines (Ramsay 1920; Robinson 1927).

Following the rediscovery of these mines in 1905, the Konia Mercury
Syndicate, Ltd. was organized to exploit the cinnabar deposits (Sharpless
1908) and operations are presently (1968) being conducted by a group
known as Itibank of Ankara, Turkey.* Production figures for 1963-1967
have averaged about 3,000 flasks of quicksilver annually (Parker 1968).

Minor deposits of mercury and cinnabar have been found in Persian
Kurdistan (Mactear 1894-1895) and in the regions of Smyrna and Ba'albek
(Schmeisser 1906).

Greece

There is some question as to whether or not the reddish mineral found at
Laurium, near Athens, by Callias about 455 B. C., was cinnabar; both Theophrastus (1956 translation) and Pliny (Plinius Secundus 1938-1963 translation) thought it was. Davies (1935) and Caley and Richards (see Theophrastus
1956 translation) have questioned this, the former suggesting haematite as
most probable. The principal product of the Laurium mines was silver, but
the frequent close association between cinnabar and argentiferous ores suggests at least a possibility that cinnabar could have been present.

The Greek Islands

Among the ancient mines described by Davies in his book on Roman
mines (1935) are those on the Island of Syra in the Cyclades in the Aegean

---

*C. A. Wendel, Minerals and Mining Attaché, U.S.A. Embassy, Ankara, Turkey. 1968:
personal communication.

Sea. He mentions prehistoric copper slag from Chalandriani (on Syra) and in the same area "an old working . . . following a thin vein of cinnabar" the date of which is uncertain. Thus there is no way of knowing whether or not cinnabar as well as copper was mined at that site in prehistoric times. That cinnabar may have been mined at a very early time is suggested by Rhousopoulos who states that several metals, including copper and mercury, were identified in shards from Syra dating from the pre-Mycenean bronze age.

Serbia

Two other early cinnabar mines are mentioned by Davies, both in Serbia. One of these is in the Kosmaj region and it was possibly worked during Roman times; the other is at Avala, about 24 kilometers south of Belgrade. It was rediscovered in 1882 during the construction of the first railway in Serbia at a site which is a mere stone's throw from the main highway between Belgrade and Niš. The finding of unglazed pottery and stone implements in the old workings has been offered as evidence that they date from the stone age (Hofmann 1886). The Avala mines have attracted attention from time to time in the 20th century (Parker 1968; Jadran in 1958; and L. Barić, former Director of Minerals Survey of Yugoslavia, personal communication, 1969).

Italy—Monte Amiata

"That cinnabar was one of the coloring materials used by prehistoric man is a known fact," wrote Mochi in 1916 in an article on prehistoric cinnabar mines in the region of Monte Amiata. The finding of two human skulls colored with cinnabar in neolithic graves called for an investigation of the source of the cinnabar and this led to Monte Amiata. Stone age implements and tools unearthed in the mines provide evidence of the ancient date of the workings. Italian neolithic skulls ". . . smeared with red ochre . . ." have been described by MacKendrick (1966) but there is no evidence that these, or those mentioned by Mochi, were subjected to chemical analysis.

Continuity in the history of mercury mines at Monte Amiata is difficult to trace. Presumably these mines were known in prehistoric times and subsequently were worked by the Etruscans (Davies 1935; Mori 1929; Bianchi Bandinelli 1925), but this leaves a gap of perhaps as much as a millennium without any records. Following the battle of Sentinum in 295 B. C. the Romans occupied all of the northern part of the Italian peninsula, including the former lands of the Etruscans where Monte Amiata is located. It is hardly conceivable that the Romans should not have known about the mercury mines, yet there are no records to show that they worked them. Pliny says nothing about mercury mines in Italy but tells about those in many other lands. The same is true of Agricola, who even includes "Britannia Scotis" among the countries where quicksilver is mined.

During late republican and early imperial times the principal source of cinnabar for the Roman market was Spain (Sisapo-Almadén). Why the Romans should have gone so far afield for this commodity is something of a mystery. A partial explanation is found in a statement by Davies (1935) to the effect that "The Etruscan mines probably closed down owing to the senate's decree forbidding mining in Italy, which must have applied almost exclusively to this region. Its date may not have been earlier than Sulla." Nef (1952) suggests late in the second century B. C. as the time of this decree. Davies' idea, based on a statement by Pliny that the prohibition against mining must have applied almost exclusively to this region, is supported by evidence of mining for gold, silver, tin, copper and iron elsewhere in Italy during the period in question. Why cinnabar was selected for special restrictions is puzzling. Agricola explains the Italian prohibition against mining on the basis of the devastation which results: injury to fields, vineyards and olive groves, pollution of streams with killing of fish, and, as a result of deforestation, the loss of birds and beasts ". . . very many of which furnish a pleasant and agreeable food for man" (Agricola 1950 translation). Agricola's comments have a distinct late-20th century ring. Perhaps the Romans were more concerned about environmental degradation in Italy than in Spain, but it is still not clear why this concern applied only to mines of quicksilver.

Some workings in the Monte Amiata area were not opened up until the beginning of the 20th century, and new deposits are still being discovered. Due principally to the output of the mines in the region of Monte Amiata, Italy vies with Spain in being the world's leading producer of mercury (Shelton 1965; Parker 1968). (See also Table III-2.)

Spain—Sisapo (Almadén)

When the Romans took Spain from the Carthaginians at the end of the Second Punic War (218-201 B. C.) among the prizes which fell into their hands were the mercury mines at Sisapo. These mines were then, and have remained to the present day, the richest known single source of cinnabar and quicksilver. The total production from the earliest times up to 1965 has been estimated to be in excess of 7 million flasks of mercury plus an unknown quantity of cinnabar that was not reduced to mercury (U.S. Bureau of Mines staff 1965).

Theophrastus (1956 translation) speaks of Iberian cinnabar, thus providing definite evidence that mines were being operated in Spain, probably only at Sisapo, at least as early as the 4th century B. C. Mellor (1922-1937, Vol. 6, p. 687) appears to stand alone in his statement that "The mines at Almadén, Greneda (sic) and Oviedo were described by Theophrastus . . . ." There is some evidence that the Sisapo mines were known and were in production several centuries before the conquest and occupation of Spain by the Romans (Strabo 1917-1933 translation).

TABLE III-2

World Production of Mercury
(flasks)

Country	1963	1964	1965	1966	1967
Bolivia (exports)	105	32[a]	52	4	100
Canada	–	73	20	–	–
Chile	613	267	428	96	184
China (mainland)	26,000	26,000	26,000	26,000	20,000
Colombia	3	3	46	84	100
Czechoslovakia	725	775	825	875	900
Italy	54,448	57,001	57,320	53,549	48,066
Japan	4,668	4,812	4,536	4,836	4,612
Mexico	17,202	12,561	19,203	22,074	23,874
Peru	3,092	3,275	3,117	3,166	2,980
Phillippines	2,651	2,496	2,384	2,443	2,612
Rumania	194	194	191	190	190
Spain	56,954	78,322	74,661	70,054	50,000
Tunisia	–	87	174	254	250
Turkey	3,042	2,615	2,755	3,420	3,500
United States	19,117	14,142	19,582	22,008	23,784
U.S.S.R.	35,000	35,000	40,000	40,000	45,000
Yugoslavia	15,838	17,318	16,419	15,896	15,890
Total[b]	239,652	254,973	267,713	264,959	242,042

SOURCE: J. G. Parker. Mercury. Preprint from 1967 Bureau of Mines' *Minerals Yearbook*. U.S. Department of the Interior. Washington: U.S. Government Printing Office. 1968.

NOTE: Compiled mostly from data available April, 1968. All 1967 figures are preliminary only. All figures for mainland China, Czechoslovakia, and U.S.S.R., as well as the 1967 figures for Columbia, Spain, Tunisia, and Turkey are estimates.

[a]Purchases by Banco Minero.
[b]Total of listed figures only; no undisclosed data included.

Pliny, in his *Historia Naturalis*, mentions a number of sources of minium, which is our cinnabar or mercuric sulfide. His list includes Athens, Colchis, Ephesus, Carmania, Ethiopia, and Hispania, particularly the region of Sisapo in Baetica. Pliny notes that Sisapo is practically the only place from which minium is exported to Rome and that this commodity is an important source of revenue to the state (Plinius Secundus 1938–1963 translation).

After Pliny, the next known written record of the cinnabar mines in Spain is that of the Arab geographer Abu-Abd-Alla-Mohamed-Al-Edrisi (commonly referred to simply as Edrisi) in the first half of the twelfth century. By this time the mines had taken on the Arab name Almadén which means "the mine." In his description of the region, Edrisi states that when he visited Almadén the mines employed some 1,000 laborers and had penetrated the earth to a depth of about 500 feet (Muḥammad ibn Muḥammad 1836–1840 translation).

During the reconquest of Spain from the Arabs a group of Cistercian monks played a particularly important role in keeping the Berber Almohades out of an area south of Toledo. These monks in 1158 formed the military-religious Order of Calatrava. As a reward for their services the monks of

Calatrava received in 1168 from King Alfonso VIII (the Noble, 1155-1214) a half interest in the Almadén mines, but the Christians suffered reverses and it was not until 1249 that the Order effectively obtained their half interest in the quicksilver mines, when Ferdinand III (1199-1252) renewed the agreement with the monks.

In 1282 Ferdinand's son Alfonso X (the Wise, 1221-1284) gave them complete proprietorship, which they retained until 1348 when Alfonso XI (1311-1350) took over control of the mines, which were then leased to private interests until the early 16th century, when the Crown again took over their management. An accurate comparison of the relative efficiency of private vs. governmental operation is not feasible, but the mines appear to have deteriorated rapidly under royal auspices, and in 1511, after less than ten years, the leasing-contract system was re-introduced.

Always important as a source of revenue for the Spanish kings, the mines of Almadén became vital during the period of imperial expansion. That the international banking firm of Fugger should have entered the picture in 1524 (See Chapter 4, below) will not surprise any student of Spanish history. By that time the crown had assumed control of the leading military-religious orders and had taken over complete ownership of the Almadén mines.

A serious fire broke out in the mines on November 18, 1550, causing such extensive damage that operations were almost entirely suspended for two years and it was not until 1555 that production returned to normal. In 1563 the Fugger negotiated the first of eight ten-year contracts which were to give them proprietorship until 1645 (Matilla Tascón 1958).

Except for a brief interruption during the French occupation (1808-1812) the Almadén mines remained under the direct control of the Spanish government from 1645 until 1835. At that time the House of Rothschild negotiated a loan to the Queen Regent Maria Cristina, with one of the conditions being that Nathan Rothschild obtain the concession to operate the mines. The full significance of this will be discussed below (Chapter 4). The Rothschilds retained some form of influence in the Almadén mines, at least intermittently, right up to the twentieth century.

During the 16th, 17th and 18th centuries a large part of the labor force of the mines was made up of slaves and convicted criminals, but in the middle of the 18th century experienced miners from Aragon were introduced and the management was placed in the hands of German technical personnel. Mining machinery has been employed since the 17th century (Hoppensack 1796; Kuss 1878).

In recognition of the unhealthful working conditions, a large hospital was built at Almadén during the 18th century and a mining school was organized at about the same time.* Miners suffering from mercurialism were treated by sweating, a form of therapy which is used at the present time (1969).

---

*A. Matilla Tascón 1968: personal communication.

An account of Almadén from the time when the Fugger control ended up to the closing years of the 18th century was written in 1796 by Hoppensack who was director of the mines for a number of years. All aspects of the mines and their operation are described in detail. He mentions a large hospital which was well-equipped and well-staffed and where fever patients were segregated from those who were hospitalized because of injuries. The slaves and convicts who worked in the mines were kept in chains, and a company of infantry was quartered in the area to maintain security.

Hoppensack noted the unhealthful working conditions and commented that many miners die in the prime of life. He was familiar with mercurial tremor and stomatitis and believed that poor food and unhygienic living conditions contributed to the poor health of the miners. To mitigate the harmful effects of exposure to mercury the workers were employed in six-hour shifts, with a month underground and two months on the surface.

Complete figures on production are given in Hoppensack's book, down to the last ounce. The output of mercury from 1542-1793 is given as 1,430,000 quintales,* 21 pounds and 13½ ounces, (of which 540,000 quintales were produced under the Fugger) the total having been supplemented to a minor degree by the neighboring mines within a few miles of the main workings.

Further details of the post-Fugger period were published by Kuss in 1878 in a work which describes the various administrative arrangements, the production methods and the types of ore found in the mines. Kuss mentions a fire which broke out in 1755 and which burned for thirty months, sharply reducing production until 1760. He gives figures for the work force as being from 2,250 to 2,500 men who work six hours a day, three days a week—the same work schedule that is in force at the present time (1969). But even the short work week, plus ventilation (natural in winter, and mechanical in summer) were not sufficient to prevent occupational poisoning; Kuss remarks that very few among the older workers manage to avoid the mercurial tremors. In characteristic style the article on mines in Diderot's *Encyclopaedia* has this to say about the situation: "Almadén—Quicksilver mine in Spain, in

---

*The standard commercial unit of mercury, the flask containing 76 pounds avoirdupois, is probably derived from the Spanish quintale of 100 troy pounds which is the equivalent of 75 pounds avoirdupois. Neumann (1904) has pointed out that the flask has not always been exactly 76 pounds, but that the Spanish flask was 34.5 kg, the Californian, Russian, and Italian flasks 34.7 kg and the Mexican flask 34.15 kg. Thus, a flask may be slightly more or less than 76 pounds, and in fact was 75 pounds in the United States up to 1927 (cf. "Mercury Potential of the U. S.," Bureau of Mines Information Circular 8252, Washington; U. S. Government Printing Office, 1965, p. 9). According to Matilla (1958), Philip II decreed in 1579 that all quicksilver sent overseas from Spain should be packed in boxes containing one quintale, since heavier units were too difficult to handle. The packing and shipping instructions issued by the Casa de la Contratación in Seville (Norte de la Contratación) in 1672 permitted three half-quintales in cases of mercury being shipped to Nueva Espana (Mexico), but limited the quantity to two half-quintales in shipments to Tierra Firme (Panama). A traditional belief among old-time mercury miners is that two quintales (or flasks) on either side for balance was a proper load for a mule.

Andalusia, which every year brings nearly two million livres to the king and ruin to many men."

Significant changes in metallurgical methods were introduced at Almadén from time to time, starting in the early part of the 17th century (Kuss 1878).

Some interesting comments, particularly on occupational safety, were written in 1954 by a Japanese metallurgist who had visited Almadén in 1939:

> This writer had the privilege of visiting the famous Almadén quicksilver mines in Spain . . . some 15 years ago, just after the Spanish civil war had ended. Almadén is a small town with a population of about 18,000, and the whole place looked very devastated, after three years of internal fighting.
>
> The machinery and facilities of the mines, too, were very dilapidated. The elevator, which took this writer and the guide down several hundred feet below the surface, was operated with worn-out hemp ropes, and the writer was in constant fear, lest the ropes might snap at any moment.
>
> But of course, things probably appear vastly different today, especially in view of the report that the U.S. is turning a considerable portion of her $80 million economic aid to Spain for improving the facilities at Almadén. And naturally, the U.S. has the priority right to acquire the mercury produced there.
>
> Who can say that, like the 'oil war' staged between the Western free nations and the communist bloc, there will not be a 'mercury war' between the same contestants (Matsumura 1961).

Recent annual production of quicksilver by the Almadén mines is shown in Table III-2.

Slovenia-Idria

Present-day Yugoslavia is the home not only of one of the earliest known mercury mines (Avala) but also one of the richest. The mines at Idria in Slovenia, along with those at Almadén and Monte Amiata, make up the "big three" of Europe. Compared with the latter two, Idria is of recent date, having been discovered during the last decade of the 15th century.

In June 1880 there was held at Idria a celebration of the three hundredth anniversary of the exclusive management of the mines by the Austrian government. As a souvenir of the event, the Mining Directorate published a ninety-page brochure which contains considerable information dealing with the history, technology, management and production of the mines (Idria. Bergdirection 1881). Having been prepared by a group of specialists who were directly connected with various aspects of the operations, this brochure might be expected to have a high degree of authenticity. This is mentioned since in a few respects the details differ from those found in some other publications.

The historical highlights (as recorded in this brochure) include the discovery of quicksilver in 1490 or 1497. The original claim was sold in 1504 to the Mining Company of St. Achazi which undertook extensive exploitation. Subsequently various misfortunes coupled with increasing expense due to constantly deeper workings, resulted in a decrease in revenue to the operators who were obliged to seek loans from the reigning prince. Ultimately it became apparent to Archduke Charles, son of Emperor Ferdinand I, that most of the financial difficulties of the several operators working the mines at that time were due to inefficiency. Therefore in 1575 he initiated negotiations with the owners to purchase all of the holdings so that the entire enterprise could be placed under a single management. The transaction was completed in 1578 and was made official by the publication on April 6, 1580 of the "Neu aufgerichten Ordnung für das Perckwerch Idria."

Continuous possession of the Idria mines by the Austrian crown was interrupted several times during the Napoleonic wars. On two occasions, one in the middle of the 18th century and the other in the early decades of the 19th, the ore appeared to have been exhausted, but renewed prospecting and the working of lower grade ore resulted in resumption of large-scale production which has continued up to the present time (1971).

Idria is said to have been known to Paracelsus, who in 1527 described the poor health and physical condition of the miners, and to Agricola who in 1530 wrote about the smelting methods then being used. He made note of the loss of teeth among the workers in the smelter (cited in Lesky 1956). Similar observations were made by Pier Andrea Mattioli (1500-1577) and recorded in a work published in 1583 (cited in Lesky 1956). At about the same time, Andrea Cesalpino (1519-1603), physician to Pope Clement VIII, wrote in his *De Metallicis* of the active exploitation of the quicksilver mines at Idria (see Michelon 1908).

The Idria mines seem to have been unusually attractive to travellers and historians. Doctor Edward Browne (1644-1708), eldest son of Sir Thomas Browne, visited Idria in 1669. In the published account of his travels he gives many details of the mines and their operation (Brown 1673; Diderot and Alembert 1751-1765).

Some interesting observations on the Idria mines were made by Johann Georg Keyssler who visited the site in 1730. Keyssler says that the mercury mines were discovered in 1497 (some say in 1490) and gives a full description of the locale and of the situation of the mines in a deep valley. He includes an account of the mining processes, the nature of the ores, the smelting and many other details. In his discussion he describes the light which appears in an inverted barometer tube when it is shaken, a phenomenon noted earlier by Hawksbee (see chapter 9, below). Although not a physician, Keyssler describes the symptoms and treatment of mercury poisoning: tremors of the hands, feet and head along with frightening grimaces, were the symptoms which seem to have impressed him most. Rats and mice which enter into the mines, he says, are poisoned in much the same manner as humans, and they die with convulsions. Poisoning is treated with steam baths similar to the

Finnish sauna which cause profuse sweating and the oozing of droplets of
mercury through the sweat glands of the skin. Syphilitics who work in the
mines are cured of their disease and if a miner holds a copper coin in his
mouth or rubs it with his fingers the surface becomes silvery.

In 1730 a barber surgeon was employed at the Idria mines and was therefore one of the first industrial surgeons (Popper 1966). Keyssler makes no
mention of such a functionary, but in 1765 the position was occupied by
Giovanni Antonio Scopoli (1723-1787) who gave his name to the alkaloid,
scopolamine.

Several accounts of the mines at Idria, in addition to those already mentioned, have been published. They include those of Johann Weichard von
Valvasor in 1689, Balthasar Hacquet in 1781 (both cited in Lesky 1956),
Peter Hitzinger in 1860 (cited in Neumann 1904), Ludwig Teleky in 1912,
and Erna Lesky in 1956. Neumann gives many details about the operations of
the Idria mines as well as about a number of other quicksilver mines in central
Europe; many of the latter are no longer being worked. Recent production
figures for Idria are given in Table III-2 (see figures for Yugoslavia).

Other European Mines

Changing national boundaries and regional designations, e.g., Carinthia,
Transylvania, etc., have created some confusion and perhaps apparent contra-
dictions in the location of various European quicksilver mines. Idria, for example, has been located in Slavonia, Austria, Italy, Slovenia, Carniola, Venice,
Friuli and Yugoslavia. There is no doubt that what has been called Germany
accounted for major production of the metal during the 17th and 18th centuries, but quicksilver production in Germany decreased during the 19th century and became insignificant in the 20th.

Of the mercury mines in France which have been mentioned from time to
time, those in Normandy near St. Lô appear to be the only ones which ever
produced significant amounts of the metal (Lemery 1686 translation; Savary
des Bruslons 1748; Diderot and Alembert 1751-1765); but the useful life of
these mines was limited to a few decades in the 18th century. The mercury
deposits at Montpellier were never seriously exploited.

Other quicksilver mines have been said to exist in Corsica, England, Norway, Portugal, Sardinia and Sweden. If, in fact, there were such mines, they
have been of little importance. The same may be said of several mines in
Africa and elsewhere listed in Table III-1.

Russia

Within the borders of the present Soviet Union, mercury deposits were
worked in Siberia as early as 1759 but these mines were shut down in 1834.

More important deposits were discovered at Nikitowka near Dnepropetrovsk (formerly Ekaterinoslav) in 1879, with major production starting in 1886. Mercury output in the U.S.S.R. in 1963 is estimated to have amounted to 35,000 flasks, which places the nation in third place, after Italy and Spain (Table III-2). The principal source of the Russian mercury is Nikitowka (Neumann 1904).

Japan

As far back as records are available, Japan has had an insufficient supply of quicksilver from its own mines. At least 90 per cent of Japanese mercury comes from the Itomuka mines in Hokkaido, the production of which was 236.5 metric tons in 1953-1954 and something over 300 metric tons in 1954-1955, a metric ton being equivalent to about 30 flasks (Matsumara 1961). Twenty years earlier it had been a mere eight metric tons (Mitsubishi Economic Research Bureau 1936) and in 1929 just one metric ton (Schumpeter 1940). Figures taken from Japanese sources differ from those published by the United States Bureau of Mines, as shown in Table III-2. The discrepancy is due to the fact that the Japanese figures represent refinery production "which includes mercury imported into Japan" while "The Bureau of Mines shows only that mercury produced from Japanese mines."[*]

Philippines

Current lists of the principal mercury-producing countries usually include the Philippines, although that country is found at or near the bottom of the lists (Shelton 1965). In terms of total world output of mercury, the Philippines have been contributing about one per cent. The principal deposits are found at Tagburos on the island of Palawan. Greatly expanded production, up to 5,000 flasks annually, has been predicted (Parker 1968).

America

Little is known about the mining of quicksilver in the New World before the time of the arrival of the Spaniards in the 16th century. There is evidence, however, that both mercury and cinnabar were known to and used by the Maya more than a thousand years earlier, but the source of these materials is unknown. Samples of red pigments from Chongoyape (Peru) dating from about 500 B. C. have been found to contain as much as 30 per cent mercury and therefore are almost certainly cinnabar.[†]

---

[*]U.S. Bureau of Mines 1970: personal communication.
[†]L. J. Goldwater. Unpublished data.

Peru

By far the most important source of quicksilver in South America, particularly during the Spanish colonial period, was the mine at Huancavelica, a small Peruvian town at one time known as the Villa Rica de Oropesa de Huancavelica. Its importance was linked to that of the silver mines of Potosí and the combination was spoken of as "two poles which support this kingdom (Peru) and that of Spain." Mercury, used in the amalgamation process of purifying silver, was held to be more important than silver in this combination since the former was virtually irreplaceable while there were other sources of the latter (Whitaker 1941).

Mercury and cinnabar were known to the natives of Peru before and at the time of the arrival of the Spaniards. Recognition by Enrique Garces of the similarity between what the Indios called "llimpi" and the Spanish "bermillon" led to the discovery of the Huancavelica mines by the Spaniards in 1563. Garces, along with Pedro Contreras, had found mercury-bearing ore at Tomaca in 1560, but this was of little importance. The discovery of mercury at Huancavelica was followed almost immediately by the shutting down of all other mercury mines in South America, possibly to facilitate the monopolistic control by the crown of Spain of all production, shipment, and sale of the important metal.

For a period of about 200 years the Huancavelica mines were operated under policies established in 1570 and 1571 by the viceroy of Peru, Francisco de Toledo. One measure was expropriation of the mines for the benefit of the Spanish crown, resulting in a prodigious increase in the royal revenue. Another was inauguration of the "mita" system, a form of forced labor under which all Indians and "mestizos" were obliged to work in the mines for periods of six months every two or three years. Responsibility for operating the mines was given by contract to the "gremio de mineros" or miners' guild (Bargalló 1955; Rivet and Arsandaux 1946).

Complete production figures for the first years of the Huancavelica mine are not available, but from 6000 to 7000 quintales of mercury have been given as the annual output during the last years of the 16th century (Bargalló 1955). Whitaker cites Rivero y Ustariz and Mendiburu as sources for complete production figures from 1571 to 1813.

For the years 1660–1679 the average annual production of mercury was about 5200 quintales and during the next 20 years between 4000 and 4500 quintales. For the first three decades of the 18th century production was scarcely more than 3000 quintales a year (Whitaker 1941), but between 1736 and 1748 the output was again brought up to the neighborhood of 5000 quintales annually (DeSola y Fuente 1748), and later up to nearly 6000 quintales. Marked fluctuations in output continued during the closing years of the 18th century and the early years of the 19th, with a generally downward trend. By 1813 the production had fallen below 200 quintales and after that no records were kept during the remainder of the Spanish regime (Whitaker 1941).

The stormy course of the Huancavelica mine was due to many causes, but principally to constant friction between the operators, the local officials and the Spanish court. Labor troubles, accidents, and mercury poisoning all contributed to the difficulties (Whitaker 1941). Peruvian independence, achieved in 1824 (or 1826) was not followed by resumption of significant mining activities at Huancavelica. Even during the 1960's the output of quicksilver amounted to not more than one-half of the maximum reached by the Spaniards (Sheldon 1965).

Mercury Mining Methods in Peru. The following account of mining methods has been provided by a veteran mining engineer, Mr. S. H. Williston:

> . . . the method of mining that was used in most of the underground mines in the Western Hemisphere was also used at Huancavelica. At the face of the drift or tunnel underground a large fire was built and the rocks adjoining were brought up to a high temperature. Water was then thrown on the face and the rapid contraction of the rock broke off spalls of the rock. Surprisingly enough, this method is sometimes still used in isolated parts of Bolivia. While the health hazard is great enough in gold and silver mines, it becomes fantastically dangerous when utilized in mercury mines. This is the reason why the life expectancy of the miners at Huancavelica was only some six months or so.*

This method of dislodging fragments of rock may well provide an explanation for the numerous occasions when disastrous fires have necessitated the shut-down of mines. It is interesting, also, that this practice is not described in any of the several detailed accounts of the various mines cited above.

Mexico

In modern times, Mexico has become an important producer of mercury, but unfortunately for Spain this was not true in colonial times. During the decade 1954–1963 the annual output has been estimated at about 20,000 flasks (Shelton 1965).

As in the case of Peru, there is evidence that the aborigines of Mexico were familiar with cinnabar, and it is believed that they used it as a cosmetic. A small mine of cinnabar was found by the Spaniards at Temascaltepec around 1580 but its output was insignificant (Bargalló 1955).

Because of the insatiable needs of the silver mines of Nueva España for mercury (1.5 kg of Hg for 1 kg of Ag) (Neumann 1904) a law was enacted in Mexico in 1609 offering rewards for the finding of significant deposits of the metal (Bargalló 1955). During the next 200 years several deposits were found but the only significant one was that at Chilapa (Chilapan) in the state of Guerrero. Even this was far from adequate and it was not until after Mexico

---

*S. H. Williston 1969: personal communication.

obtained its independence (1824) that really rich mercury mines were discovered. These are at Guadalcazar, Huitzuco, Zacatecas, and Nueva Potosí, and of these, Guadalcazar is by far the most important.

In 1968 the Mexican government reduced export and production taxes hoping thereby to reduce smuggling and to obtain more complete reports on production (Parker 1968).

United States

A memorandum from the Spanish governor of the Province of California to the Commandante of Santa Barbara, written in 1796, contains the first hint of the existence of veins of mercury-bearing ore within the confines of the present United States of America (see Discovery in California prior to 1860). Apparently nothing was done at that time to pursue this discovery. Similar lack of zeal was shown by the Mexican governors when deposits of cinnabar were found in the New Almaden region in Santa Clara County in 1834 or earlier. These ore bodies, which have been described as "fabulously rich" were further investigated in 1845, but large-scale mining was not undertaken until 1850, when California was admitted to the Union. Intensive operations were carried on until 1890, by which time depletion of the richest ore bodies had begun. In spite of this, the United States led the world in mercury production during the latter half of the 19th and the early years of the 20th century (Neumann 1904; Parker 1968). From then until the present, smaller operations utilizing lower grade ores have continued at New Almaden. The plot of Bret Harte's "The Story of a Mine" suggests that it was inspired by the early history of New Almaden.

Of second importance among California's mercury mines are those of the New Idria area in San Benito and Fresno Counties. Production in these started in 1854 and has been going on more or less continuously since that time (U.S. Bureau of Mines Staff 1965).

The discovery of gold in California in 1848 and subsequent gold discoveries in various parts of the West up to the 1870's created a huge demand for mercury to be used for purification of the precious metal by amalgamation. This in turn stimulated a search for quicksilver deposits to supplement the supply available from New Almaden and New Idria. Altogether, in California, mercury ores have been found in 41 districts, of which twenty have been classified as "major" (U.S. Bureau of Mines Staff 1965).

Compared with the output of the mines in California, that of the other states is of minor importance. Production figures for the period 1850–1961 show that California contributed more than 70 percent of all mercury mined in this country. Only Texas, Oregon and Nevada, in that order, have made significant contributions to the supply of mercury and the remaining 11 states in which mercury has been found have been responsible for less than 3 percent of the United States output.

Canada

Prior to 1969, mercury production in Canada never reached significant proportions except during the war years, 1940-1944, when output reached a total of about 53,000 flasks for the period (Parker 1968). Steps were taken by the Consolidated Mining and Smelting Company of Canada, Ltd. during 1968 to re-open the quicksilver mines at Pinchi Lake in British Columbia, and active production began early in 1969. Projected output had been estimated to reach about 20,000 flasks annually and early results indicate that the goal is being achieved.* If this rate of production can be maintained, Canada will become one of the leading producers of mercury.

Extraction

Physical Properties of Mercury

"Of all the metals, quicksilver is probably the most easily recovered from the ore, as it can be volatilized at a comparatively low temperature, and thus separated from nearly all other substances that might be present in the ore" (Duschak and Schuette 1925).

Quicksilver has a boiling point of 357.3°C. At any temperature above its boiling point mercury will exert a pressure greater than atmospheric so that when any ore which contains free quicksilver is heated to above 357.3°C the vaporized metal will force its way out through crevices in the ore. For mercury present as crystalline mercuric sulfide (cinnabar) temperatures above 580°C are required (Duschak and Schuette 1925). The liberation of mercury vapor from mercuric sulfide is enhanced by the introduction of oxygen, resulting in the formation of sulfur dioxide along with the mercury vapor. This reaction is of practical importance, since cinnabar ore roasted in a retort with little or no air must be heated to temperatures above 580°C in order to separate the mercury and sulfur. Common practice is to add lime (CaO) to the charge to provide oxygen.

History of Extraction of Mercury

Pliny, Dioscorides, and Vitruvius understood the principle of recovering quicksilver by distillation and condensation. Theophrastus did not; but he did describe a separation by amalgamation. His process consists of grinding cinnabar with vinegar in a copper mortar with a copper pestle. According to Caley and Richards (Theophrastus 1956 translation) this represents the earliest account of the separation of quicksilver from the sulfide ore and also the first description of "isolating a metal from one of its compounds." Because doubt

*R. Weiss 1970: personal communication.

had been expressed as to the practicability of this method, Bailey performed the experiment and found that copper turnings were amalgamated, the reaction taking place slowly at ordinary temperatures and being speeded up by heat.

Two methods for extracting quicksilver from cinnabar are described by Pliny (Natural History, Book XXXIII, 123). One is the vinegar-copper process which he probably got from Theophrastus, and the other is the application of heat. The latter, which depends on vaporization and condensation of the metal, has been used since the days of Pliny up to the present, with various technical refinements from time to time. This process is, in effect, a form of distillation. Pliny probably learned of the distillation method from Dioscorides. Vitruvius (Book VII, viii) describes the extraction of quicksilver from cinnabar by means of heat.

For a period of about 1500 years after Pliny, nothing of significance was added to the literature on the extraction of quicksilver, perhaps because the method was so simple and efficacious.

Figure III-1 Apparatus Described by Biringuccio for Distilling Mercury from Its Ore. 1A. Chambers for condensing mercury vapor on to branches. 1B. Pots for distilling mercury from ore. 1C. Earthenware pots with glazer covers on which the mercury vapor condensed. In the pots illustrated in 1B and 1C, the ore was covered with sand or ashes on which the condensed mercury came to rest and from which it was separated by washing. 1D. Distilling bells; mercury vapor condensed and flowed out of the apparatus through the long "beaks." (From *Pirotechnia*, by Biringuccio. 1942 tranlsation MIT Press. By permission of the Metallurgical Society of the American Institute of Mining, Metallurgical, and Petroleum Engineers.)

In his *Pirotechnia*, first published in 1540, Vannucio Biringuccio (1480-1539) describes several methods of preparing quicksilver and accompanies his descriptions with illustrations (Figs. III-1 to III-4). The first step in all of the methods is grinding and washing the ore. If there is a great deal of rock, the ground ore is placed in pots which are set in a furnace and the space between the pots and the roof of the furnace is filled with green tree branches.

When a fire is lighted under the pots, "the quicksilver feels the warmth of the fire and tries to flee from it as its opposite, it rises in evaporation and issues from the mouth of the vessels. Feeling a certain moist coolness which the leaves of those branches have like itself, it runs out and attaches itself to them." When distillation is complete the fire is allowed to go out and the furnace to cool. At this point the furnace is opened and the branches shaken to dislodge the adherent quicksilver. The distilled metal is then recovered from the smooth floor of the furnace.

Four of the five methods for the extraction of quicksilver described by Agricola (1494-1555) in the *De Re Metallica* closely resemble those of Biringuccio. The fifth method, actually the first to be given by Agricola, is distillation *per descensum*, which does not appear in the *Pirotechnia* of Biringuccio. According to Neumann the method of extraction used at Almadén from the time of the Moors until the end of the 16th century was Agricola's first method. In the Hoover translation (1950) it is as follows:

> ... pots ... of earthenware, having a narrow bottom and a wide mouth ... are nearly filled with crushed ore, which ... is covered with ashes to a depth of two digits and tamped in. The pots are covered with lids a digit thick, and they are smeared over on the inside with liquid litharge [lead monoxide, probably as a paste], and on the lid are placed heavy stones. The pots are set on the furnace, and the ore is heated and similarly exhales quicksilver, which fleeing from the heat takes refuge in the lid; on congealing there it falls back into the ashes, from which, when washed, the quicksilver is collected.

At Idria, the extraction method used until 1530 was quite crude—the so-called Haufenbrennen or "heap-burning" (Idria. Bergdirection 1881). Great piles of ore and wood were built up in layers and the wood set afire. After the fire had burned out, the remaining debris was scattered and whatever quicksilver had formed into pools was gathered up. Much was lost by evaporation and by being mixed with the residual ashes. About 1530, Agricola's method of distillation *per descensum* was introduced, and the practice of roasting the ore with lime was initiated in 1580. The use of iron retorts described by Edward Browne when he visited Idria in 1669, was begun in 1641 or possibly a few years later (Neumann 1904). This method resulted in reducing the losses of vaporized mercury but it had two important disadvantages: the retorts were expensive and they could process only a relatively small amount of ore.

While the Idria mines were continuing to use their primitive methods of extraction, significant improvements were being made elsewhere. Of prime importance was the introduction of the cupola or furnace process devised by Lope de Saavedra Barba in 1633 and described in his book which was published in 1640 (cited in Kuss 1878). Barba was at that time director of the Huancavelica mines and he found that his new method permitted the treatment of much larger batches of ore than could be handled with the older procedures. Barba was succeeded at Huancavelica by Juan Alonso Bustamonte who, in 1647, presented a memorandum to King Philip IV describing the advantages of the Barba method. Shortly thereafter this process was introduced at Almadén and became known as the "metodo de Almadén," while the furnaces were called "hornos Bustamonte" (Bargalló 1955). Some of the original Bustamonte furnaces at Almadén remained in operation into the 20th century. Although credit for the invention was rightfully given to Barba in 1667, the name of Bustamonte became firmly attached to the furnaces. When the first installations of this type were constructed at Idria by Hauptmann and Poll in 1750 they were called "Bustamonteofen" (Neumann 1904).

All of the extraction methods described above are time-consuming "batch" processes which require loading, heating, cooling, and finally recovery of the liberated quicksilver and then a repetition of the cycle. Continuous extraction methods were introduced at Idria by Alberti in 1842 and by Hähner in 1849 but these proved to be unsuccessful and were abandoned after a few years (Neumann 1904). Ultimately the engineering problems were solved so that continuous processes could be widely adopted.

Considerable information on extraction methods up to the beginning of the 20th century is given by Neumann and the subject was comprehensively reviewed by Spirek in 1906 (cited in Duschak and Schuette 1925). Additional data have been published by Duschak and Schuette with particular emphasis on practices in the United States.

Modern Methods

A description of modern practice, as given by J. E. Shelton (1965) of the U.S. Bureau of Mines is as follows:

Milling of mercury ore consists of crushing, sometimes followed by screening. The principal purpose of these operations is to reduce the material to a size required for furnacing; mercury minerals crushing easier and finer than the gangue rock can be upgraded by screening and rejecting the larger pieces of low-grade or barren materials. The material also can be upgraded by hand sorting. Gravity concentration by jigs and tables has been attempted but with little success due to excessive losses in slimes. Concentration of mercury minerals by flotation is efficient and produces a high-grade concentrate. Despite the necessity of fine grinding and an ample water supply, the treatment of low-grade ores by flotation before roasting or leaching is advantageous.

Mercury is extracted from ore and concentrate by heating in retorts or furnaces to liberate the metal as a vapor, followed by cooling of the vapor and collection of the condensed mercury. Retorts are inexpensive installations for small operations and require only simple firing and condensing equipment. They are best adapted to operations treating 500 pounds to 5 tons per day of high-grade sorted ore. One of the most objectionable and costly features of retorts is the manual charging and removal of material.

For larger operations, either rotary or multiple-hearth furnaces with mechanical feeding and discharging devices are preferred. It is unnecessary to size the feed, which may range up to 3 inches for rotary and 1.25 inches for multiple-hearth furnaces. Standard furnace capacities range from 10 to

100 tons per day with larger sizes to suit requirements. Mercury-laden gases pass from the furnaces through dust collectors into the condensers, where the vapor is cooled and the mercury is collected. Final traces of mercury in the gases from the condensers are removed in washers, and the stripped gases are discharged through a stack into the atmosphere. The mercury from the condenser always contains some dirt and soot, which are separated by hoeing the mixture with lime. Mercury from this operation is clean and sufficiently pure for marketing.

Mercury also can be leached from its ores and concentrates with a solution of sodium sulfide and sodium hydroxide and recovered as the metal by precipitation with aluminum or by electrolysis [Von Bernewitz 1937]. Leaching of mercury ores has not been practiced extensively because of reagent-consuming constituents in some ores, irregularity in compositions of ores, and the cost of fine grinding. However, recent studies have indicated that some of these objections can be overcome by concentrating the ore by flotation and leaching the resulting concentrate.

Some experts believe that flotation should be used only with coarse cinnabar ore as losses are great when fine-grained cinnabar is treated in this way.*

History of Cinnabar Extraction and Purification

All available evidence indicates that cinnabar was widely used long before quicksilver, mainly because of its coloring properties. This created an interest in the purification of cinnabar ores for the purpose of producing more brilliant colors. Only later did the process of separating cinnabar from its gangue become important as a step in improving the efficiency of quicksilver extraction. The flotation process is a common way of accomplishing this, being based on the differences in specific gravity of the different minerals present in any sample of ore. Elutriation, used to concentrate the material that sinks in water rather than that which floats, is another application of the same principle.

What has been described as "the earliest account of the process of separating a pure mineral from its associated impurities" was the use of elutriation by Theophrastus (1956 translation) (who credits the process to Kallias) to purify cinnabar ore taken from the mines near Ephesus. Through repeated treatment with water the heavier cinnabar would sink more rapidly than the lighter impurities, thus affording purification or at least concentration of the cinnabar.

Vitruvius (Book VIII, 9) in his account of the preparation of vermilion, describes a combination of elutriation and heating of cinnabar ore. The repeated washing, he notes, enhances the color, but the subsequent heating results in some loss of its "natural" powers. He thus anticipates the modern process of concentration of the ore before the distillation of its quicksilver although it is clear that his intention was to produce good vermilion and not quicksilver.

---

*S. H. Williston 1969: personal communication.

One method of preparing cinnabar is given by Agricola in Book IX of *De natura fossilium*. This consists of combining one part of ground sulfur and two parts of quicksilver in a shallow dish and subjecting the mixture to heat. In Book X there is described a procedure for making "minium" from cinnabar. This involves crushing and pulverizing the cinnabar ore, followed by screening and washing. The resulting minium is, in fact, a refined cinnabar.

Vermilion manufacture was started at Idria at first by simple pulverization of pure cinnabar ore and later by sublimation of this material. The product was inferior to that made by the Dutch and by the Venetians and, therefore, found few markets. Concerted attempts to improve the quality of the vermilion were made in 1681, 1726, 1740 and on several subsequent occasions. From time to time first class vermilion was produced but this phase of the work at Idria was not commercially important at any time up to 1880 (Idria. Bergdirection 1881).

REFERENCES

Acts 13:51.

Acts 14:12.

Agricola, G. 1558. *De veteribus et novis metallis*. Basel.

Agricola, G. 1950. *De re metallica*. Trans. from the 1st Latin ed. of 1556 by Herbert C. Hoover and Lou Henry Hoover. New York: Dover Publications.

Agricola, G. 1955. *De natura fossilium*. Trans. from the 1st Latin ed. of 1546 by Mark Chance Bandy and Jean A. Bandy. New York: Geological Society of America.

Bailey, K. C. 1929. *The Elder Pliny's chapters on chemical subjects*. Part I. London: E. Arnold and Co.

Bargalló, M. 1955. *La minería y la metalurgia en la América Española durante la epoca colonial*. Mexico: Fondo de Cultura Económica.

Bianchi Bandinelli, R. 1925. Richerche archeologiche e topographiche su chiusi e el suo territorio in Etá Etrusca. *Monumenti Antichi* 30:429.

Biondi, C. 1934. Mercury mines. In *Occupation and health*, International Labour Office, vol. 2. Geneva: International Labour Office.

Biringuccio, R. 1942. *Pirotechnia*. Trans. Cyril S. Smith and Martha T. Gnudi from the first edition of 1540. Cambridge: M.I.T. Press.

Born, I. von. 1772. *Index fossilium*. Prague.

Brown, R. 1673. *A brief account of some travels*. London.

Davies, O. 1935. *Roman mines in Europe*. Oxford: Clarendon Press.

Dewhurst, K. 1963. *John Locke (1632-1704): Physician and philosopher*. London: Wellcome Historical Library.

Diderot, D., and Alembert, J. le R. d'. 1751-65. *Encyclopédie, ou Dictionnaire raisonné des sciences, des arts et des métiers, par une société de gens de lettres*. 17 vols. Paris.

Dioscorides, P. 1934. *The Greek herbal*. Englished by John Goodyer, A. D. 1655. Edited and first printed, A. D. 1933, by Robert Gunther. Oxford: At the University Press.

Discovery in California prior to 1860. 1953. *Calif. J. Mines Geol.* 49 (Suppl.):1-144.

Duschak, L. H., and Schuette, C. N. 1925. *The metallurgy of quicksilver*. Washington: U.S. Government Printing Office.

Gowland, W. 1920. Silver in Roman and earlier times: I. Prehistoric and proto-historic times. *Archaeologia* 69:121-160.

Haüy, R. J. 1801. *Traité de minéralogie*. 5 vols. Paris.

Hintze, C. 1904. *Handbuch der mineralogie*. Vol. 1. Leipzig.

Hofmann, R. 1886. *Der Quecksilberbergbau Avala in Servien*.

Hoppensack, J. M. 1796. *Ueber den Bergbau in Spanien ueberhaupt und den Quecksilber-Bergbau zu Almadén ins besondere*. Weimar.

Idria. Bergdirection. 1881. *Das K. K. Quecksilberwerk zu Idria in Krain.* Herausgegeben von der K. K. Bergdirection zu Idria. [Introd., etc., M. V. Lippold]. Vienna.

Jadranin, D. 1958. Mercury deposit Suplja Stena. In *Proceedings of the Geological Convention of Yugoslavia*, 1958, Belgrade.

Keyssler, J. G. 1751. *Neuste Reisen durch Deutschland, Boehmen, Ungarn, die Schweiz, Italien und Lothringen.* Hanover.

Kuss, H. 1878. Mémoire sur les mines et usines d'Almadén. *Ann. Mines Mém.* 13:39–141.

Lemery, N. 1686. *A course of chymistry.* 2d ed., enl. and trans. from the 5th ed. in the French by W. Harris. London.

Lesky, E. 1956. *Arbeitsmedizin in 18. Jahrhundert, Werksarzt und Arbeiter in Quecksilberbergwerk Idria.* Vienna: Verlag des Notringes der wissenschaftlichen Verbande Oesterreichs.

Lucas, J. A. H. 1804–13. *Tableau méthodique des espèces minérales.* 2 vols. Paris.

MacKendrick, P. 1966. *The mute stones speak.* New York: New American Library.

Mactear, J. 1894–1895. Some notes on Persian mining and metallurgy. *Trans. Inst. Mining Met.* 3:2–39.

Matilla Tascón, A. 1958. Desde la epoca romana hasta el año 1645. In *Historia de las minas de Almadén*, vol. 1. Madrid: Graficas Osca, S. A.

Matsumura, Y. 1961. *Japanese economic growth, 1945–1960.* Tokyo: Tokyo News Service.

Mellor, J. W. 1922–1937. *A comprehensive treatise on inorganic and theoretical chemistry.* 16 vols. London: Longmans, Green and Co.

Mendiburu, M. de. 1931–1938. *Diccionario histórico-biográfico del Perú.* 2d ed. 19 vols. Lima: Evaristo San Cristóval.

Michelon, E. G. J. 1908. *Histoire pharmacotechnique et du mercure à travers les siècles.* Ph. D. dissertation, University of Bordeaux. Tours: Delis Frères.

Mitsubishi Economic Research Bureau. 1936. *Japanese trade and industry: present and future.* London: Macmillan and Co.

Mochi, A. 1916. Indizi di miniere preistoriche di cinabro nella regione dell'Amiata. *Boll. Paletnol. Ital.* 1:5–12.

Mori, A. 1929. Amiata, Monte. In *Enciclopedia Italiana*, vol. 2. Milan: Instituto Giovanni Treccani.

Movers, F. C. 1850. *Die Phoenizier.* 2 vols. Berlin.

Muḥammad ibn Muḥammad, al-Idrisi. 1836–1840. *Géographie d'Édrisi.* Traduit de l'arabe en français par P. Amedée Jaubert. 2 vols. Paris.

Nef, J. U. 1952. Mining and metallurgy in mediaeval civilization. In *Cambridge economic history of Europe*, ed. M. Postan and E. E. Rich, vol. 2. Cambridge: At the University Press.

Neumann, B. 1904. *Die Metalle.* Halle a S.: Wilhelm Knapp.

Parker, J. G. 1968. Mercury. In *1967 Minerals Yearbook*, U.S. Department of the Interior, Bureau of Mines. Washington: U.S. Government Printing Office.

Plinius Secundus, C. 1929. *The elder Pliny's chapters on chemical subjects.* Ed. and trans. K. C. Bailey, part 1. London: E. Arnold and Co.

Plinius Secundus, C. 1938–1963. *Natural history.* Trans. H. Rackham, W. H. S. Jones and D. E. Eichholz. 10 vols. London: W. Heinemann.

Popper, L. 1966. *The development of occupational hygiene in Austria.* Specially printed for the 15th International Congress for Industrial Medicine, September 1966, Vienna.

Postlethwayt, M. 1774. *The universal dictionary of trade and commerce.* 4th ed. vol. 2. London.

Quintero y Atauri, P. 1928. *Compendio de la historia de Cádiz.* Cádiz: M. Alverez.

Ramsay, W. M. 1920. Military operations on the north front of Mount Taurus. *J. Hellenic Studies* 40:88–112.

Rawlinson, G. 1889. *History of Phoenicia.* London: Longmans, Green and Co.

Rhousopoulos, O. A. 1909. Beitrag zum Thema über die chemischen Kentnisse der alten Griechen. In *Beitrage aus der Geschichte der Chemie*, ed. Paul Diergart, pp. 178 ff. Leipzig: Deuticke.

Rivero y Ustariz, M. E. de. 1857. Memoria sobre la mina de Azogue de Huancavelica y la de Chonta. In *Colección de memorias científicas, agrícolas e industriales*, 2 vols. Brussels.

Rivet, P., and Arsandaux, H. 1946. La métallurgie en Amérique précolombienne. *Trav. Mém. Inst. Ethnol. (Univ. Paris)* 39:6.

Robinson, D. M. 1927. The discovery of a prehistoric site at Sizima. *Am. J. Archaeol.* 31:26–50.

Savary des Bruslons, J. 1748. *Dictionnaire universel de commerce.* New ed. 5 vols. Paris.

Schmeisser, C. 1906. Bodenschatze und Bergbau Kleinasiens. *Z. Prakt. Geol.* 14:186–196.

Schumpeter, E. B., ed. 1940. *The industrialization of Japan and Manchukuo, 1930–40.* New York: Macmillan Co.

Sharpless, F. F. 1908. Mercury mines at Konia, Asia Minor. *Eng. Mining J.* 86(13): 601–603.

Shelton, J. E. 1965. Mercury. In *Mineral facts and problems*, U.S. Department of the Interior, Bureau of Mines. Washington: U.S. Government Printing Office.

Sola y Fuente, G. de. 1748. *Relación de Guancavelica.* Lima.

Strabo. 1917–33. *The geography of Strabo.* Trans. Horace Leonard Jones. 8 vols. New York: G. P. Putnam's Sons.

Te-Kong Tong. 1967. The tortoise shell which set off a mighty chain reaction. *Columbia Library Columns* 16:11–18.

Teleky, L. 1912. *Die gewerbliche Quecksilbervergiftung.* Berlin: A. Seydel.

Theophilus. 1961. *Schedula diversarum artium.* Trans. and ed. by C. R. Dodwell. London: Thomas Nelson and Sons.

Theophrastus. 1956. *Theophrastus on stones.* Introd., trans., and commentary by Earle R. Caley and John F. C. Richards. Columbus: Ohio State University Press.

Theophrastus. 1965. *De lapidibus.* Ed., with introd., trans., and commentary by D. E. Eichholz. Oxford: Oxford University Press.

Turner, W. H. 1908. Quicksilver mining in foreign countries. *The Mineral Industry During 1908.* 17:746.

U.S. Bureau of Mines Staff. 1965. *Mercury potential of the United States.* Washington: U.S. Government Printing Office.

Vitruvius Pollio. 1914. *Vitruvius: The ten books on architecture.* Trans. Morris Hicky Morgan. Cambridge: Harvard University Press.

Von Bernewitz, M. W. 1937. Occurrence and treatment of mercury ore at small mines. *U. S. Bur. Mines Inform. Circ. 6966.*

Whitaker, A. P. 1941. *The Huancavelica mercury mine.* Cambridge: Harvard University Press.

Zippe, F. X. M. 1857. *Geschichte der Metalle.* Vienna.

ADDITIONAL READINGS

Hartendorp, A. V. H. 1958, 1961. *History of industry and trade of the Philippines.* 2 vols. Manila: Philippine Education Co.

Huke, R. E. 1963. *Shadows on the Land: An economic geography of the Philippines.* Manila: Bookmark.

Commerce, Trade, and Finance

Some of the uncertainties about knowledge and use of mercury and cinnabar in early times may be resolved, at least partially, by a study of what is known about their distribution. Access to a material, and interest in its use, may support a belief that it actually was used; and, conversely, absence of evidence pointing to availability weakens any such contention.

Ancient Trade and Commerce

Specific information on commerce in quicksilver and cinnabar in ancient Sumer, Egypt, India, China and elsewhere has not been recorded, but evidence of trade and commerce in general is plentiful. Egypt is known to have had many contacts with the Aegean islands and with peoples in the Tigris valley in pre-dynastic times, long before the Phoenicians began (c. 1500 B. C.) to build their trading empire. Furthermore, there was a close trade connection between the Tigris and the northwestern provinces of India at least as early as 3000 B. C. (Partington 1935). Indirect evidence of a flourishing trade all over the Mediterranean Sea is found in the numerous early references to piracy. Unless there were many ships carrying valuable cargoes professional pirates could not have made a living.

Along with this early commerce by water there were also important overland trade routes connecting all parts of the civilized world, from China to the western parts of Europe. Channels of distribution for many commodities, including cinnabar and quicksilver, certainly existed as long ago as the third millennium B. C. Clear evidence of Chinese trade with neighboring states at least as early as 800 B. C. has been adduced by Britton (1935).

Another possible, but conjectural, source of trade in quicksilver and cinnabar in the ancient world was the Hittites. Not only have these people been credited with being the earliest distributors of iron in the Mediterranean

58

world and the East (Breasted 1916, p. 239), but "Above all, it was the Hittites who controlled the mines of Asia Minor which supplied the ancient world with silver, copper, lead, and perhaps also tin" (Sayce 1910). Quicksilver mines are known to have existed within their territories, and quicksilver may have been distributed along with other metals.

Except for the detailed accounts given by Pliny of the shipment of cinnabar from Spain to Rome, there is practically no information on commerce in this commodity for the time prior to Pliny's day. Nothing is known, for example, of the source of the cinnabar discovered at Mohenjo-Daro (Marshall 1931, Vol. 2, p. 691) nor how it found its way to that site. No local source is presently known, but this does not necessarily mean that none existed four or five thousand years ago. One may speculate on the possibility of shipment from China but there is no evidence to support this notion except the existence of trade routes between the two countries. Similarly, there is no way of knowing whether the red mercurial pigment (presumably cinnabar) found on the island of Syra (Davies 1935) and dating from the pre-Mycenean bronze age was of local origin or imported. Whether or not the ancient Egyptians imported cinnabar for use as a pigment is a mooted question which will be discussed elsewhere (p. 75). Channels of distribution through which the Assyrians obtained mercurial compounds (if, in fact, they did so) are not known, although mines producing cinnabar and mercury, located in Asia Minor, and dating from Hittite times, have been described (Sharpless 1908).

Certainly the Phoenicians were among the earliest traders in the Mediterranean basin; they were interested in trading in anything that might yield a profit. These enterprising pioneers had established a trading post at Gades (modern Cádiz) in Spain by the seventh or eighth century B. C., and the date may have been several hundred years earlier (Lorimer 1950; Bosch-Gimpera 1929; Castro y Rossi 1858; Livermore 1960; Contenau 1949; Moscati 1966; Rawlinson 1889; and Movers 1850). Thus there is the possibility that they had access to the products of the mercury mines at Sisapo (Almadén).

While both cinnabar and quicksilver were known to Aristotle and Theophrastus in the Athens of the fourth century B. C., the source of these materials is far from certain. Attic silver mines at Laurium also might have been the source for some cinnabar for the Athenians, but it seems more likely that cinnabar and quicksilver were imported from overseas, probably from Spain, a source known to Theophrastus. Firm evidence that "miltos" was shipped to Athens from the island of Ceos is found in records of a treaty governing this trade (Michell 1940) but it is not known whether the Cean miltos was cinnabar or something else.

Mercury and Cinnabar Trade after Ancient Times

Brief mention of first century commerce in cinnabar is made by Strabo, the geographer, who lists "*miltos*" along with grain, wine, and kermes (a red dye made from insects) among the exports of Turditania (Spain). Vitruvius at this time called attention to the common practice of adulterating the ore with lime and described a test for determining the purity of any batch of cinnabar.

Pliny the Elder in the first century stated that the mines at Sisapo were most important to the economy of the Roman nation, that smelting and refining of the ore was not permitted at the mines and that production was limited under strict State control. The raw ore was shipped to Rome under seal and there purified and sold at a price fixed by law at 70 sesterces per pound (libra). The import quota was 2,000 pounds annually.

Little else is known about the production and distribution of quicksilver and cinnabar during the days of the Roman Empire and for some thousand years thereafter. According to Nef, a decline in output began in Europe in the 3rd century and continued for hundreds of years, until the 15th century. In all probability the use of cinnabar as a pigment continued through the middle ages. Illumination of medieval manuscripts was dependent on the availability of, and hence trade in, cinnabar. The word "miniature" is derived from "miniate" which in turn comes from minium or vermilion. Although the original Latin meaning of minium as mercuric sulfide (cinnabar) was changed to red lead by Theophilus in the 11th century, this conversion was not widely adopted until three or four hundred years later. Merrifield interprets minium as red lead, but this must have been a matter of conjecture since neither she nor anyone else seems to have performed chemical analyses on samples of pigments from illuminations. Quite possibly both cinnabar and lead oxide were used. Modern non-destructive analytical methods could easily be applied to a solution of this question.

Use of quicksilver by alchemists is known to have persisted from as far back as the third century. Figures on demand and consumption are not available, but the amount was sufficient to lead to the edict by Diocletian that all alchemical writings be destroyed (Chap. 2). This may have forced the purveyors of quicksilver to operate clandestinely. At any rate there are no contemporary records of the sale and distribution of the metal.

In addition to the alchemical uses of mercury metal, *miltos*, which may have been, and *kinnabari* which definitely was mercury sulfide, found a place in medicine in the time of Hippocrates (Chap. 14). Obviously, various forms of the mineral must have been available through some channels of commerce, but exact information on this is not available.

The Almadén Mines

Edrisi has recorded that the Arabs, during their occupation of southern Spain, operated the quicksilver mines of Sisapo (Almadén) (see Muhammad ibn Muhammad, al Idrisi) and shipped both mercury and cinnabar throughout the world (Matilla Tascón 1958). This may be a bit of hyperbole but it is not unreasonable to conclude that they distributed these products to numerous points around the Mediterranean basin which, in fact, constituted "the world" in the 11th century.

Eleventh century lists of commodities transported from West to East between 1045 and 1096 include quicksilver, with Spain as the source (Goitein, 1967).

The military-religious order of Calatrava (Chap. 3, above) obtained an interest in the Almadén mines during the 13th century. The Grand Master of the Order may or may not have been familiar with the Roman policy of monopolistic control of production and sale of cinnabar but in any case in 1286 he petitioned King Sancho IV (1257–1295) for this kind of protection, but got no action. A flourishing trade with foreign merchants had already developed at that time. Ferdinand IV (1285–1312) became king in 1295 and undoubtedly was preoccupied initially with other matters, such as trying to rid the country of Moors, so it was not until 1308 that he acted on the petition, decreeing that no quicksilver or cinnabar could be sold within the kingdom without the permission of the Order of Calatrava (Matilla Tascón 1958). This re-established the monopolistic policy inaugurated by the Romans and forms a link in a chain which has continued in one form or other to the present day.

Near Eastern Trade

Specific mention of quicksilver as an article of commerce in the 14th century is found in *The Book of Duarte Barbosa*, the author, Barbosa, being a Portuguese geographer and Magellan's brother-in-law. Barbosa describes the year-round shipping activities by which trade was carried on between India and Arabia, with roots, herbs and drugs being brought to Jidda (the port of Mecca) by sea and the returning ships carrying large quantities of copper, quicksilver, and other commodities to Calicut (in Southwest India). Quicksilver was also transhipped at Aden destined for other countries bordering on the Red Sea but no information is given by Barbosa as to its source. The extensive use of mercurials in Indian medicine and alchemy during the period covered by Barbosa's account may have been one reason for this active trade (Partington 1965).

Additional accounts of trade in mercury and cinnabar in the 14th century are found in the ledgers of Francesco Balducci Pegolotti, an agent of the Florentine banking House of Bardi. The frequent appearance of "argento vivo" and "cinabro" on Pegolotti's lists provides further evidence that these commodities were distributed as items of trade to all parts of the Mediterranean basin as well as to many other points in Europe. A third form of mercury, corrosive sublimate (argento silimato), was also distributed during the first half of the 14th century, with retail sales being a monopoly of the spicers guild. This association of drugs and spices survived into modern times in the familiar "Droguerie et Épicerie" in France. In Britain the apothecaries became divorced from the grocers during the reign of James I.

Chinese Trade

Trade relations between Portugal and China were established early in the 16th century. One Giovanni da Empoli, an Italian in the service of Portugal, wrote from Cochin on November 15, 1515 that "There come from there . . . much white alum and good vermillions . . ." (Chang 1934). The latter could have been cinnabar, although mines yielding the ore are not presently known in that area. Possibly its origin, if in fact it was cinnabar, was China itself. Other records show that China imported *vermilion* and *quicksilver* (Chang 1934). On the other hand, the famous Chinese alchemist, Ko Hung is said to have asked his emperor to send him to Kou-lou because cinnabar, which he needed for his experiments, could be obtained there from Cochin-China (Sarton 1927–1948).

That vermilion should have been exported from and imported into China by Portuguese traders during the same period of time would be difficult to understand were it not for the fact that the ports involved were separated by a considerable distance. The exports were shipped from Cochin while the imports probably entered through Canton. Transport routes between the latter and the sources of quicksilver and cinnabar in the interior of China may not have been established in the 16th century. The simultaneous exportation from and importation into China of cinnabar was known in the nineteenth and early part of the twentieth centuries (Morse 1920).

Role of Quicksilver in Spanish History

Spanish economic, and, therefore, political history for a period of about three hundred years may fairly be said to have been written in quicksilver. Earlier events had foreshadowed the vital role the metal was to play in Spain from the beginning of the 16th to the end of the 18th centuries: the Phoenician and Arab trade, the Roman monopoly, the reward given by Alfonso VIII to the monks of Calatrava and the subsequent control of the Almadén mines by Alfonso XI. The last of these events, taking place in 1348, inaugurated a period of royal control which was to last well into the 19th century. The security afforded by the quicksilver of Almadén repeatedly stood between the Spanish government and bankruptcy (Ehrenberg 1928).

Charles I (1500–1558) became King of Spain in 1516 and, as Charles V, Holy Roman Emperor in 1519. His campaign for election to the latter post cost him some 850,000 gold florins which he raised largely through loans from the great Welser and Fugger banking interests, his indebtedness to the latter in 1523 being more than 400,000 florins (Ehrenberg 1928; Livermore 1960; Clough et al. 1965). The Fugger were not at all timid about pressing the emperor for payment, but the royal treasury was far from able to satisfy the debt. As an alternative, the Fugger, in 1524, were given a lease on the mines at Almadén as well as on other mining properties and farm lands. Although they paid rent to the king it was presumed that the revenues from the mines and farms would be far in excess of the rents. This must, in fact, have been a profitable arrangement since the Fugger family maintained it intermittently for more than a century and accepted it from a series of kings as a means of repayment for their loans. Thus it is clear that quicksilver served as the principal underpinning for the financial needs of Spain during an era of frequent military adventures and of great territorial expansion in America.

Important as quicksilver was to the Spanish economy during the first half of the 16th century, it became even more so following the discovery of the "bonanza" of silver at Taxco in Nueva España (Mexico) in 1542 and at Potosí in Peru in 1545. The finding of additional deposits of silver at Zacatecas and elsewhere in the New World during later years further increased the need for and importance of mercury (Bargalló 1955). Prior to the discovery of quicksilver at Huancavelica all of the metal that was needed for the amalgamation of silver had to be shipped from Spain and even afterward there were times when the Peruvian mercury had to be supplemented by supplies from the mother country. The royal monopoly on Spanish mercury and the insistence by the king that he derive his prescribed benefits from the sale of that which was mined at Huancavelica often resulted in artificial shortages and in complicated shipping arrangements. According to de Launay the high prices which the refiners were forced to pay for quicksilver was one of the demoralizing actions in relation to mining and refining of precious metals which contributed to the decline of Spain in the 16th century.

Exercise of the royal prerogatives was not the only complicating factor in the distribution of quicksilver in the New World. Transportation from the mines to points of use was far from simple even when both were located in the same country, Peru. Huancavelica, the producer, and Potosí, the user, are about a thousand miles apart "as the crow flies," but in between are found some of the highest mountain ranges on the South American continent (Fig. IV-1). Mules, alpacas, and llamas were the only means available for transporting heavy burdens. But fortunately for the Spaniards, the Incas had constructed a network of roads, many of them paved, which have been described as being not inferior to those of the Romans (Bargalló 1955). In the words of Hernando Pizarro, "The mountain road is something to see, for in truth, in all christendom in such rugged country, such beautiful roads have never been seen, for the most part paved. All the valleys have bridges of stone or wood. . . ."

Despite such paved highways, mines were not always easily accessible. In the case of the quicksilver deposits of Huancavelica, they were located about three miles from the town and at an elevation of about a thousand feet above it. The refineries were at the edge of the town and the ore was carried to it in baskets on the backs of alpacas and llamas (Whitaker 1941). Fortunately it was downhill.

Of the mercury itself, a small amount was used in nearby silver mines and another minor quantity was sent to Lima, but most of the product was

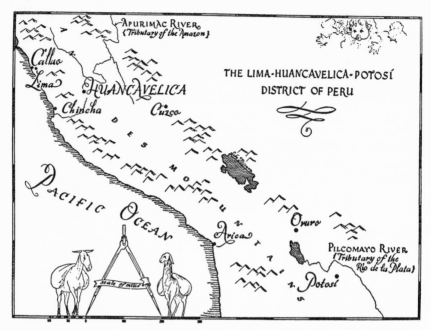

Figure IV-1 Map of Huancavelica-Potosi Area in Peru. The two cities, some 1000 miles apart, are separated by high mountains, but excellent Inca-built roads covered much of the distance. (From *The Huancavelica Mercury Mine* by A. P. Whitaker. Cambridge, Massachusetts: Harvard University Press. 1941.

shipped to Potosí. The usual route was overland to the port of Chincha, by sea to Arica and then by pack-animal to the silver mines at Potosí. In the early days the quicksilver was packed in sheepskin pouches in much the same manner as was done at Almadén in Spain. In the 18th century the use of iron flasks with contents of 100 quintales (75 lbs. av.) was introduced, again in accordance with the practice in the mother country. This probably represents the introduction into America of the unit of measurement which has become the universal standard (Bargalló 1955).

As early as 1503 an organization was established to regulate the trade between Spain and the colonies in America, the Casa de la Contratación set up in Seville by order of Queen Isabella of Castile (Guerra 1966; Haring 1918). At first the principal precious metal shipped from America to Spain was gold, but discovery of important deposits of silver in Mexico and the fabulous wealth of the Potosí silver mines in Peru soon relegated gold to secondary importance to that of silver.

Recovery of silver by the mercury amalgamation "patio" process was first practiced in Mexico in 1554, having been introduced by Bartholomé de Medina and his German colleague Gaspar Lomann (Bargalló 1955; Guerra 1966). The process is described by Mr. S. H. Williston:

Very briefly, this consisted of roasting, grinding, screening, and then the addition to the ground material of vinegar, and water, mercury bichloride, a little copper sulfate and a very considerable amount of mercury. This material was spread out in an open patio and horses or mules, sometimes with attached rollers, were driven around and around the patio. The mercury, which had collected almost all of the silver, was then separated out and retorted leaving the silver sponge behind.*

It is interesting to note that the use of vinegar goes back at least to the time of Theophrastus.

From 1554 until 1557 urgent appeals to the Spanish crown for the shipment of mercury to Mexico were without avail. (This was the period in which Charles I abdicated and Philip II ascended the Spanish throne and therefore a time of preoccupation with affairs at home.) Finally, in 1558, Philip II ordered that mercury be shipped from Almadén to Mexico and early in 1559 he instructed the operator of the mines, Ambrosio Rótulo, to send to New Spain, through the Casa de la Contratación, the entire output of the mines except for 50 or 60 quintales. The amounts available were far from adequate to meet the needs in Mexico. In order to meet the deficiency the Casa de la Contratación purchased mercury outside of Spain, from Germany, Idria and other sources (Matilla Tascón 1958).

From time to time, efforts were made to supplement the mercury sent from Europe to Mexico with that mined at Huancavelica in Peru and continued intermittently until 1657 when it was finally realized that the cost and the difficulties of transporting the mercury from Peru to Mexico made the effort impractical. In 1751, however, an accident at the Almadén mines resulted in drastic reduction of output and once again it became necessary to send quicksilver from Peru to Mexico (Whitaker 1941).

The French Trade in Mercury

Savary noted that the Spanish government prohibited the export of quicksilver and cinnabar to other European countries; therefore most of the quicksilver used in France came from Hungary and Idria. That from Hungary was shipped by way of Vienna and Holland. Dutch traders had an agreement with those in Germany to buy all of the metal not needed in the latter country. The quicksilver was transported in lamb-skin sacks which were packed in kegs. England also furnished France with a small amount of mercury which was shipped in heavy glass flasks of different sizes and weights. For purposes of import duty, quicksilver brought into France was classified with drugs and spices and was taxed at the rate of 100 sols per hundredweight.

British Trade in Quicksilver

Evidence that quicksilver was known in Scotland in the early days of the 16th century is found in the ledgers of Stirling Castle which are said to record

---

*S. H. Williston 1969: personal communication.

the purchase of a supply for James IV in 1502 at four shillings a pound (Read 1947). Unless the mercury came from the "lost" mines which Agricola said were present in "Britannia Scotis" it must have been imported, and this may be the first record since the time of Pliny in which the price of mercury is given.

Information on 18th century trade in quicksilver and cinnabar as seen by British eyes can be found in the *Lex Mercatoria Rediviva* of Wyndham Beawes. He records that Dutch and English traders take quicksilver to Constantinople and that several European nations provide the territories of the Great Mogul with vermilion as well as with other commodities. Quicksilver and vermilion are exported from China and specifically from Canton by the East India Company. Apparently a surplus was brought to Britain, as the metal was exported from there to France and to Holland. The demands of the American mines for quicksilver are reflected in the absence of this material from lists of exports from Spain to the British Isles.

Details about the manner in which quicksilver (argentum sublimatum) and cinnabrium were packed for shipping are given by Beawes and also the amount of tares (packing) that are allowed by British customs authorities on different containers. The accepted tare for tubs in which vermilion was shipped was 36 lb. . . . "But the Officers, having been dissatisfied with the above Allowance, have on some Occasions tared the tubs, and found them on an Average only to deserve 24 lb." Apparently there is nothing new about trying to beat the customs officers.

Another example of 18th century British views on the mercury trade is found in the *Universal Dictionary of Trade and Commerce* of Malachy Postlethwayt (1707?-1767).*

Mercury and International Finance

Diderot's caustic comment on the quicksilver mine at Almadén that "it brings to the king nearly two million pounds every year, and ruin to many men" can be applied with equal cogency to the effects on Spanish domestic economy of the influx of silver and gold from America during large parts of the 16th and 17th centuries. The use of quicksilver in the extraction of the

---

*The arrangement of the dictionary is such that information on trade and commerce in quicksilver is submerged in masses of facts that are totally unrelated to the subject. There is a detailed account of the quicksilver trade between Old and New Spain in which Postlethwayt emphasizes the strict regulations which are in effect to guarantee that the king receives his revenues from this trade. He is so impressed with the system that he calls it ". . . perhaps, the best established commerce in the world." Some of the details are given in an entry entitled "Azoga Ships" and others under "Mercury." The latter includes a long account of the extraction of mercury from wild purslane (see Chapter I, above).

Azoga ships, incidentally, are of considerable historical importance, having been responsible for the defeat of the Spanish fleet in the second battle of Cape St. Vincent in 1797, a turning point in European and world history (Lloyd 1963).

precious metals resulted in more efficient production and thus contributed to the economic disruption which was felt most acutely in Spain but also had repercussions in other parts of Europe (Rickard 1932). Bullion from the overseas empire was a necessity for the Spanish kings and their governments, but the effect on the economy was serious inflation with drastic increases in the prices of merchandise and corresponding decreases in the purchasing power of money. Not only did this cause great hardship at home but it also placed Spain at a disadvantage in foreign trade (Guerra 1966; Parry 1964).

The Fugger were not the only bankers who used quicksilver as an instrument of international finance during the 16th century. Their control of the Almadén mines was interrupted from 1533 to 1537 by the proprietorship of Bartholomew Belser and it had been challenged a few years earlier by the Hochstetter interests. Mercury was to be the undoing of the latter (Ehrenberg 1928).

Between 1511 and 1517 the Hochstetter had gotten control of a major part of the output of the copper and silver mines of the Tyrol. Their mining holdings were extended by 1523 to include a monopoly on the production and sale of mercury from the mines of Idria and Bohemia. As the monks of Calatrava before and the Rothschilds after, they aimed at a wider scope for their monopoly on quicksilver. Their obvious main target was Almadén, but the less important mercury mines in Hungary also engaged their attention. In anticipation of monopolistic control of the price of mercury, the Hochstetter invested about 200,000 florins in the metal, expecting to sell it at an inflated price. At the same time they became involved in large-scale operations in grain for England in exchange for English cloth. This deal was to be made through the Gresham brothers who had recommended the financiers to Cardinal Wolsey. But everything went wrong for the Hochstetter. The shipment of grain, which was aboard ships in the Netherlands, was seized by the government and could not be delivered to England. In an effort to extricate themselves from their difficulties, the Hochstetter agreed as a sort of bribe, to make a loan of 200,000 gulden to the Court of Brussels (the current seat of government), but being short of cash they put up about 350,000 pounds of mercury and 60,000 pounds of cinnabar which they were holding. The government turned these over to an agent to sell, but due, at least in part, to a surplus in the supply from Almadén, the amount realized was only 126,000 gulden instead of the required 200,000. The downfall of the Hochstetter was completed by their failure to get control of Almadén and with it, the ability to fix the price of quicksilver.

Mercury appeared again in a major international deal during the closing years of the 17th century. The locale was Amsterdam, which had become the financial capital of Europe. In 1695 Emperor Leopold I went to the Amsterdam money market seeking a loan of 1,500,000 gulden offering as security a quantity of quicksilver from the mines at Idria. The transaction was handled by the Amsterdam banking firm of Dentz, which considered the collateral to be adequate, and the Emperor got the money he needed (Ehrenberg 1928).

Throughout most of the 18th century, for all practical purposes, the output of Almadén and Idria completely dominated the market. This situation prevailed also during the early decades of the 19th century, a fact which did not escape the notice of the House of Rothschild. The ancient idea of a mercury monopoly once more came to life, this time in the minds of real masters.

Some time prior to 1835, with the assistance of Metternich, Solomon Rothschild of the Austrian branch of the House obtained control of the Idria mercury mines from the Austrian government. If the Rothschilds should be able also to gain control of the Almadén mines they would then have a monopoly on this important commodity.

The disturbed political situation in Spain created ideal circumstances for a typical Rothschild maneuver, the position of the Queen Regent Maria Cristina having been challenged by the Pretender, Don Carlos. Both sides in this contest needed money and both approached the House of Rothschild for a loan. With an eye on the Almadén mines, Nathan Rothschild of the London branch decided to place his money on the Queen Regent since she was the ruling monarch at the time. In what outwardly appeared to be competitive bidding, Nathan Rothschild was awarded the contract to operate the mercury mines. The agreement was signed by Nathan's son Lionel and the Spanish minister of Finance, Count José Maria Toreno, on February 21, 1835. The Queen Regent was so pleased that she, too, signed the contract and made Lionel a member of the Order of Isabella the Catholic (Morton 1962; Corti 1928).

The subsequent history of the Rothschild mercury monopoly is not entirely clear. Neumann calls it a cartel (*Vertrag*) and says it was dissolved in 1901. Large scale production of mercury in Italy, Russia and the United States during the second half of the 19th century may have introduced sufficient competition to render the cartel ineffective. A new cartel known as Mercurio Europeo was organized in 1928 by Spanish and Italian producers to control production, distribution and sale at a time when world stocks of quicksilver were in excess of demand. More than 80 percent of world production was controlled by the interests which formed the combine. By agreement, 55 percent of sales were allocated to Spain and 45 percent to Italy. In January 1950, following the purchase by the United States government of large quantities of mercury by means of counterpart funds in Italy, the Mercurio Europeo cartel was dissolved (Shelton 1965).

In 1958 Italian producers set up an organization known as Mercurio Italiano to regulate the sale of mercury from the principal mines in Italy. This agreement lasted for five years and was then discontinued because it was in conflict with the antimonopoly rules of the European Common Market (Shelton 1965). Because of the traditional secretiveness of the House of Rothschild in their business dealings it is not easy to determine what part, if any, they played in these manipulations of the mercury market, but there is evidence that they had important interests in Italian and Spanish mercury after World War I (Engel 1967) and in the Almadén mines as recently as 1961 (Schveitzer 1961).

In view of the conflicting published statements about the Rothschild interests in mercury, an inquiry was directed to their London office in 1969. The following reply was received from Edmund de Rothschild, with gracious consent for its publication:

> I have looked through some of the archives here, and there are many references to the Almadén mines and the quicksilver contracts with the Spanish Government. In 1831 the Almadén quicksilver mines were mortgaged by the Spanish Government as security for the due payment of the interest of their loans, and in 1832 control of the mines passed to the Rothschilds. Baron Solomon acquired the Idria quicksilver mine from the Austrian Government about this time. There is not much else about the Idria mine, but with regard to the Almadén mines in 1850 N. M. Rothschild & Sons renewed their contract with the Spanish Government at the rate of $70 per quintal of quicksilver. This was in the face of competition from another house at $54¼ per quintal. In 1870 an agreement was reached whereby the produce of the Almadén quicksilver mines was consigned to both N. M. Rothschild & Sons and de Rothschild Frères in Paris for 30 years.

During the past fifty or sixty years the price of mercury has been subject to marked fluctuations in world markets and in the United States (Fig. IV-2). Based on "constant dollars" the price in the United States fell below $100 per flask in 1920-21 and again in 1948-50. Between 1910 and 1961 the high point was $411 per flask in 1940. Limited supplies in 1965 resulted in a rise

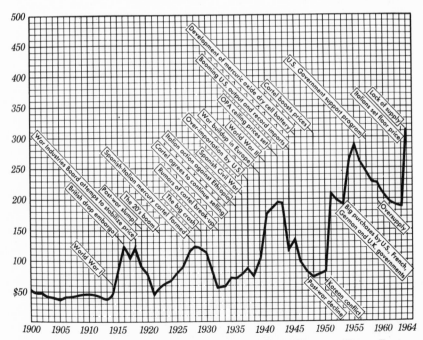

Figure IV-2 The "Why's" of Mercury's Erratic Price History (From *Engineering and Mining J.* 166[2]: 141-143, 1965.)

in price to $500 per flask and prompted the United States government to offer stockpiled mercury at $430. (Correcting these last two figures to "constant dollars" somewhat reduces the price.) The market quotation for mercury in September 1968 was $550 per flask (in 1968 dollars).

REFERENCES

Barbosa, D. 1918-21. *The book of Duarte Barbosa: An account of the countries bordering on the Indian Ocean.* Trans. from the Portuguese text and ed. and annot. by Manuel Longworth Dames. 2 vols. London: Hakluyt Society.

Bargalló, M. 1955. *La minería y la metalurgia en la América Española durante la epoca colonial.* Mexico: Fondo de Cultura Económica.

Beawes, W. 1761. *Lex mercatoria rediviva* or *The merchant's directory.* 2d ed. London.

Bosch-Gimpera, P. 1929. Fragen der Chronologie der Phoenizischen Kolonization in Spanien. *Klio* 22:345-68.

Breasted, J. H. 1916. *Ancient times: A history of the early world.* Boston: Ginn and Co.

Britton, R. S. 1935. Chinese interstate intercourse before 700 B. C. *Am. J. Intern. Law* 29:616-35.

Castro y Rossi, A. de. 1858. *Historia de Cádiz y su Provincia.* Cádiz.

Chang, T'ien-Tse. 1934. *Sino-Portuguese trade from 1514 to 1644.* Leyden: Late E. J. Brill.

Clough, S. B.; Hicks, D. L.; Bradenburg, D. J.; and Gay, P. 1965. *A history of the western world: Early modern times.* Boston: D. C. Heath and Co.

Contenau, G. 1949. *La civilization phénicienne.* Paris: Payot.

Corti, E. C. 1928. *The reign of the House of Rothschild.* Trans. from the German by Brian and Beatrix Lunn. New York: Cosmopolitan Book Corporation.

Davies, O. 1935. *Roman mines in Europe.* Oxford: Clarendon Press.

Ehrenberg, R. 1928. *Capital and finance in the age of the Renaissance: A study of the Fugger and their connections.* Trans. from the German by H. M. Lucas. London: Jonathan Cape.

Engel, G. T. 1967. Mercury. In *Kirk-Othmer Encyclopedia of chemical technology*, 2d ed., vol. 13, pp. 224-25. New York: John Wiley and Sons.

Goitein, S. D. 1967. *A Mediterranean society.* Economic Foundations, vol. 1. Berkeley: University of California Press.

Guerra, F. 1966. Drugs from the Indies and the political economy of the sixteenth century. In *Materia medica in the sixteenth century*, Analecta Medico Historica, vol. 1. Oxford: Pergamon Press.

Haring, C. H. 1918. *Trade and navigation between Spain and the Indies in the time of the Hapsburgs.* Cambridge: Harvard University Press.

Launay, L. de. 1908. *La conquête minérale.* Paris: E. Flammarion.

Livermore, H. 1960. *A history of Spain.* New York: Grove Press.

Lloyd, C. 1963. *St. Vincent and Camperdown.* London: B. T. Batsford.

Lorimer, H. L. 1950. *Homer and the monuments.* London: Macmillan and Co.

Marshall, Sir John, ed. 1931. *Mohenjo-Daro and the Indus civilization.* Vol. 2. London: Arthur Probsthain.

Matilla Tascón, A. 1958. Desde la epoca romana hasta el año 1645. In *Historia de las minas de Almadén*, vol. 1. Madrid: Graficas Osca, S. A.

Merrifield, M. P. 1849. *Original treatises on the art of painting.* London.

Michell, H. 1940. *The economics of ancient Greece.* New York: Macmillan Co.

Morse, H. B. 1920. *The trade and administration of China.* 3d ed., rev. London: Longmans, Green and Co.

Morton, F. 1962. *The Rothschilds: A family portrait.* New York: Atheneum.

Moscati, S. 1966. *Die Phoeniker.* Zurich: Kindler Verlag.

Movers, F. C. 1850. *Die Phoenizier*, Vol. 2, Part 2. Berlin.

Muḥammad ibn Muḥammad, al-Idrisi. 1836-40. *Géographie d'Edrisi.* Traduit de l'arabe en français par P. Amedée Jaubert. 2 vols. Paris.

Nef, J. U. 1952. Mining and metallurgy in mediaeval civilization. In *Cambridge economic history of Europe*, vol. 2, ed. M. Postan and E. E. Rich. Cambridge: At the University Press.

Neumann, B. 1904. *Die Metalle.* Halle a S.: Wilhelm Knapp.

Parry, J. H. 1964. *The age of reconnaissance.* New York: New American Library.

Partington, J. R. 1935. *Origins and development of applied chemistry.* London: Longmans, Green and Co.

Partington, J. R. 1965. *A short history of chemistry.* 3d ed. New York: Harper and Row, Harper Torchbooks.

Pegolotti, F. B. 1936. *La practica della mercatura.* Ed. Allan Evans. Cambridge, Mass.: Medieval Society of America.

Plinius Secundus, C. 1938-63. *Natural history.* Trans. by H. Rackham, W. H. S. Jones and D. E. Eichholz. 10 vols. London: W. Heinemann.

Postlethwayt, M. 1774. *The universal dictionary of trade and commerce.* 4th ed. 2 vols. London.

Rawlinson, G. 1889. *History of Phoenicia.* London: Longmans, Green and Co.

Read, J. 1947. *Humor and humanism in chemistry.* London: G. Bell and Sons.

Rickard, T. A. 1932. *Man and his metals: A history of mining in relation to the development of civilization.* New York: McGraw-Hill Book Co.

Sarton, G. 1927-48. *Introduction to the history of science.* 3 vols. in 5. Baltimore: Williams and Wilkins.

Savary des Bruslons, J. 1748. *Dictionnaire universel de commerce.* New ed. 5 vols. Paris.

Sayce, A. H. 1910. In introductory note to *The land of the Hittites* by J. Garstang. London: Constable and Co.

Schaefer, E. 1947. *História y Organización del Consejo y de la Casa de la Contratación de las Indias.* Seville: Centro de Estudios de História de América.

Schveitzer, M. N., ed. 1961. *Hachette world guides: Spain.* Paris: Hachette.

Sharpless, F. F. 1908. Mercury mines at Konia, Asia Minor. *Eng. Mining J.* 86(13):601-3.

Shelton, J. E. 1965. Mercury. In *Mineral facts and problems*, U. S. Department of the Interior, Bureau of Mines. Washington: U. S. Government Printing Office.

Strabo, 1917-33. *The geography of Strabo.* Trans. Horace Leonard Jones. 8 vols. New York: G. P. Putnam's Sons.

Theophrastus. 1956. *Theophrastus on stones.* Introd., trans. and commentary by Earle R. Caley and John F. C. Richards. Columbus: Ohio State University Press.

Theophrastus. 1965. *De lapidibus.* Ed., with introd., trans., and commentary by D. E. Eichholz. Oxford: Oxford University Press.

Vitruvius Pollio. 1914. *Vitruvius: The ten books on architecture.* Trans. Morris Hicky Morgan. Cambridge: Harvard University Press.

Whitaker, A. P. 1941. *The Huancavelica mercury mine.* Cambridge: Harvard University Press.

ADDITIONAL READINGS

Robinson, D. M. 1906. Ancient Sinope. *Amer. J. Philol.* 27:125-153.

Veitia Linage, J. de. 1945. *Norte de la contratación de las Indias Occidentales.* Buenos Aires: Comisión Argentina de Fomento Inter-americano.

5

Early Knowledge and Uses

Cinnabar probably was known to and used by man earlier than mercury, which came from it. In some early societies cinnabar was used as a coloring material; in others it may have served as a valuable commodity in trade and in still others it may have been nothing more than a curiosity.

Certainly one of the earliest known uses of cinnabar was its application to human bones by neolithic man. Götze (1924–1932, Vol. 14, p. 538) has described human bones, including a skull, found in Italy, the red color of which he claims was identified on chemical analysis to be cinnabar. As is true in most reports dealing with archaeochemistry, no details on analytical procedures are given. The finding of a large piece of cinnabar ore in a bronze age grave at Marcella in Portugal is also mentioned by Götze, but there is no suggestion of why it was there. Perhaps it had some religious or magical significance.

The Near East

Assyria

Interpretation of the Sumerian IM.KAL.GUG by Thompson (1936) to mean vapor of cinnabar, therefore quicksilver, has been offered as evidence that this material was known to and used by the Assyrians. Partington has questioned this on the basis that cinnabar has not been found in Assyrian or Babylonian remains; and Oppenheim,* too, has disputed Thompson's interpretation; but as Sigerist points out the Assyrians were energetic merchants and traders, and had established a colony at Kanish in northern Asia Minor as early as 1900 B. C. This could very well have made available to them cinnabar

---

* A. L. Oppenheim 1969: personal communication.

and quicksilver from mines in what later was known as Cappadocia. The same source of ore could also have been available to the Hittites.

Indus Valley

Archaeological explorations at Mohenjo-Daro and Harappa have revealed remains of a Sumerian type civilization which at first was thought to date from as early as the fourth (Marshall 1931, vol. 2) but which now is believed to be no earlier than the third millennium B. C. (Wheeler 1966). Cinnabar was reported to be among the findings at Mohenjo-Daro (Marshall 1931), but neither its source nor its uses are known.

Except for extensive use in alchemy and medicine, quicksilver does not figure prominently in early Indian history. This may be due, at least in part, to the fact that mercury mines apparently were unknown in India. On the other hand, if it is true that cinnabar was found at Mohenjo-Daro as reported, the Indians would have been among the first peoples to be familiar with this compound.

Indian technology of the Vedic Period (1500-600 B. C.) is described in the Arthashastra of Kautilya. In his discussion of this work, P. C. Ray (1956) suggests that gold, silver, copper, iron, tin, lead, and mercury ores were imported from neighboring countries, and sulfide of mercury (hingul) possibly from China.

The custom among Hindu women of wearing a red circular dot on the forehead is well known. Less familiar is the fact that at one time it was customary for the groom as well as the bride to adorn the forehead with red paint, the usual procedure being to place a streak of red across the scalp at the hair line and sometimes to color the eyebrows. These practices are supposedly a part of the marriage rites and the red color a blood symbol (Wunderlich 1925). Perhaps it would be insulting and irreverent to suggest that cinnabar could be effective in treating nits and lice, favorite haunts of which are along the hair line and in the eyebrows. Beautification of the soles of the feet by painting them with cinnabar, another use of the pigment, is more difficult to explain. Possibly it was related to the Chinese alchemical belief that this practice would impart unusual powers.

Phoenicia

The Phoenicians may have handled cinnabar as an item of trade as far back as the 12th century B. C. or even earlier. There is no evidence of any use of quicksilver by the early Phoenicians themselves nor of cinnabar other than for trading purposes.

Palestine

According to Partington, "There is no indication that the metal mercury is ever mentioned in the Bible or was known to the old Hebrews. . ."

(Partington 1935). This view has not been challenged although what is probably a fanciful suggestion has been made that the stone of Tharshish was
cinnabar from the mines of Almadén (Haupt 1902). Partington concedes that
this is a possibility, but most historians give little credence to the theory. His
statement that "mercury is said to be mentioned (with sulphur and a crucible) in the Talmud . . ." with a reference to the Rodkinson translation,
proved, on investigation, to be only partially correct. The passage in question
has to do with the disposition of rubble remaining after demolition of monuments to the god Mercurius. The pious rabbis decided that these stones were
suitable only for filling privies—a rather unique form of crucible and an un-
conventional source of sulfur.

Egypt

Repeatedly it has been stated in the literature that the oldest known
specimen of mercury metal is that which was found in a tomb at Kurna, dating from the 16th to 15th century B. C. (See Colopinto 1964; King 1957;
Mellor 1922-37; Egenhoff 1953; Abramowitz 1934; Marshall 1931; Almqvist
1928; Stillman 1960; Thompson 1923; Schelenz 1904; Lippmann 1919;
Davies 1935; and Strunz 1906.) If this story were related only by writers with
unimpressive archaeological credentials, it might not have to be taken seriously. But when Sir John Marshall, Oliver Davies, J. W. Mellor, J. M. Stillman,
and several well-known German scholars subscribe to the authenticity of the
"mercury of Kurna," a careful examination of the matter is called for.

For 20th century writers the source of the Kurna story appears to have
been an anonymous article published in the *Pharmazeutische Zeitung* in 1890.
This article describes several vessels containing mercury which had been do-
nated to the Berlin Ethnological Museum by Schliemann and which supposedly had been found in Kurna, near Thebes, in a tomb of the 18th-19th
dynasties, therefore dating from about 1500 B. C. As Schliemann himself had
done no archaeological work in that location, the flasks, if they existed, must
have been given to him by someone who found them at Kurna (or somewhere
else). These objects are not known to G. R. Meyer, the Director of Museums
in East Berlin (personal communication, 1968).

Schelenz in 1904 repeats the Schliemann story with a slight variation. He
refers to a single vessel and suggests that it might have been used for transporting quicksilver from Spain to Arabia. Mellor (1922-1937, Vol. 4, p. 695)
complicates the picture still more by stating that "According to S. Seligmann,
the discovery by H. Schliemann of a small vessel full of mercury in a grave at
Kurna, and estimated to belong to the 16th or 15th century B. C., shows that
this element must have been known for a very long time." The statement is
supported by a reference to S. Seligmann, *Der böse Blick* (Berlin: 1910). This
work, however, contains no mention of Schliemann nor of Kurna. Since
Mellor's 16-volume treatise is widely used for reference purposes, it is quite
probable that many recent writers have gotten the Kurna story from that
source, although Mellor obviously was in error.

Partington (1935, p. 84) comments on this question:

> Mercury . . . does not appear to have been known in early Egypt. That found in tombs in glazed earthenware bottles like pineapples is supposed to have been introduced by the Arabs, and the small bottle containing mercury which Schliemann brought back from Egypt, where it is said to have been found in a grave of the 16th–15th centuries B. C. at Qurnah, is probably of the same origin. These small bottles, also nuts, quills, etc., containing mercury, are used in the East as amulets.

In this passage Partington correctly cites Schelenz and Seligmann. It seems possible that Mellor's incorrect reading of these references led to his error.

Vessels answering to the descriptions given by Partington and by the author of the article in the *Pharmazeutische Zeitung* (1890) were the subject of attention by several scholars during the latter part of the 19th century and later. (See Wilkinson 1878; Chester 1871; Clermont–Ganneau 1885; Perrott and Chipiez 1882-1914, vol 3; Saulcy 1874; Sarre 1925; Riis et al. 1957; Ettinghausen 1965; and Goitein 1967.)

Many, perhaps most, Egyptologists have been of the opinion that neither quicksilver nor cinnabar was known in Egypt prior to Hellenistic times (fourth century B. C.).

Analyses of pigments from an 18th dynasty Theban tomb were performed by Ure and described in this way: "The red pigment obtained by washing the coloured stone in a tomb of the kings with a wet sponge, and evaporating the liquid to dryness, when treated with water, evinces the presence of a glutinous gummy matter. It dissolves readily, in a great measure, in muriatic acid, and affords muriates of iron and alumina. It is merely red earthy bole" (Ure 1878). One does not have to be an experienced analytical chemist to recognize the inadequacy of this test procedure. Had mercury been present it could easily have been lost in the evaporating process and, furthermore, there is no evidence that any test for mercury was performed.

Colors in tombs at Tell El Amarna are discussed by Petrie, but unfortunately no details of any proper analyses are mentioned.

Fragments of painted limestone from the tomb of Perneb (c. 2650 B. C.) now in the Metropolitan Museum of Art in New York were analyzed by Toch in 1918. Without giving any actual analytical data on methods or specific results, Toch states that the reds were hematite and not ochre since they contain more than 50 percent of iron oxide. Nothing in his report effectively eliminates the possibility that cinnabar might have been present in addition to hematite.

One bit of evidence that the Egyptians may have known and used cinnabar before Greek times is found in an article (Herausgeber 1826) which deals with some material taken from a tomb in the Valley of the Kings, which Professor John R. Harris (personal communication, 1969) believes may have been that of Seti I. This would give the objects a date of about 1300 B. C. Chemical analysis revealed the presence of cinnabar mixed with iron oxide. The material in question is described as a fragment of plaster about two inches square and a half inch thick which came from a limestone column with fresco painting found by Belzoni in a tomb at Biban el Moluk. The analytical method by which mercury was identified can hardly be described as super-sensitive, so that the report of a "geringen Menge Zinnober" must, in fact, have been more than what today would be called a trace.

Mérimée (father of Prosper), in commenting on the colors used in paintings in a Theban tomb, says that for the most part the reds were ochre but that it is not impossible that vermilion also was used. He points out that the Egyptians could have gotten cinnabar from India where it was known even in ancient times.

Red pigment scraped from the outer surface of a shred from Tell el Amarna, of the New Kingdom period, and analyzed at Columbia University by a sensitive photometric method (Jacobs and Goldwater 1961) showed nine parts per million of mercury. A red-colored fragment of Ptolemaic mummy cartonnage contained 2,200 parts per million of mercury. These findings indicate the presence of a red, mercuric pigment which must be either cinnabar or mercuric oxide,* assuming that there had been no accidental contamination over the centuries.

To conclude from these meagre reports that the Egyptians of the time of Seti I, or earlier, knew and used cinnabar in their wall paintings or for decorating pottery might not be justified. On the other hand, the pharaohs almost certainly had access to this pigment and were surely interested in using the best available coloring materials. Furthermore, they are known to have imported other products from great distances. All of this suggests that the question of the Egyptians' knowledge and use of cinnabar should be kept open until further studies can be performed using modern analytical methods.

The Aegean World and Rome

As was true of other peoples in the Mediterranean basin and elsewhere, the Greeks almost certainly knew and used cinnabar much earlier than they knew quicksilver. In all probability the former antedates the latter by as much as 1500 years.

Rhousopoulos, while serving on the staff of the Acropolis Museum between 1887 and 1909, studied objects found on the island of Syra. His special interest was the chemical analysis of ancient pigments and his specimens dated back as far as the pre-Mycenian bronze age, that is, 2500 to 1600 B. C.

One of the specimens reported on in detail by Rhousopoulos came from a tomb on Syra dating from approximately 2000 B. C. His analyses showed the presence of three metals: copper, iron and quicksilver, and he reports, "The definite presence of quicksilver was repeatedly demonstrated." He also ana-

---

* L. J. Goldwater and J. Herndon 1969. Unpublished data.

lyzed material taken from painted statues in the Acropolis Museum dating from the sixth century B. C., and reports that the more beautiful and rarer reds were cinnabar. He makes a special point that pigments containing manganese and quicksilver were used. Unfortunately Rhousopoulos's laboratory records have not been preserved and consequently his analytical techniques cannot be checked.*

Analyses of bulk pigments by Caley showed that cinnabar was used in Greece during the fifth and third centuries B. C. and that other red pigments employed were hematite (iron oxide) and realgar (sulphide of arsenic). The use of cinnabar in the fourth century B. C. was demonstrated by Midgley (cited in Caley 1946). Aristotle's use of the term "argyros chytos" establishes the fact that quicksilver was known in fourth century Athens, but his writings say very little about uses to which the metal was put.

In addition to being the first to describe the purification of cinnabar, Theophrastus was also first in stating that quicksilver had some practical use (1956 and 1965 translations). Unfortunately surviving works give no indication of what the use might have been, but his description of the purification process may be " . . . the earliest account of the process of separating a pure mineral from its associated impurities" (Caley, 1946). The purpose of purifying cinnabar was to produce a superior pigment suitable for use by artists. Theophrastus states that ". . . use is made of the washings floating above, particularly as a wall paint . . ." (1965 translation) an early allusion to the use of an industrial by-product. It is conceivable that this residual material, from which most of the precious cinnabar had been elutriated, was enriched with ochre or hematite to intensify the color, both of these adulterants being relatively inexpensive.

Herodotus says that in ancient times all ships were painted with a red material which he calls "miltos" (1922 translation) and Hasebroek has suggested that Cean miltos was used as an anti-fouling paint on warships. If this theory is correct, the material must have been cinnabar which would be more effective than iron oxide. The superior rating given by Theophrastus to the miltos from Ceos supports the idea that it might have been cinnabar.

An interesting use for miltos, described by Aristophanes, is for coloring ropes used by officers to round up laggard citizens who were slow to assemble for meetings on the Pnyx. By stretching the colored rope across the agora and walking toward the meeting place the officers performed a sort of round-up. Those who moved slowly would be touched by the rope which would shed some of the red color on their clothes, thus leaving a "brand" which would subject them to shame and ridicule.

Several uses for quicksilver and cinnabar are described by Vitruvius, not all of them in his special field of architecture. He prefaces his remarks on the subject by telling how cinnabar exudes droplets of quicksilver when it is

---

* W. Willis 1969: personal investigation and communication.

pounded with a hammer and how the heated ore gives off vapors which on condensation form quicksilver. His discussion of the weight of mercury anticipates the later system adopted in Spain which is the basis for the present-day standard flask of approximately 76 pounds. When Vitruvius relates that a stone weighing 100 pounds will float on quicksilver while a piece of gold weighing a scruple (approximately 1.3 grams) will sink he is enunciating the principle of specific gravity and using mercury to prove it experimentally. He also describes uses of a more practical nature, e.g., for gilding and, by amalgamation, for recovering gold from embroidery on worn-out clothing.

The process for making first-class vermilion pigment from cinnabar as described by Vitruvius is more complicated than that of Theophrastus and seems to result in a less satisfactory product. The ore is not only ground and washed repeatedly but also is heated. Vitruvius was aware that this resulted in the loss of quicksilver as well as impairing the lasting qualities, particularly when the pigment is used on the exteriors of buildings. By covering the vermilion with a coat of Punic wax and oil the deterioration caused by exposure to the elements could be prevented.

Pliny indicates two uses for argentum vivum and minium (cinnabar): in amalgamation for the purification of gold (Book 33, line 99) and in copper and silver gilding (Book 33, lines 66, 100). Cinnabar is used principally for painting, but other applications are of interest.

In one passage, Pliny states that minium is used in writing books and that ". . . it makes a brighter lettering for inscriptions on a wall or on marble even in tombs" (Book 33, line 122). The special, possibly religious, significance of minium is brought out in his account of the custom of painting the face of Jupiter's statue with minium on holidays and of the wide use of this pigment in celebrating triumphs, it being ". . . one of the first duties of the Censors . . . to place a contract for painting Jupiter . . . " with minium (ibid.). Another interesting note by Pliny which suggests religious significance for minium is his mention, in describing the Sphinx of Egypt, that ". . . from a feeling of veneration, the face of the monster is coloured red" (Book 36, line 17).

Dioscorides was familiar with both cinnabar and quicksilver, and he gives directions for preparing the latter from the former. His method of making a superior red pigment is similar to that of Vitruvius in that it involves heating. He speaks of the use of cinnabar in art and for ". . . the sumptuous adorning of walls . . . " (Book 5, Chap. 109). His medical uses and toxicological ideas are described in Chapter 14.

A serious error is made by Dioscorides when he states that quicksilver ". . . is kept in glassen, or leaden, or tinnen, or silver vessels, for it eats through all other matter, and makes it run out" (Book 5, Chap. 110). This strongly suggests that his statement is not based on personal observation for he would have found that quicksilver will readily dissolve lead, tin and silver.

Additional details on the use of cinnabar and other red pigments in Greece and Rome are given by John, Lenz, Blümner, Donner, and Kroll. Quite naturally there is considerable duplication of coverage among these exhaustive works.

The Far East

China

That the Chinese have for many centuries assigned special virtues to red pigments is well known. An ancient Chinese word for cinnabar was tan-sha (Tsien [Ch'ien] 1962). According to Needham, "tan" which meant red, was a great witch word in Chinese alchemy and mineralogy, having signified cinnabar as far back as it can be traced.

One of the earliest uses of cinnabar was in making red ink. Literally thousands of so-called oracle bones used in divination have been found (Tong 1967) with inscriptions in red ink which has been shown to contain cinnabar (Britton 1937). A particularly rich source of old oracle bones was discovered in 1914 in the Anyang district of Honan province in central China, and these bones have been traced to the Shang-Yin dynasty (1751-1112 B. C.) (Tong 1967). The use of cinnabar ink may even go back for a considerable time before that period. The durability of cinnabar and the consequent survival of the inscriptions on oracle bones have played an important part in the reconstruction of early events in Chinese history. The writing on the oracle bones of the Shang-Yin dynasty confirmed the history of that period written by Ssu-ma Ch'ien (ca. 154-86 B. C.) (cited by Tong). Because of its superior qualities cinnabar ink was originally restricted to imperial use (Britton 1937), a practice which is said to have been enforced in Constantinople ca. 470 A. D. (Sarton 1927-1948, Vol 1; Laufer 1926). By the time of the Eastern Chou dynasty (770-265 B. C.) silk had largely replaced bone and shells as a writing surface but cinnabar ink continued in use (Tsien [Ch'ien] 1962).

A totally different use of cinnabar is suggested by Cheng in his description of a Shang tomb: "The ground is stained red probably as the result of the use of cinnabar in preserving the body of the dead." He does not indicate that analyses of the red pigment were actually performed, but he does show that preservation was not very good when he says that there was in the tomb ". . . some white powdery material, probably the remaining traces of a skeleton . . . ." Cheng's theory about the use of cinnabar to preserve the body of the dead receives support from an account by Lord Redesdale of burial practices in Japan: "The rich and noble . . . are partially preserved from decay by filling the nose, ears, and mouth with vermilion. In the case of the very wealthy, the coffin is completely filled with vermilion" (Mitford 1915).

Cinnabar ink is known to have been used during the Han Dynasty (second century B. C.- second century A. D.) and in fact has almost certainly been in continuous use in China right up to the present time (Tsien [Ch'ien] 1962). It has been reported that parts of the Great Wall of China were painted red,

and while there is no information as to whether or not this was cinnabar, it is
not entirely unlikely that it was.

A mercury still dating from the Han Period provides evidence that the
metal itself, as well as its sulfide, was known in China at least as early as that
time. Needham and Wang Ling (Vol. 4) mention the tomb of Chhin Shih
Huang Ti which is described in the Shih Chi as having a relief map showing the
course of the Chiang (Yangtze) and Ho (Yellow) rivers as well as other water
courses and the sea—all imitated by means of flowing quicksilver with a mechanism for producing circulation. This is said to date from about 210 B. C.
which is within the period of the Han Dynasty. Other Chinese mechanical devices using quicksilver described by Needham are a perpetual motion machine,
a form of clepsydra (a clock based on the regulated flow of liquid through a
small opening) in which mercury is used instead of water, and an armillary
sphere (an instrument with an arrangement of rings to show relationships of
various celestial circles) for which mercury provided the motive power.

Mirror making was known in China at least as early as the fourth century
A. D. and possibly as far back as the second century B. C. The Taoists are
said to have had a sort of patron saint of mirror polishers. The reflecting surface was made by depositing a tin-mercury amalgam on a smooth sheet of
bronze, the process thus being similar to that of later times. Some of the
emperors had special mirrors made which were said to show the internal
viscera and to reveal dangerous thoughts of imperial concubines (Needham and
Wang Ling, Vol. 4).

Chinese stamp ink, the familiar red ink used for seals, contains cinnabar as
its coloring material. According to tradition, the red seal for personal use
originated during the latter part of the Chou dynasty (ca. 1100–255 B. C.)
and for exclusive imperial use in the Ch'in dynasty (255–206 B. C.) (Li
Ch'iao P'ing 1948). The oil base was made from either castor bean, sesame
seed, rape seed, or tea seed. To this was added pulverized cinnabar (vermilion)
and sometimes powdered coral, pearl or other ingredients. Finally, moxa punk
was added and the entire batch was mixed and ground to form a smooth, uniform product. Moxa (*Artemisia indica, A. chinensis, and A. moxa*) was also
used in making stamp pads. Marco Polo (1254?–1324?) states that this
vermilion stamp ink was used by Kublai Khan for placing the royal signature
on paper currency. Cinnabar may thus be said to have been part of some of
the earliest printing inks.

Much of the early Chinese technology of quicksilver comes from (the)
T'ien kung K'ai wu. The manufacture of vermilion, the most important of the
pigments, is described in detail, with instructions both for its synthesis from
quicksilver and sulfur and for purification of the naturally occurring ore (Li
Ch'iao P'ing 1948).

Quite as familiar as the vermilion stamp ink is the rich, red pigment used in
decorating Chinese furniture and as a solid color on all kinds of ornate wooden
objects such as boxes and plates. This is cinnabar. Its use in this way is of relatively recent origin, most examples dating from the 18th century or later. The

large exhibit of "cinnabar ware" in the British Museum contains one object from the early part of the 15th century (Yung Lo Period), but all the rest are of a much later time.

Japan

Cinnabar has been found in the red colors used in Japanese wall painting at least as far back as the third or fourth century A. D. and has been shown to have had similar use from that time through the 16th century (Yamasaki 1964). There is a possibility that this pigment may have been used for decorating pottery very much earlier, in fact as early as the Jomon period which ended in the third century B. C.

The Americas

Peru

Cinnabar certainly, and quicksilver probably, were known to the natives of Peru in pre-Columbian times. A sample of a brilliant red powder found at Changoyape and dating from no later than 500 B. C. analyzed at Columbia University by a sensitive photometric technique (Jacobs et al. 1960) contained about 32 percent of mercury. There can be little doubt that this material is cinnabar.

The practice of body painting and tattooing is known to have been followed by the pre-Inca peoples as early as 200 B. C. (Montell 1929) and this may have been one of the early uses of cinnabar. Red was a favorite color as it was supposed to frighten away evil spirits and demons, and also to frighten enemies in battle. Body painting was practiced as a part of the puberty rites of young girls and sometimes merely for cosmetic purposes. Red pigments retained their popularity into and through the Inca period, and there is strong evidence that cinnabar was widely used (Montell 1929; Whitaker 1941), not only for coloring but also for divination and other religious purposes (Jiménez de la Espada 1965). One historian of the early part of the 17th century states that the Inca kings had obtained quicksilver and admired its properties but that because of its toxic nature they had prohibited its mining (Bargalló 1955).

Decorative arts of northern Peru in Chimu times (A. D. 1200–1400) made use of cinnabar in a manner resembling the cloisonné work of Europe (Easby 1966). The techniques ". . . included painting with . . . cinnabar, encrusting with turquoise, and dividing the surface of a metal object with small partitions and filling the spaces with cinnabar. . . ."

Middle America

Both quicksilver and cinnabar from pre-Columbian times were found at Copán (Honduras) by Maudslay during the latter part of the 19th century.

Similar findings later were described by Gann, but there is a striking resemblance between Gann's description and that of Maudslay. A vial containing about 90 grams of mercury originating at Copán and now in the Field Museum of Natural History in Chicago may possibly be the specimen described by Maudslay and Gann.*

Mexico

According to Clavigero, quicksilver mines at Chilipan were exploited by the inhabitants of Mexcio in pre-Columbian times. The Mexicans may have used mercury in amalgamating gold and silver (Rivet and Arsandaux 1946).

REFERENCES

Abramowitz, E. W. 1934. Historical points of interest on the mode of action and ill effects of mercury. *Bull. N. Y. Acad. Med.* 10:695–705.
Almqvist, A. 1928. Quecksilberschädigungen. In *Handbuch der Haut und Geschlechtskrankheiten.* Vol. 18, Berlin: Springer.
Aristophanes *Acharnians* 22.
Aristophanes *Eccleziarae* 378.
Bargalló, M. 1955. *La minería y la metalurgia en la América Española durante la epoca colonial.* Mexico: Fondo de Cultura Económica.
Blümner, H. 1884. *Technologie und Terminologie der Gewerbe und Künste bei Griechen und Römern.* Leipzig.
Britton, R. S. 1937. Oracle-bone color pigments. *Harvard J. Asiatic Studies* 2:1–3.
Caley, E. R. 1946. Ancient Greek pigments. *J. Chem. Ed.* 23:314–16.
Cheng, Te-k'un. 1960. *Shang China.* Archaeology in China, vol. 2. Cambridge: W. Heffer and Sons.
Chester, G. 1871. *The recovery of Jerusalem.* Edited by Walter Morrison. London.
Clavijero, F. J. 1780–81. *Storia antica del Messico cavata da' migliori storici spagnuoli, e da' manoscritti, e dalle pitture antiche degl'Indiani.* 4 vols. Cesena: G. Biasini.
Clermont-Ganneau, C. 1885. *Les fraudes archéologiques en Palestine.* Paris.
Colopinto, L. 1964. Breve storia del mercurio. *Ann. Med. Navale (Roma)* 69:314–17.
Davies, O. 1935. *Roman mines in Europe.* Oxford: Clarendon Press.
Dioscorides, P. 1934. *The Greek herbal.* Englished by John Goodyear, A. D. 1655. Edited and first printed, A. D. 1933, by Robert T. Gunther. Oxford: Published for the Author at the University Press.
Donner, O. 1869. *Die erhalten antiken Wandmalereien in technischer Beziehung.* Leipzig.
Easby, D. T., Jr. 1966. Early metallurgy in the New World. *Sci. Am.* 214:72–83.
Egenhoff, E. L. 1953. De argento vivo. Supplement to the *California Journal of Mines and Geology*, October, 1953.
Ettinghausen, R. 1965. The uses of spherico-conical vessels in the Muslim East. *J. Near Eastern Studies* 24:218–29.
Gann, T. 1931. *The history of the Maya.* New York: Chas. Scribner and Son.
Götze, A. 1924–32. In *Reallexikon der Vorgeschichte,* ed. M. Ebert, 15 vols. Berlin: W. de Gruyter and Co.
Goitein, S. D. 1967. *A Mediterranean society.* Economic Foundations, vol. 1. Berkeley: University of California Press.
Hasebroek, J. 1933. *Trade and politics in ancient Greece.* Trans. L. M. Fraser and D. C. MacGregor. London: G. Bell and Sons.
Haupt, P. 1902. Die "Steine von Tarsis." *Mitteilungen zur Geschichte und Naturwissenschaften* 1:386–87.

---

* D. Collier 1970: personal communication.

Herausbeger [pseud.]. 1826. Untersuchung einiger Farben, und der Decke auf welche sie aufgetraten waren welche aus einem alten ägyptischen Grabmal erhalten wurden. *Mag. Pharm.* 14:41–62.

Herodotus. 1922. *Herodotus.* Trans. A. D. Godley. New York: G. P. Putnam's Sons.

Jacobs, M. B., and Goldwater, L. J. 1961. Ultramicrodetermination of mercury in apples. *Food Technol.* 15:357–60.

Jacobs, M. B.; Goldwater, L. J.; and Gilbert, H. 1961. Ultramicrodetermination of mercury in blood. *Am. Ind. Hyg. Assoc. J.* 22:276–79.

Jacobs, M. B.; Yamaguchi, S.; Goldwater, L. J.; and Gilbert, H. 1960. Microdetermination of mercury in blood. *Am. Ind. Hyg. Assoc. J.* 21:475–80.

Jiménez de la Espada, M. 1965. *Relaciones geográficas de Indias: Perú.* Vol. 2. Madrid: Ediciones Atlas.

John, J. F. 1836. *Die Malerei der Alten.* Berlin.

King, C. V. 1957. Mercury: Its scientific history and its role in physical chemistry and electrochemistry. *Ann. N. Y. Acad. Sci.* 65:360–68.

Kroll, W. 1894–1962. In *Paulys Real-Encyclopädie der classischen Altertumswissenschaft,* by A. F. von Pauly, ed. G. Wissowa, vol. 15, pt. 2, cols. 1848–54. Stuttgart: J. B. Metzler.

Laufer, B. 1926. In *Printing ink,* ed. F. B. Wilborg. New York: Harper and Brothers.

Lenz, H. O. 1861. *Mineralogie der alten Griechen und Römer.* Gotha.

Li Ch'iao P'ing. 1948. *The chemical arts of old China.* Easton, Pa.: Journal of Chemical Education.

Lippmann, E. O. von. 1919. *Entstehung und Ausbreitung der Alchemie.* Berlin: Springer.

Lucas, A. 1962. *Ancient Egyptian materials and industries.* 4th ed., rev., and enl. J. R. Harris. London: Edmond Arnold.

Marshall, Sir J., ed. 1931. *Mohenjo-Daro and the Indus civilization.* Vol. 2. London: Arthur Probsthain.

Maudslay, A. P. 1889–1902. Archaeology. In *Biologia centrali-americana;* or, *Contributions to the knowledge of the fauna and flora of Mexico and Central America,* ed. F. D. Godman and O. Salvin. 9 vols. in 6. London.

Mellor, J. R. 1922–37. *A comprehensive treatise on inorganic and theoretical chemistry.* 16 vols. London: Longmans, Green and Co.

Mérimée, J. F. L. 1826. Sur la préparation et l'emploi des couleurs, des vernis et des émaux, dans l'ancienne Égypte. In *Catalogue raisonné et historique des antiquités decouvertes en Égypte,* by J. Passalacqua, pp. 258–59.

Mitford, A. B. F. Baron Redesdale. 1915. *Tales of old Japan.* London: Macmillan Co.

Montell, G. 1929. *Dress and ornaments in ancient Peru.* Göteborg: Elanders.

Needham, J., and Wang Ling. 1954–1962. *Science and civilization in China.* 4 vols. Cambridge: Cambridge University Press.

Partington, J. R. 1935. *Origins and development of applied chemistry.* London: Longmans, Green and Co.

Perrott, G., and Chipiez, C. 1882–1914. *Histoire de l'art dans l'antiquité.* 10 vols. Paris: Hachette.

Petrie, W. M. F. 1894. *Tell El Amarna.* London.

Pharm. Z. 1890. 35:630.

Plinius Secundus, C. 1938–63. *Natural history.* Trans. by H. Rackham, W. H. S. Jones and D. E. Eichholz. 10 vols. London: W. Heinemann.

Polo, M. 1929. *The book of Ser Marco Polo.* Trans. and ed. Sir Henry Yule. Third ed., by H. Cordier. 2 vols. London: J. Murray.

Ray, P. 1956. *History of chemistry in ancient and medieval India,* incorporating *The history of Hindu chemistry* by A. P. C. Ray. Calcutta: Indian Chemical Society.

Rhousopoulos, O. A. 1909. Beitrag zum Thema über die chemischen Kentnisse der alten Griechen. In *Beiträge aus der Geschichte der Chemie,* ed. Paul Diergart, pp. 178 ff. Leipzig: Deuticke.

Riis, P. J.; Paulsen, V.; and Hammershaimb, E. 1957. *Hama: Fouilles et recherches, 1931–1938.* Vol. 4, pt. 2. Copenhagen: National Museum.

Rivet, P., and Arsandaux, H. 1946. La métallurgie en Amérique précolombienne. *Trav. Mém. Inst. Ethnol. (Univ. Paris)* 39:6.

Sarre, F. 1925. *Keramik und andere Kleinfunde der Islamische Zeit von Baalbek.* Sonderabdruck aus dem III. Bande der Ergebnisse der Ausgruben und Untersuchungen in den Jahren 1898 bis 1905. Berlin: W. de Gruyter.

Sarton, G. 1927-48. *Introduction to the history of science.* 3 vols. in 5. Baltimore: Williams and Wilkins.

Saulcy, F. de. 1874. Note sur des projectiles à main, creux et en terre cuite, de fabrication arabe. *Mém. Soc. Nat. Antiq. France* 35:18-34.

Schelenz, H. 1904. *Geschichte der Pharmazie.* Berlin: Springer.

Seligmann, S. 1910. *Der böse Blick und Verwandtes.* 2 vols. Berlin: H. Barsdorf.

Sigerist, H. E. 1967. *A history of medicine: I. Primitive and archaic medicine.* New York: Oxford University Press, Galaxy Book.

Stillman, J. M. 1960. *The story of alchemy and early chemistry.* New York: Dover Publications.

Strunz, F. 1906. *Über die Vorgeschichte und die Anfänge der Chemie.* Leipzig: F. Deuticke.

Theophrastus. 1956. *Theophrastus on stones.* Introd., trans., and commentary by Earle R. Caley and John F. C. Richards. Columbus: Ohio State University Press.

Theophrastus. 1965. *De lapidibus.* Ed., with introd., trans., and commentary by D. E. Eichholz. Oxford: Oxford University Press.

Thompson, C. J. S. 1923. *Poison mysteries in history, romance and crime.* London: Scientific Press.

Thompson, R. C. 1936. *Dictionary of Assyrian chemistry and geology.* Oxford: Clarendon Press.

Toch, M. 1918. The pigments of the tomb of Perneb, *J. Ind. Eng. Chem.* 10:118-19.

Tong, Te-Kong. 1967. The tortoise shell which set off a mighty chain reaction. *Columbia Library Columns* 16:11-18.

Tsien (Ch'ien), Tsuen-Hsuin. 1962. *Written on bamboo and silk.* Chicago: University of Chicago Press.

Ure, A. 1878. In *Manners and customs of the ancient Egyptians,* edited by J. G. Wilkinson, vol. 2. London.

Vitruvius Pollio. 1931-34. *On architecture.* Ed. from the Harleian manuscript 2767 and trans. by Frank Granger. 2 vols. New York: G. P. Putnam's Sons.

Wheeler, Sir Mortimer. 1966. *Civilizations of the Indus Valley and beyond.* London: Thames and Hudson.

Whitaker, A. P. 1941. *The Huancavelica mercury mine.* Cambridge: Harvard University Press.

Wilkinson, Sir J. G. 1878. *The manners and customs of the ancient Egyptians.* New ed. rev. by Samuel Birch. London.

Wunderlich, E. 1925. *Die Bedeutung der roten Farbe im Kultus der Griechen und Römer.* Giessen: Töpelmann.

Yamasaki, K. 1964. Pigments employed in old paintings in Japan. In *Archeological chemistry: A symposium,* ed. M. Levey, pp. 347-65. Philadelphia: University of Pennsylvania Press.

ADDITIONAL READINGS

Agricola, G. 1955. *De natura fossilium.* Trans. from the 1st Latin ed. of 1546 by Mark Chance Bandy and Jean A. Bandy. New York: Geological Society of America.

Aristotle. *De anima.* 1. 3.

Caley, E. R. 1928. Mercury and its compounds in ancient times. *J. Chem. Ed.* 5:419-24.

Gann, T. 1926. *Ancient cities and modern tribes.* London: Duckworth.

Haeser, H. 1875-82. *Lehrbuch der Geschichte der Medicin und der epidemischen Krankheiten.* 3rd. rev. ed. 3 vols. Jena.

Homer. *Iliad* 2.

Homer. *Odyssey* 9.

MacKendrick, P. 1966. *The mute stones speak.* New York: New American Library, Mentor Book.

Maspero, Sir G. 1926. *Manual of Egyptian archaeology.* Transl. and enl. Agnes S. Johns. 6th English ed. New York: G. P. Putnam's Sons.

Michell, H. 1940. *The economics of ancient Greece.* New York: Macmillan Co.
Mochi, A. 1916. Indizi di miniere preistoriche di cinabro nella regione dell'Amiata. *Boll. Paletnol. Ital.* 1:5–12.
Torr, C. 1894. *Ancient Ships.* Cambridge: Cambridge University Press.
Watson, W. 1966. *Early civilization in China.* London: Thames and Hudson.

Medieval Knowledge and Uses

Although quicksilver was important during the middle ages chiefly in alchemy and medicine, it was not unknown in medieval science and technology. One craftsman, Theophilus the Monk, who probably lived during the latter part of the 11th or early part of the 12th century, wrote of mercury in his book, *De Diversis Artibus*. This is the work of a skilled and experienced master, containing, among other things, detailed descriptions of a number of metallurgical techniques, including cupellation, refining, smelting and casting. The third and longest of the three books in Theophilus's work, is concerned primarily with metals and devotes several chapters to processes in which quicksilver is used.

A chapter on the milling of gold describes how one part of gold is mixed with eight parts of vivum argentum. Another chapter describes the recovery of gold by means of amalgamation, and three chapters are devoted to various types of gilding. Of the latter, one is concerned with putting a quicksilver finish on tin jugs.

The first book of *De Diversis Artibus* deals mainly with the preparation and use of colors in painting, anticipating by four centuries the famous work of Cennino Cennini. Of particular interest is the statement that minium is prepared from "plumbeas tabulas" while cenobrium is made from sulfur and vivum argentum. This provides conclusive proof that the metamorphosis of minium from mercury to lead had been completed, in spite of Pliny and Isidore.

Alfonso X

King Alfonso X of Castile (1221-1284), commonly known as El Sabio—the Wise—was responsible for the construction of one of the very few mechanical devices of the middle ages in which quicksilver was used. Historians of

86

science and technology are familiar with King Alfonso X through his *Libros del Saber de Astronomía,* compiled for him by a group of scholars about 1275. One of the books, "Libro del Relogio dell Argen Vivo" (Book of the Quicksilver Clock) gives a detailed description of how the clock is constructed, complete with working drawings and the method by which it is to be used as the motive power for the operation of an astrolabe. The moving parts are designed to make one complete rotation in a day and a night, "ni mas ni menos" (Fig. VI-1). (About 300 years later [c. 1570] Tycho Brahe measured time ". . . either by direct astronomical observation of the positions of particular stars or by weighing the mercury delivered by a device based on the principle of the water clock" [Usher 1929].)

Arabian Writings

Among the important Near Eastern writers on mercury during the Middle Ages was Jabir ibn Hayyan (Geber, Djeber) who, according to Holmyard, may have lived during the 8th and 9th centuries, or who may have been a mythical

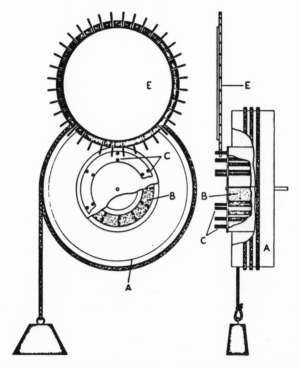

Figure VI–1 Mercury Clock of Alfonso X, c. 1275. A, weight-driven drum; B, mercury container divided by perforated partitions; C, studs that turn the dial E. (From *A History of Mechanical Inventions* by A. P. Usher. Copyright 1929 by McGraw-Hill Book Co. Used with permission of McGraw-Hill Book Co.

being, as maintained by Multhauf. In either case, a large body of Arabic al-chemical and chemical literature does exist under the name of Jabir, but the discussions of quicksilver are primarily alchemical. Al-Razi (Rhazes), reputed to have lived from 825 or 826 to 925, also made some major contributions to the Arabic literature on mercury and its compounds.

Among the works translated from Arabic to Latin by Gerard of Cremona (1114-1187) was one known as "Rasis de aluminibus et salibus." This book may or may not have been written by Al-Razi, but it ". . . is indeed much in-debted to him" (Multhauf 1966). In this important work mercury and its compounds receive more attention than any other metal, occupying thirteen of the seventy chapters. Of particular significance, according to Multhauf, is its introduction of ". . . the important artificial substance, corrosive subli-mate." While most of the Arabic writings dealing with mercury were either medical or alchemical, a few uses for other purposes have been recorded. Amalgamation for the recovery of gold was seen by the geographer al-Edrisi when he visited Wadi-Allaqi and Safala in central Africa in 1154 (Muhammad ibn Muhammad, Vol. I, pp. 42, 67). He describes how the auriferous sands were washed with water and then mixed with quicksilver; the resulting amalgam was then heated, driving off the mercury and leaving the gold.

A story is told of a novel use of quicksilver by the Caliph Abderrahman (ruled 912-961) for the purpose of impressing the guests at his palace in Cordoba, not far from Almadén. The caliph's device consisted of a large shell-like basin of porphyry mounted on rockers and filled with mercury. The walls of the pavilion in which the basin was placed were finished in ebony and ivory of high polish so that when the sun's rays struck them there was a blinding re-flection. When the guests had assembled in the pavilion, the basin was made to rock in such a way that the reflected light would create an illusion that the building was a ship being tossed on a stormy sea (Zippe 1857).

Leonardo

Somewhat later another devotee of practical jokes, Leonardo da Vinci (1452-1519), ". . . a man of very high spirit . . .," found a use for quicksilver. The story, as related by Vasari (1511-1574), runs as follows:

> When Leo X became pope, Leonardo went to Rome with Duke Giuliano de'Medici. A gardener of the Belvedere one day brought in a curious lizard for which Leonardo made wings from the skins of other lizards. In these wings he put quicksilver, so that, when the animal walked, the wings moved with a tremulous motion. He then made eyes, horns, and a beard for the creature, which he tamed and kept in a cage. He showed it to his visitors, and all who saw it ran away terrified.

Excluding alchemical and medical uses, medieval European information on quicksilver, as on most technical matters, is found principally in the writings of of the compilers and encyclopedists. To avoid useless repetition, four have been chosen for consideration: Isidore of Seville (c. 560-636), Vincent of

Beauvais (c. 1190–1264), Bartholomaeus Anglicus (fl. c. 1260) and Albertus Magnus (c. 1200–1280).

Isidore of Seville

Isidore's major work, the *Etymologies,* has been called one of the outstanding feats of scholarship of all time. It achieved great popularity and was one of the most widely read books for some thousand years after its publication, serving as a model for later encyclopedists through the 13th century. Prior to the appearance of Isidore's book, Pliny's *Natural History* had been the most widely used reference work, but the *Etymologies* soon supplanted it (Stahl 1962). Much of Isidore's material (despite Crombie's statement to the contrary) was based on Pliny, whether directly or through Solinus (Albert, 1967 translation, p. 266).

Seven metals are described by Isidore: gold, silver, copper, electrum, tin, lead and iron. Argentum vivum is looked upon as a special kind of silver. It can be distilled from minium but is also found in the oldest excrement of sewers and in the slime of cesspools, he says (XIX, xvi).

The section on colors in the *Etymologies* includes a discussion of red pigments, including minium, cinnabar, rubrica, ochra and Sinopis. Some of the wording is practically a verbatim reiteration of Pliny (XIX, xvii).

An additional reference to cinnabar is found in that section of Isidore's book which describes certain customs and distinguishing marks of various peoples: "we may see the curls of the Germans . . . cinnabar of the Goths, and scars of the Britons" (XIX, xxiii,7).

Vincent of Beauvais

Possibly there is some justification for Sarton's condemnation (Vol. 2, pp. 929–930) of Vincent of Beauvais as having contributed nothing new or original and for his ". . . backwardness . . . with regard to the more scientific subjects," but it is unfair to ignore the comprehensive nature and positive values of Vincent's *Speculum Quadruplex.* In the *Speculum Naturale*, Vincent's discussions of quicksilver, minium and cinnabar are typical of the pains to which he went to gather and record all the information (and misinformation) he could find. According to Stillman, more than 300 authors are quoted. This in itself is sufficient to justify an important place for Vincent in the history of science. Had the works of the authorities whom he cited and those of his contemporary encyclopedists failed to survive, the *Speculum Quadriplex* might have become the most important book of the middle ages.

In his chapters on argentum vivum (Book 7, Chaps. 61–65) Vincent quotes from Isidore, Pliny, Avicenna, Constantine, Platearius, Dioscorides, Aristotle, and Galen as well as works, not identified by him, *de vaporibus, de natura rerum* (probably Bede) and *de aluminibus et salibus* (possibly Rhazes). The medical uses mentioned are those of the authorities he cites and involve exter-

Rgentum viuf. Cyfdoze. Ar
gent vif eft ainfi dit pource quil
trenche les matieres efquelles
il eft mis/auffi il eft luyfant et q
court toufiours. On fe treuue p
cfpeciales metaufp:ou es four/
naifes dargent adßerat auy parois par petite⁵
goutes. On fe treuue auffi bien fouuent en fie
tes bien vieilles des latrines ou en fymon des
puys.

Figure VI-2 An Early Pictorial Representation (Possibly the First Known) of
Quicksilver. Ca. 1500. (From *Hortus Sanitatis. Le Jardin de Santé.* Paris:
Antoin Vérard [not later than 1501/02].)

nal applications in the form of ointments. The toxic effects of mercury are
described and treatment outlined, including the familiar goat's milk, wine,
wormwood, and hyssop. The dire effects of mercury vapors on man and
"reptilia" are not overlooked.

Vincent's chapter on minium (Book 7, Chap. 101) could well be an example
of the "backwardness with regard to the more scientific subjects" deprecated

by Sarton. Although Vincent quotes or cites Theophrastus and Pliny he confuses matters by combining minium and ochra in a single chapter. His discussion must refer to minium as he says it comes from Ephesus and Hispania and that if it is distilled it gives argentum vivum. Vincent probably was not aware of the change in the meaning of minium so clearly brought out by Theophilus some two hundred years earlier and thus he inadvertently contributed to the confusion which has persisted up into the 20th century.

Bartholomaeus Anglicus

The inclusion of quicksilver by Bartholomaeus Anglicus in his book entitled *De Lapidibus Preciosis* does not necessarily imply that he considered it to be in the same category as gold and silver, for in the same chapter he treats of many other metals, including such common ones as iron and lead. Classifying mercury as a "stone" was common practice in the 13th century; by some writers all minerals were called stones (lapides).

Credit for determining that mercury is an element is usually given to Lavoisier, but Bartholomaeus in a sense anticipated him by about 400 years when he wrote: "Quicke silver is matter of all metal, and therefore in respect of them it is a simple element" (Book 16, Chap. 8). He was aware of its toxic properties in that "The smoke thereof is most grevous to men that ben therby. For it bredeth the palsey, and quaking, shakynge, neshynge [softening, O.E.D.] of the synewes. If it be take in at the mouthe, or do in the ere, it thyrleth [pierces] and sleeth [slays] the members."

Bartholomaeus values quicksilver next after gold but gives no special uses for it other than as a medicine to heal wounds and prevent infection, his wording on this point being similar to that of Dioscorides on ochra.

Minium and cinobrium are discussed by Bartholomaeus in his book *De Coloribus* (Book 19, Chaps. 26, 27). His account of the former, in toto is: "Minium is a red colour, and the Grekes found the matter thereof in Ephysum: In spayne is more suche Pigmente than in other londes, as Isidore sayth." His chapter De Cinobrio starts with "Sinobrium hyghte Cinabrin amonge the Grekes . . ." and ends with the familiar "As Isidore sayth."

Albertus Magnus

Quicksilver appears frequently in Albertus Magnus's *Book of Minerals* (see Albert 1967 translation), and in Book IV it is found on nearly every page. Completely without warning, Albert changes back and forth between the chemical and alchemical meanings of the word, which suggests that to him there was no difference. This practice is a reflection of the many influences under which he worked ranging from Aristotle to Hermes Trismegistus. On the one hand, Albertus exhibits a keen scientific sense when he writes that quicksilver may be found "live" or may have to be extracted from its ores by roasting, and that minium (cinnabar) can be made by subliming it with sulfur, and on the other hand he ends his chapter on quicksilver in alchemical jargon, saying that "All this has been said so that it may be understood that Quicksilver is nothing but the matter [female principle] in metals, since it undoubtedly suffers complete dissolution by means of sharp waters, either natural or artificial" (Albert, p. 208).

Very little is said by Albertus on practical uses for quicksilver, but he does mention that it is a kind of poison and therefore can cause loosening of the sinews and paralysis and that ". . . it kills lice and nits and other things that are produced from filth in the pores" (Albert, p. 207).

Albertus' *Book of Stones* contains a chapter (Book II, Tractate ii, Chap. 19) entitled Varach, which is another name for *sanguis draconis*. Albertus quotes "Aristotle" (pseudo-Aristotle, the alchemist) as saying that it is a stone, ". . . but some medical men say that it is the juice of a certain plant." In Book III, Tractate i, Chapter 10, Albertus mentions mines in Goslar [Saxony] in which ". . . quicksilver is found running out . . . ."

There is ample evidence that quicksilver and cinnabar were well known in medieval times. It would be supererogatory to cull additional proof from the writings of other encyclopedists. Roger Bacon (?1214–1292) knew argentum vivum and used it in his alchemical experiments. To search through the voluminous writings of men such as Bede (?673–735) and Grosseteste (c. 1170–1253) and the lesser known encyclopedists to see what they say about mercury would result in nothing new.

Evidence of the persistence of medieval ideas about mercury up to the beginning of the 16th century is found in the *Hortus Sanitatis* of Antoine Vérard, published around 1500. Of particular interest are the woodcuts, one of which is perhaps the earliest to depict the appearance of quicksilver (Fig. VI-2).

REFERENCES

Albert, Saint, Bishop of Ratisbon, surnamed the Great. 1967. *Book of minerals*. Trans. Dorothy Wyckoff. Oxford: Clarendon Press.

Alphonso X, King of Castile and León, surnamed the Wise. 1836–67. *Libros del saber de astronomía del Rey D. Alfonso X de Castilla*. Copilados, anotados y comentados por D. M. Rico y Sinobas. 5 vols. Madrid.

Bacon, Roger. 1920. *Secretum secretorum*. Ed. Robt. Steele. Oxford: Clarendon Press.

Bartholomaeus, Anglicus. 1535. *De proprietatibus rerum*. London.

Cennini, C. 1933. *Il libro dell'arte, the craftsman's handbook*. Trans. from the Italian by D. V. Thompson, Jr. New Haven: Yale University Press.

Crombie, A. C. 1959. *Medieval and early modern science*. 2d ed. Garden City, N.Y.: Doubleday Anchor Books.

Holmyard, E. J. 1968. *Alchemy*. Harmondsworth, Middlesex: Penguin Books.

Isidore, Saint, Archbishop of Seville. 1911. *Etymologiarum sive originum libri XX*. Ed. W. M. Lindsay. 2 vols. Oxford.

Muḥammad ibn Muḥammad, al-Idrisi. 1836–40. *Géographie d'Edrisi*. Traduit de l'arabe en français par P. Amedée Jaubert. 2 vols. Paris.

Multhauf, R. P. 1966. *The origins of chemistry*. London: Oldbourne.

Sarton, G. 1927–48. *Introduction to the history of science*. 3 vols in 5. Baltimore: Williams and Wilkins Co.

Stahl, W. H. 1962. *Roman science*. Madison: University of Wisconsin Press.

Stillman, J. M. 1960. *The story of alchemy and early chemistry*. New York: Dover Publications.

Theophilus. 1961. *De diversis artibus.* Translated from the Latin with notes and intro-
duction by C. R. Dodwell. London: Thomas Nelson and Sons.
Usher, A. P. 1929. *A history of mechanical inventions.* New York: McGraw-Hill Book Co.
Vasari, G. 1946. *Lives of the artists.* Abridged and ed. by Betty Burroughs. New York:
Simon and Schuster.
Vincentius, Bellovacensis. 1964–65. *Speculum quadruplex.* Facsimile of the 1624, Duaci,
edition. Graz: Akademische Druck- u. Verlagsanstalt.
Walsh, J. J. 1924. *The thirteenth: Greatest of centuries.* New York: Catholic Summer
School Press.
Zippe, F. X. M. 1857. *Geschichte der metalle.* Vienna.

ADDITIONAL READINGS

Edelstein, L. 1967. *Ancient Medicine: Selected Papers of Ludwig Edelstein.* Ed. Owsei
Temkin and C. Lilian Temkin. Baltimore: Johns Hopkins Press.
Steele, R. R. 1929. Practical chemistry in the twelfth century: Rasis de aluminibus et
salibus. *Isis* 12:10–48.

7

Mercury in the History of Chemistry

Historians of science in the mid-twentieth century frequently use the word "paradigm" in describing a system of beliefs, and have pointed out that old paradigms do not easily give way to new. The transition from alchemy to chemistry illustrates this thesis very well with mercury responsible for much of the difficulty and at the same time serving as a link between the old and the new. Without mercury, in both the literal and alchemical senses, alchemy may not have existed at all, or, if it did exist, it would have had to assume a form entirely different from that which actually developed. It is difficult to imagine what course chemistry would have taken had there been no alchemy and even more difficult to conceive of what would have happened had there been no mercury.

Before the new chemistry could come into its own it was essential to take steps to clear away the accumulated accretions of more than a thousand years of alchemy. This was clearly recognized well before the end of the 17th century, as exemplified by Lemery's famous *Cours de Chymie*, the first edition of which was published in 1675. In the words of Walter Harris who translated Lemery's Fifth Edition into English, this treatise

> . . . is not writ after the usual way of ordinary Chymists; it has none of the bombastick expressions, nor ridiculous pretences, none of the Melancholick Dreams, and wretched Enthusiasms, none of the palpable falsities, and even impossibilities wherewith the common rate of Chymical Books has been stuff'd heretofore. The Author is no Believer in that great stumbling-block, the Mystery of Projection, nor at all addicted to the Transmutation, or rather Adulteration of Metals, . . . he may be said to have Purified and Refined Chymistry from the many dregs and feculencies, which by other mens over-refining, and over-curious diligence it had been tainted with before (Lemery 1686 translation, Introduction).

94

Similar views were expressed early in the 18th century by Joseph (sic)
(= John) Freind (1675-1728), Professor of Chemistry at Oxford who wrote
in 1712: "Chemistry has made a very laudable progress in Experiments; but
we may justly complain, that little Advances have been made towards the Ex-
plication of 'em. . . No body has brought more Light into this Art than Mr.
Boyle . . . who nevertheless has not so much laid a new Foundation of Chemistry as he has thrown down the old" (cited in Butterfield 1965).

Mercury, as one of the cornerstones of the hermetic art, figured prominently in the process of demolition but promptly showed its versatility by becoming an essential tool in the hands of those who were building the new
science.

Among the first to join the revolutionary movement was Paracelsus (1493-
1541). He forcefully rejected the reigning Aristotelian concept of the four
"elements," earth, air, fire and water, but with equal strength espoused the
alchemical *tria prima* of mercury, sulfur and salt (Pagel 1958). Possibly because of his personal idiosyncrasies, his teachings made little immediate impact, but eventually many of his ideas prevailed.

Seventeenth Century

Nearly a century had passed before the first significant post-Paracelsian
step was taken, this one by Johann Baptista van Helmont (1579- 1644), the
most prominent chemist of his time. Van Helmont refused to accept the elemental nature not only of the four principles of Aristotle but also of the *tria
prima* of Paracelsus and the alchemists. Paradoxically, however, he believed in
alchemical transmutation and claimed to have accomplished the "magnum
opus" himself by means of the philosopher's stone; he even named his son
Franciscus Mercurius. Known best, perhaps, for his pioneering studies on the
gastric juice, van Helmont made a major contribution to chemical theory
when he demonstrated the principle of the indestructibility of matter. This
he did by an experiment in which he boiled quicksilver in oil of vitriol (sulfuric acid) thus producing a precipitate from which the original mercury
could be quantitatively recovered (Partington 1965).

The first real broadside in what might be called the chemical Hundred
Years War was fired by Robert Boyle (1627-1691) with the publication in
1661 of *The Sceptical Chymist: or Chymico-Physical Doubts & Paradoxes,
touching the Spagyrist's Principles commonly call'd Hypostatical, as They are
wont to be Propos'd and Defended by the Generality of Alchymists.* One of
the principal objectives of the *Sceptical Chymist* was to attack the idea that
either the four "principles" of Aristotle or the *tria prima* of the alchemists
should be considered elements. To do this Boyle created his own definition of
principles or elements as being "Those primary and simple bodies of which
the mixt ones are said to be composed, and into which they are ultimately re-

solved." To bolster his argument he cited some of the properties of quicksilver, thus using mercury in its literal sense to destroy "mercury" as an alchemical essence or principle.

Boyle is probably best known from the "law" (announced in 1662) which bears his name: at constant temperatures the volume of a confined gas is inversely proportional to the pressure. (Boyle's experiments were done with air. The word "gas" was first employed by van Helmont about 1630). The experiments which led to the formulation of Boyle's law involved the use of mercury. In proving the law for pressures greater than atmospheric, Boyle employed a U-tube containing the metal, and for pressures less than atmospheric he used a straight glass tube in which air was confined above mercury. It is, of course, possible that some liquid other than mercury might have been used in demonstrating the pressure-volume relationships of gases, but the fact remains that quicksilver was Boyle's choice, as it was of Edmé Mariotte (1620–1684) who independently discovered the pressure-volume relationship in 1676 or 1679 (Magie 1935).

Problems related to combustion had long puzzled the chemists. In particular they were at a loss to explain the increase in weight of metals that had been subjected to high temperatures such as in calcination or roasting. Boyle was unable to solve this problem. He found that when mercury was calcined in air, it formed a red precipitate which he believed could be reduced to free mercury if greater heat could be applied. He was not sure whether this was a purely mechanical process, or whether it involved the association of some "penetrating igneous particles" with the "mercurial corpuscles." This may be interpreted as at least a hint of the existence of what later was to be identified as oxygen.

In other experiments of lesser significance, Boyle used corrosive sublimate to produce anhydrous cuprous chloride and bismuth chloride. These were examples of the same type of double decomposition previously described by Glauber.

Several additional developments in 17th century chemistry depended on the use of quicksilver. They are noteworthy principally because they show that many of the most distinguished chemists of the period employed the metal in their experiments. The "spiritus fumans Libavi" of Andreas Libavius (1540?–1616), which was tin tetrachloride, was made from tin and mercuric chloride. Johann Rudolph Glauber (1604–1670) of Glauber's salt fame, demonstrated the phenomenon of double decomposition, using corrosive sublimate ($HgCl_2$) and stibnite (antimony sulfide, Sb_2S_3) to produce antimony chloride ($SbCl_3$). Similarly he made arsenic chloride ($AsCl_3$) from mercuric chloride and arsenic trisulfide (Partington 1965). Glauber was an admirer of Paracelsus and a student of his writings, accepting the *tria prima* but substituting the term "water" for the "mercury" of Paracelsus. In distinguishing between the true mercury and "mercury" as one of the three principles (a distinction also made by other contemporary chemists), Glauber contributed to the downfall of alchemical doctrine.

Robert Hooke's work with thermometers and barometers (Chapter 8, below) represents only one aspect of his experimental activities in which quicksilver was used. He was a man of many parts and he even challenged Sir Isaac Newton's claim to priority in formulating the laws of universal gravitation. For a time, Hooke (1635–1703) served as Boyle's assistant and was extremely active in the Royal Society during its formative years, in both of which connections he had occasion to employ mercury in various experiments. In the preface to his best known work, *Micrographia*, Hooke describes in great detail the construction of an instrument for ". . . discovering the effluvia of the Earth mixt with the Air." This instrument employed a column of mercury ". . . by means of which contrivance, every the least variation of the height of the Mercury will be exceedingly visible by the motion to and fro of the Small Index. . . ." This is but one of several devices based on the Torricellian barometer that were designed by Hooke and used in a number of experiments. Mercury was also used by Hooke in his studies of capillary action, reflection and refraction of light, and other phenomena. It appears again and again throughout *Micrographia*, sometimes as "mercury," sometimes as "Quicksilver" and sometimes represented as the alchemical symbol ☿. Minium, cinnabar, sublimate, and vermilion also make their appearance in Hooke's discussions of colors.

Despite the significant strides toward a new chemistry made during the latter part of the 17th century, alchemical ideas about mercury were far from dead. This can be illustrated by a passage from a work by John Mayow (1641–1679), his *Tractatus Quinque Medico-Physici*: "First then, in the birth of plants, the nitro-aerial spirit or Mercury when set in motion by the impulse of solar rays, descends in virtue of its very penetrating nature into the depths of the earth and attacks there its most bitter enemy terrestrial sulphur, firmly united with fixed salt and nearly hidden and buried in its embrace . . . ." Words such as these, in spite of recognition of the mercury-sulphur relationship, might be dismissed as the fanciful ramblings of a "puffer" were it not that Mayow was recognized as a competent chemist whose experiments on combustion and respiration were of sufficient merit to prompt Robert Hooke to nominate him for Fellowship in the Royal Society. Along with Hooke and Boyle, he made up the trio known as the "Oxford Chemists."

Further evidence of the entrenched position of "essential mercury" can be found in the writings of Johann Kunkel (1630–1703) and Johann Joachim Becher (1635–1682), both of whom are important in the history of chemistry. Kunkel served as court alchemist to Elector John George of Saxony and Frederick William of Saxony and he believed that all metals contained "mercury." This did not prevent him from making at least one important contribution to scientific chemistry by being one of two independent discoverers of phosphorus. Best known as an early proponent of the Theory of Phlogiston, Becher believed that all matter was composed of air, water and three earths, one of which was "mercurial." In effect, his three earths were equivalent to the sulfur, salt, and mercury of the alchemists (Partington 1965).

Mercury must have enjoyed a great vogue in Restoration England since biographers of Charles II have noted that the king started experimenting with it shortly after his return from exile (Bryant 1931). Exposure to the vapors of mercury in the course of his research has even been suggested as being at least partly responsible for the king's final illness (Wohlbarsht and Sax 1961). Charles had at his court Nicasius le Febure (Nicholas Le Fèvre), whose title was Royal Professor in Chymistry to His Majesty of England, and Apothecary in Ordinary to His Honourable Household. He was a Fellow of the Royal Society and published *A Compleat Body of Chymistry* in London, 1670. The king's interest in research has been described in this way:

> Within the Palace walls were the Privy Gardens . . . the bowling green . . . the little Physic Garden, where he culled herbs for his laboratory. The latter, which he had made beneath his Closet, was a constant delight to him. Here his chemist, Le Febvre [sic] presided over the chemical glasses which so puzzled Pepys, and his successor, Williams, made the famous King's Drops, which not only cured the sick but, surreptitiously dropped by the invaluable Chiffinch into the wine of guests, made cunning tongues babble secrets. The King loved to work here, making researches into the heart of things more exact and curious than the nebulous nature of political life afforded, dissecting, composing cordials or endeavoring to fix mercury (Bryant 1931).

Eighteenth Century

Mercury's contributions to the chemistry of the 17th century, while important, were not spectacular. In the 18th century they were both important and spectacular. Without mercury some of the most significant advances in chemistry would have been delayed for years and possibly for centuries.

Surely one of the most important developments, perhaps the most important single development, in the history of chemistry was the discovery of oxygen. Observations which for many years had baffled the chemists, found, in oxygen, a sound, scientific explanation. Throughout the train of events that led to the discovery of oxygen, including the final isolation and identification of the element, mercury occupied a prominent position. Full appreciation of the importance of this discovery and the role played by mercury requires some familiarity with what went before.

As early as about 1300 A. D. it had been observed by the so-called pseudo-Geber that calcination of tin and of lead added to the weight of these metals (see Stillman 1960). Toward the end of the 15th century Eck von Sulzbach (fl. c. 1500) described an experiment in which six pounds of mercury-silver amalgam were heated for a period of eight days. At the end of this time the weight of the samples had increased to nine pounds (Stillman 1960). (Such a marked gain in weight cannot be explained on the basis of oxidation alone since the oxides of neither mercury nor silver have atomic weights of 50 percent more than the parent metal. Perhaps some of the material from the vessels in which the heating was done came off with the sample, or perhaps the calcined material took up moisture before it was weighed. In any case, a substantial gain in weight had been observed.) Biringuccio (1480–1538) in the early part of the 16th century had noted that the calcination of lead in a furnace leads to a substantial increase in weight of the material. There is evidence that Jerome Cardan (1501–1576) also knew that the weight of metals increased on calcination (Partington 1965). Somewhat later, but antedating Boyle, Jean Rey (ca. 1583–1645) and later Otto Tachenius (died ca. 1700) had made similar observations (Partington 1965; Stillman 1960; Rey 1904).

During the latter part of the 16th, and throughout the 17th century, a number of chemists had attempted to explain the observed increase in weight of metals which had been subjected to calcination. Various theories were propounded to explain these increases in weight. To test some of these theories, Boyle heated tin and mercury in sealed flasks, but still found a slight addition to the weight of his samples. Boyle was convinced that something in the fire was the source of the augmentation, thus overlooking the possible role of the air contained in the closed, but not evacuated, vessels. The wide acceptance of Boyle's theory of "igneous Corpuscles" undoubtedly contributed to the delay of a century before the true function of air in combustion was understood. The work of Scheele and Priestley, both of whom used mercury, was to provide the solution.

The bulk of Boyle's work on combustion was reported in 1673; the experiments of Scheele and of Priestley (leading to the discovery of oxygen) were performed almost exactly 100 years later. In the meantime the phlogiston theory commanded wide support in the chemical world. The two principal proponents of this theory were Johann Joachim Becher and Georg Ernst Stahl (1660?–1734), both German. Phlogiston was an hypothetical substance, essence, or principle presumed to be present in all inflammable bodies, and to be given off during combustion. It was supposed to have "negative weight," a property which accounted for gain in weight in a body which was calcined— negative weight was lost. The phlogiston theory had wide acceptance as the most reasonable explanation of this phenomenon until Lavoisier showed that combustion was dependent on oxidation.

The Discovery of Oxygen

At the end of the 17th century and in fact well into the 18th very little was known about gases. The word "gas" had been coined by van Helmont, probably around 1630, derived from the Greek "chaos" as a general term for air. Van Helmont described a "gas sylvestre" now recognized as carbon dioxide, and "gas pingue" which may have been either impure hydrogen or methane (marsh gas). The term "gas" was not employed by any of van Helmont's immediate successors but was reintroduced by Lavoisier about 150 years later (ca. 1780). Boyle's use of mercury in studying the compressibility of air has been mentioned.

Real progress began during the third decade of the 18th century, with the work of Hales, followed by that of Black, Cavendish, Scheele, Priestley, and Lavoisier. As will be seen, mercury figured prominently in many of the most significant advances.

Although trained for the clergy, Stephen Hales (1677-1761) found science more interesting than religion and moved from the pulpit to the laboratory. In one set of experiments, Hales (1727) placed fermenting peas in a bottle over mercury and sealed a vertical glass tube into the bottle, with its lower end below the level of the mercury. The gases generated by the fermentation of the peas forced the mercury upward into the glass tube, thus permitting measurement of the pressure. Throughout his work, the mercury gauge served as one of Hales' most useful instruments.

Hales is best known to physicians and to physiologists as the first to measure blood pressure (Chapter 8). Had he used mercury in this work, the results would surely have been even more significant than they were in leading to a useful sphygmomanometer.

Joseph Black (1728-1799) used mercury in his experiments which led to the concepts of specific heat and latent heat.

The fact that some gases are soluble in water was not fully appreciated by the early chemists. This was undoubtedly responsible for many of the difficulties they encountered in interpreting their observations. Henry Cavendish (1731-1810) appears to have been the first to collect gases over mercury, thus eliminating losses through solution in water. Later, Gay-Lussac (1778-1850) used this technique in his epoch-making studies on gases; and adoption of the technique by Joseph Priestley (1733-1804) was one of two uses of mercury which led him to the discovery of oxygen. There is some question as to whether Priestley or Carl Wilhelm Scheele (1742-1786) was the first to discover oxygen. Both used mercury in their crucial experiments.

The process by which Scheele produced "fire air" (oxygen) involved heating mercuric oxide which united with "phlogiston" (ϕ) to give mercury and "fire air." The reaction may be expressed:

$$\text{calx of Hg} + (\phi + \text{fire air}) = (\text{calx of Hg} + \phi) + \text{fire air}$$

This reaction had been suggested by Bayen (1725-1798) in 1774 and, interestingly, resulted in the discovery of oxygen within the framework of the phlogiston theory. Scheele also found that "fire air" could be produced by heating mercuric and mercurous nitrates, mercurous carbonate, and several other metals.

Like Hales, Priestley, who emigrated to America in 1794, was trained for the ministry and later became interested in chemistry. Priestley's early work on gases was facilitated by a ready supply of carbon dioxide from a brewery located close to his home near Leeds. He is credited with the invention of soda water, but investigated many gases in addition to CO_2. His use of Cavendish's technique of collection over mercury permitted him to study some gases which are soluble in water and which otherwise could not have been readily isolated.

An essential feature of Priestley's experimental armamentarium was the pneumatic trough, which he used in the collection of gases (Fig. VII-1). For gases that are soluble in water, the trough was filled with mercury. Using this device, Priestley prepared not only oxygen, but also sulfur dioxide, silicon fluoride, carbon monoxide, nitrogen oxides, ammonia (alkaline air) and hydrochloric acid gas (acid air). In his preparation of oxygen he used red oxide of mercury which was heated by means of a convex lens (burning glass) in vessels inverted over mercury. The resulting gas he called "dephlogisticated air."

Although Robert Boyle is sometimes called the father of modern chemistry, the honor might more appropriately be bestowed on Antoine Laurent Lavoisier (1745-1794). Involved in public affairs as well as in scientific work, Lavoisier was guillotined during the Reign of Terror along with other members of the *Ferme Générale.*

Figure VII-1 Priestley's Pneumatic Trough and other Apparatus. (Photo. The Science Museum, London.)

Priestley visited Lavoisier in Paris in 1774 and told him of his preparation
of "dephlogisticated air" from mercuric oxide. Lavoisier quickly grasped the
significance of Priestley's observations and shortly after the latter's departure
from Paris set about to repeat the experiments on the oxides of mercury. He
produced a gas which could support respiration and combustion and which
combined with metals. This he first called *principe oxygine* (1777). The name
underwent a series of modifications through 1786: *principe oxygine, oxygène,
oxigène* and finally back to *oxygène* (Duveen and Klickstein 1954).

Disputatious anglophiles have accused Lavoisier of not giving suitable credit
to Priestley for having told him of his use of red oxide of mercury in generating
the gas which eventually was called oxygen. It is not known whether or not
Lavoisier was aware of the work of his countryman Pierre Bayen (1725-1798)
who, in 1774 or 1775 published an article describing experiments with
mercuric oxide. In this article, Bayen stated that when mercury is calcined
it does not lose phlogiston but combines with a gas (which, of course, was
oxygen) and the increase in weight was thus explained. He did not know the
nature of the gas but his theory seems a clear rejection of phlogiston several
years before the phlogiston theory was proven false by Lavoisier (Weeks 1934).

The use of mercury in his experiments with oxygen was only one of several
important investigations in which Lavoisier used the metal or its compounds.
His isolation of nitrogen,* which at first he called *atmosphére mofette* (me-
phetic air) and later *azote,* involved the use of mercury and mercuric oxide. The
name nitrogen was given to this gas by Jean Chaptal (1756-1832) in 1790. In
experiments leading to determination of the composition of sulfuric acid and
of carbonic acid, Lavoisier made use of mercury (Duveen and Klickstein
1954). In his experiments on air Lavoisier was the first to distinguish mercury as an element.

It took many men many years to discover oxygen, and with it the key to
understanding combustion and other forms of oxidation. This may well be
considered the single most important discovery in the history of chemistry.
Possibly, even probably, this *could* all have come about without mercury. The
fact is that mercury, in one form or another, was a central feature in many of
the experiments which led to the discovery of oxygen. This event signals the
opening of the era of modern chemistry and the end of alchemy. Of comparable importance was the demise of the phlogiston theory, which occurred
at about the same time and as a direct result of these developments.

Other Important Advances in Which Mercury Figured

Mercury's importance in the history of chemistry did not end with the discovery of oxygen and the major corollaries: an understanding of combustion
and respiration, and the downfall of phlogiston and alchemy.

---

*Scheele, Cavendish, and Priestley also discovered nitrogen at about this time, but Daniel
Rutherford (1749-1819) had described it in a dissertation on experiments performed in
1772.

General Law of Equivalent Weights

For a time, development of the theory or law of equivalent weights of acids and bases was attributed to Carl Friedrich Wenzel (1740-1793). This was the result of a misconception on the part of Berzelius, who later corrected the error. Wenzel had reacted silver chloride with mercuric sulfide, producing silver sulfide and mercuric chloride. In his results he found uncombined residues resulting from this double decomposition, thus failing to demonstrate equivalence. It was, in fact, Jeremias Benjamin Richter (1762-1807), a mathematician who had studied under Kant, who, around 1795, first proved the law of reciprocal proportions or equivalents (Partington 1965).

Discovery of Potassium, Sodium, and Ammonium

Among the many studies performed by Sir Humphry Davy (1778-1829) were included *"Electro-chemical researches on the decomposition of the earths...."* Davy's early efforts to produce electrolysis of the alkaline earths were not entirely successful. (Berzelius had had a similar experience [see Jorpes 1966]) until a mercury electrode was used. Mercury was placed in a vessel and the electrolyte, an alkaline earth in solution, was overlaid. Passage of an electrical current to the mercury cathode electrolysed the solution, the metallic component forming an amalgam. On distilling off the mercury, the alkaline earth in pure form remained. (This procedure is used today in the chlor-alkali industry.)

Avogadro's Hypothesis

The hypothesis that equal volumes of all gases, at the same temperature and pressure, contain the same number of molecules (atoms) was advanced in 1811 by Amedeo Avogadro (1776-1856). Eventually this hypothesis gained universal acceptance, but originally it was rejected by Berzelius and by other distinguished contemporary chemists. In part, the objections were based on studies by Jean Baptiste André Dumas (1800-1884) of the vapor densities of mercury, sulfur, phosphorus and arsenic. Dumas did not realize that he was dealing with Hg, S_6, P_4 and As_4 (Foundations 1923). Eventually Avogadro was vindicated, however, largely through the work of his countryman, Stanislao Cannizzaro (1826-1910). Mercury, in the hands of Dumas, had contributed to the rejection of Avogadro's Hypothesis, but it reversed its role in Cannizzaro's experiments.

Organic Synthesis

It was not only in the field of inorganic chemistry but also in the newly developing science of organic chemistry that mercury contributed to important progress. Up until the first quarter of the 19th century it had been believed that organic substances depended for their formation on some mysterious "vital force." In 1828, Friederich Wöhler (1800-1882), with the help of mercury and mercuric oxide, converted ammonium cyanate to urea, thus for the first time creating an organic compound in the laboratory, and showing that no "vital force" was needed. This served as the basis for an entire new era in chemistry (Partington, 1965).

A synthesis of an organic mercurial compound was achieved in 1852 by Edward Frankland (1825–1899), who reacted methyl iodide with mercury to form crystalline methyl mercuric iodide. Earlier, Frankland, using mercury in his experiments, had propounded the important concept of combining power, later known as valence or valency.

Working primarily with mercury, calomel, and corrosive sublimate, Guillaume-François Rouelle (1703–1770) was the first to distinguish between acid, neutral, and basic salts. He was also the first to consider calomel and similar insoluble compounds as salts.

Discovery of New Elements

Mercury's role in the discovery or identification of new elements has been impressive: in addition to oxygen, nitrogen, and the alkaline earths mentioned above, mercury contributed directly or indirectly to the discovery of twenty-two elements (Weeks 1934).

Miscellaneous Discoveries

Thomas Graham (1805–1869) made many contributions to basic chemistry, and mercury played an essential part in much of his work. His experi-

TABLE VII-1

**Elements in the Discovery of Which
Mercury Played a Direct Part**

Element	Date	Discoverer
Oxygen	1773	Scheele (reported in 1777)
	1774	Priestley
Chromium	1797	Vauquelin
Palladium	1803	W. H. Wollaston
Calcium	1808	Davy
Barium	1808	Davy
Strontium	1808	Davy
Magnesium	1808	Davy
Aluminum	1825	Oersted
Argon	1894–5	Rayleigh and Ramsay
Helium	1895	Ramsay
Krypton	1898	Ramsay and Travers
Xenon	1899	Ramsay and Travers
Neon	1900	Ramsay and Travers

**Elements in the Discovery of Which Mercury Played
an Indirect Role or Was Used in Preliminary
Work Which Led to the Discovery**

Nitrogen	Zirconium	Chlorine
Hydrogen	Thorium	Titanium
Iodine	Rhenium	Cerium

ments on the diffusion of gases led to the formulation of Graham's Law—that the rate of diffusion of gases is inversely proportional to the square root of density. A pneumatic trough containing mercury was an important part of Graham's experimental equipment. In studies which led to the discovery of dialysis, Graham made use of the Sprengel mercury pump (p. 125), an instrument which he also employed in his research on osmosis (Graham 1876; Thorpe 1911).

Radicals—groups of atoms which can enter into chemical reactions as a combination and not lose their identity, thus behaving much like individual atoms—commanded the attention of most of the prominent early 19th century chemists; and mercury has been important in studying these groups. An important clue to the understanding of the behavior of radicals was provided by Seebeck (1770-1831) who used a mercury electrode in the electrolysis of ammonium carbonate. Similarities in the behavior of the ammonium radical and of sodium in forming amalgams with mercury led to the application of Seebeck's methods in further studies on radicals by Berzelius, Davy, Gay-Lussac, and others (Berry 1968). More recently, sodium-mercury amalgams have been used in elucidating the theory of acids and bases.

Colloidal suspensions of selenium, gold, and mercury in water were used by Westgren in 1915 in studies of Brownian movement and of Avogadro's constant (cited in Berry 1968). This work, incidentally, provided validation for Einstein's formula for the diffusion of colloidal particles.

Among physicians, mercury's best known application in analytical chemistry is in Nessler's reagent, which is an alkaline solution of the double iodide of mercury and potassium ($HgI_2 \cdot 2KI$). The reaction of Nessler's reagent with ammonia to form the yellow-colored Millon's base is used in a variety of clinical and other tests for nitrogenous bodies. Numerous other uses for mercury and mercury compounds are common in analytical chemistry.

Mercury was particularly useful to Rayleigh, Ramsay, and Travers in their work which led to the discovery of the rare gases in the atmosphere during the closing years of the 19th century (Travers 1928). In the discovery of argon, first of the rare gases to be identified, Rayleigh and Ramsay used the Sprengel mercury pump and also employed mercury to produce circulation of the gas being studied.

Argon was discovered in 1894 (or 1895), followed by helium in 1895, krypton in 1898, xenon in 1899 and neon in 1900. The mercury trough and other uses for mercury figured prominently in all of this work.

The preceding account covers the most important advances in chemistry in which mercury played a prominent or essential role. Additional uses in chemistry, too numerous to mention, are to be found (Bergman 1788-91 translation; Proust 1797; Gay-Lussac 1814, 1815, 1816a, 1816b; Hoff 1887, 1929; Arrhenius 1887; Burndy Library 1955; Ihde 1964; Leicester and Klickstein 1952; Leicester 1956; Muir 1907; Lowry 1926).

REFERENCES

Arrhenius, S. 1887. On the dissociation of substances in aqueous solution. *Z. Physik. Chem.* 1:631-48.

Bergman, T. O. 1788-91. *Physical and chemical essays.* Trans. Edmund Cullen. 3 vols. Edinburgh: J & J. Fairbairn.

Berry, A. J. 1968. *From classical to modern chemistry.* New York: Dover Publications.

Biringuccio, R. 1942. *Pirotechnia.* Trans. Cyril S. Smith and Martha T. Gnudi from the first edition of 1540. Cambridge: M. I. T. Press.

Black, J. 1910. *Experiments upon magnesia alba, quicklime, and some other alcaline substances.* Alembic Club Reprint, No. 1. Edinburgh: Alembic Club.

Boyle, R. 1772. *Works.* Ed. Thomas Birch. 6 vols. London. J. & F. Rivington.

Boyle, R. 1910. *The skeptical chymist.* Original edition published in London, 1661. London: J. M. Dent and Sons.

Bryant, A. 1931. *King Charles II.* London: Longmans, Green and Co.

Burndy Library, Norwalk, Conn. 1955. *Heralds of science.* With notes by Bern Dibner. Norwalk.

Butterfield, H. 1965. *The origins of modern science.* New York: Macmillan Co.

Cannizzaro, S. 1910. *Sketch of a course in chemical philosophy (1858).* Alembic Club Reprint, No. 18. Edinburgh: Alembic Club.

Cavendish, H. 1926. *Experiments on air.* Alembic Club Reprint, No. 3. Edinburgh: Alembic Club.

Davy, H. 1935. *The decomposition of the fixed alkalies and alkaline earths.* Alembic Club Reprint, No. 6. Edinburgh: Alembic Club.

Duveen, D. I., and Klickstein, H. S. 1954. *A bibliography of the works of Antoine Laurent Lavoisier.* London: Dawson and Sons and Weil.

Foundations of the molecular theory, comprising papers and extracts by John Dalton, Joseph-Louis Gay-Lussac and Amedeo Avogadro (1808-1811). 1923. Alembic Club Reprint, No. 4. Edinburgh: Alembic Club.

Frankland, E. 1852. On a new series of organic bodies containing metals. *Phil. Trans.* 142:417-44.

Gay-Lussac, J. L. 1814. Mémoire sur l'iode. *Ann. Chim.* 91:5-160.

Gay-Lussac, J. L. 1815. Recherches sur l'acide prussique. *Ann. Chim.* 95:136-230.

Gay-Lussac, J. L. 1816a. Experiments on prussic acid. *Ann. Phil.* 7:350-64.

Gay-Lussac, J. L. 1816b. Experiments on prussic acid. *Ann. Phil.* 8:37-53.

Graham, T. 1876. *Chemical and physical researches.* Edinburgh.

Hales, S. 1727. *Vegetable staticks.* London.

Hoff, J. H. van't. 1887. Die Rolle des osmotischen Druckes in der Analogie zwischen Lösungen und Gasen. *Z. Physik. Chem.* 1:481-508.

Hoff, J. H. van't. 1929. In *Foundations of the theory of dilute solutions.* Alembic Club Reprint, No. 19. Edinburgh: Alembic Club.

Hooke, R. 1962. *Micrographia or Some physiological descriptions of minute bodies made by magnifying glasses with observations and inquiries thereupon.* Facsimile of 1665, London, edition. New York: Dover Publications.

Ihde, A. 1964. *The development of modern chemistry.* New York: Harper and Row.

Jorpes, J. E. 1966. *Jac. Berzelius: His life and work.* Trans. from the Swedish manuscript by Barbara Steele. Stockholm: Almqvist and Wiksell.

Kuhn, T. S. 1962. *The structure of scientific revolutions.* Chicago: The University of Chicago Press.

Leicester, H. M., and Klickstein, H. S. 1952. *A source book in chemistry.* New York: McGraw-Hill Book Co.

Leicester, H. M. 1956. *The historical background of chemistry.* New York: John Wiley and Sons.

Lemery, N. 1686. *A course of chymistry.* 2nd ed., enl. and trans. Walter Harris, M. D. from the 5th French ed. London.

Lowry, T. M. 1926. *Historical introduction to chemistry.* London: Macmillan and Co.

Magie, W. F. 1935. *A source book in physics.* New York: McGraw-Hill Book Co.

Mayow, J. 1907. *Medico-physical works.* Trans. of the 1674 text by A. C. Brown and L. Dobbin. Alembic Club Reprint, No. 17. Edinburgh: Alembic Club.

Muir, M. M. P. 1907. *A history of chemical theories and laws.* New York: J. Wiley and Sons.

Pagel, W. 1958. *Paracelsus.* Basel: S. Karger.

Partington, J. R. 1965. *A short history of chemistry.* third ed. New York: Harper and Row, Harper Torchbooks.

Priestley, J. 1776. *Experiments and observations on different kinds of air.* 2nd ed. 3 vols. London.

Proust, J. L. 1797. Recherches sur le bleu de Prusse. *Ann. Chim.* 23:85.

Rey, J. 1904. *On an enquiry into the cause wherefore tin and lead increase in weight on calcination (1630).* Alembic Club Reprint, No. 11. Edinburgh: Alembic Club.

Rouelle, G. F. 1744. Mém. sur les sels neutres, dans lequel on propose une division méthodique de ces sels etc. *Mém. Acad. Roy. Sci.* pp. 353–64.

Rouelle, G. F. 1754. Mém. sur les sels neutres, dans lequel on fait connaître deux nouvelles classes de ces sels etc. *Mém. Acad. Roy. Sci.* pp. 572–88.

Scheele, C. W. 1901. *The chemical essays.* Trans. J. Murray from the 1786, London, *Transactions of the Academy of Sciences at Stockholm.* London: Scott, Greenwood and Co.

Stillman, J. M. 1960. *The story of alchemy and early chemistry.* New York: Dover Publications.

Thorpe, E. 1911. *Essays in historical chemistry.* 3rd ed. London: Macmillan and Co.

Travers, M. W. 1928. *The discovery of the rare gases.* London: Edward Arnold and Co.

Weeks, M. E. 1934. *The discovery of the elements.* 2nd ed., rev. Easton, Pa.: Journal of Chemical Education.

Wöhler, F. 1828. Ueber Kunstliche Bildung von Harnstoff. *Ann. Physik. Chem.* 12:253–56.

Wohlbarsht, M. L., and Sax, D. S. 1961. Charles II, a royal martyr. *Notes Records Roy. Soc. London.* 16:154.

ADDITIONAL READINGS

Hales, S. 1731–33. *Statical essays.* 2 vols. London.

Jaffe, B. 1949. *Crucibles: The story of chemistry.* Rev. and enl. New York: Hutchinson's Scientific and Technical Publications.

Barometers, Thermometers, and Sphygmometers

Barometers

Any scientist in the early part of the 17th century was faced with a number of obstacles peculiar to that period. For one thing, many of the views of Aristotle (B. C. 384-322) had been incorporated into Christian dogma, and observations which seemed to run counter to Aristotle's principles were very likely to be considered heretical. The unhappy confrontation between Galileo Galilei (1564-1642) and the church authorities is the best known example of what happened when science challenged accepted belief.

Galileo was intimately concerned with some of the theoretical and practical considerations which led to the development of the barometer. He accepted the Aristotelian belief that air had no weight, but disagreed with Aristotle's dictum that a vacuum was logically impossible. Several contemporary natural philosophers shared with Galileo the belief that the existence of a vacuum was possible, the best known among these being Isaac Beeckman (1588-1673) in the Netherlands, Giovanni Battista Baliani (1582-1666) in Genoa, and Rafael Magiotti (1579-1658) in Rome (Middleton 1964).

Prior to Torricelli's famous experiment, probably in 1641, a young mathematician-astronomer, Gasparo Berti (of whose life little is known) had demonstrated that the atmosphere had weight and that a vacuum was possible. His apparatus consisted of a lead tube about 12 meters long, the lower end of which was immersed in a vat of water. The top of the tube was fitted into a globe of glass, with a continuous column of water from the globe into the vat. The lower end of the tube was fitted with a stop-cock below the water level of the vat. When this cock was opened the water in the tube descended, but did not completely run out, the level in the tube coming to rest at a fixed point. This showed that air had weight. When the globe at the top of the tube was opened, air rushed in, indicating that a vacuum had been created.

There is no evidence that Evangelista Torricelli (1608-1647) actually witnessed Berti's experiment; more likely he heard about it from his friend Magiotti who had been present. At any rate, Torricelli began to think about designing an instrument capable of measuring changes in the pressure of air. With advice from Galileo and technical assistance from Vincenzio Viviani (1622-1703), Torricelli performed the crucial experiment (probably in June 1644) which served as the basis for the mercury barometer. One major modification of the Berti apparatus was his substitution of mercury for water, an important practical improvement, since it permitted him to use a tube one meter, rather than 12 meters, in length.

Aristotle's ideas on the impossibility of a vacuum had now been demolished. Because of the religious climate in Italy at the time, and with Galileo's experience with the Holy Office in 1633 not forgotten, Torricelli's work was not given wide publicity in Italy. (The idea of a vacuum was anathema to the church.) It has been reported that the first repetition of Torricelli's experiment was performed by or for Cardinal Giovanni Carlo de'Medici, in February 1645, but kept secret for several years (Middleton 1964).

Because of the intellectual climate in France, French scientists did not suffer from the same handicaps which inhibited their Italian colleagues. No lesser personages than René Descartes (1596-1650) and the youthful Blaise Pascal (1623-1662) were among those who became involved in measuring atmospheric pressure. Pascal demonstrated a difference in the height of the mercury column in a Torricelli tube when measurements were made at Clermont and at the higher elevation on Puy-de-Dôme, and Descartes applied a measuring scale to the instrument, thereby converting what had been more properly called a "baroscope" into a barometer (Descartes, Vol. 1, pp. 205-8).

When word of the Torricelli experiment reached England it caused great excitement and engaged the attention of such notables as Robert Boyle (1627-1691) and Sir Christopher Wren (1632-1723).

Early interest in the use of the mercury barometer for meteorologic purposes is found in a letter from Robert Hooke to Boyle, October 6, 1664, in which the writer notes that observation of the "baroscopical index" may be used to predict the weather. In a second letter, on December 13, 1664 Hooke says "I have lately observed many circumstances in the height of the mercurial cylinder, which do very much cross my former observations. . . ." Thus, Hooke questioned the reliability of barometric readings to predict weather. George Sinclair, however, described a graduated weather glass marked to indicate the type of weather to be expected with different heights of the mercury (Middleton 1964).

Over the years there have been many modifications and adaptations of the mercury barometer, and new patents are being issued right up to the present day (Brombacher 1960). An important basic tool in the physical and biological sciences since the time of its birth, the mercury barometer and its close

relatives have recently assumed new importance in medicine with the application of hyperbaric states in the treatment of certain diseases.

Important as the basic function of a mercury barometer may be in measuring atmospheric pressures, this use of the instrument is overshadowed by developments arising from the application of physical principles which underlie
barometry. Demonstration of the fact that a vacuum could exist laid the
foundation for the entire range of scientific and technological advances based
on evacuated vessels, i.e., the vacuum tube and everything which sprang from
it. Torricelli's use of mercury in the 17th century set the stage for the technology of electronics in the 20th.

Thermometers

Second to the mercury barometer in the date of its perfection, but not in
its importance, is the mercury thermometer. The story of the birth of the
latter is much more complex than that of the former, including doubts on
paternity. Torricelli's siring of the mercury barometer has not been seriously
questioned, but there have been numerous contestants for the honor of in-
venting the mercury thermometer. Claims, denials, and counter-claims have
resulted in a literature as confusing as it is prodigious. Furthermore, quicksilver did not achieve success in thermometry without a long struggle.

The mercury thermometer as it is known today represents a late step in a
series of events which began some 2000 years ago. Hero of Alexandria and
Philo of Byzantium, both of whom lived at about the beginning of the Christian era, knew that air expands when heated, and the latter is said to have constructed a device which demonstrated this phenomenon (Middleton 1966;
Wolf 1935). This was the first thermoscope. It is possible that Galileo was
familiar with the observations of Hero and Philo and from them got the idea
of a thermoscope. Many historians credit Galileo in 1592 with being the first
in early modern times to suggest a means for demonstrating the expansion of
heated air by the displacement of water (Bolton 1900).

A thermoscope merely shows that expansion takes place when a gas or
liquid is heated; it becomes a thermometer when a measuring scale or "meter"
is attached to it. This was first done around 1610 but it is by no means clear
by whom. Leading contenders for the honor of taking a step which in effect
amounts to designing the first thermometer are Galileo himself, Sanctorius
(1561-1636), Robert Fludd (1574-1637), and Cornelius Drebbel (1572-
1634). In any case, mercury had not yet become involved.

Galileo's thermoscope, as just mentioned, demonstrated the expansion of
air by the displacement of water. (See Fig. VIII-1.) In 1632, Jean Rey (ca.
1583-1645) reversed the roles of air and water and thus became the first to
use the displacement of a gas by a liquid as a means of thermometry (Singer
1959; Burkhardt 1902).

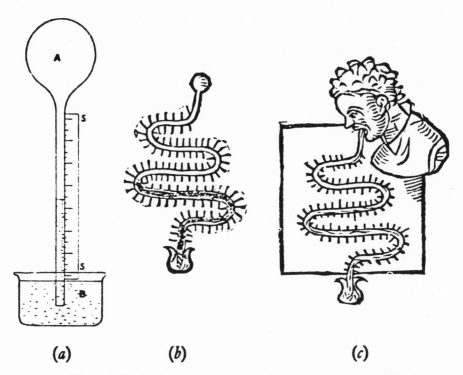

(a)   **(b)**   **(c)**

Figure VIII-1 Galileo's Thermometer and an Adaptation of it. (a) Galileo's
thermometer was inaccurate since it acted both as a thermometer (when a rise
in temperature forced the liquid [Galileo used water] down in the tube) and
as a barometer (when a fall in atmospheric pressure had the same result). A is
a bulb containing air. B is the liquid into which the tube is inserted. (b).
Sanctorius' adaptation of Galileo's thermometer. (c) A patient using the
Sanctorius thermometer. (From *A Short History of Medicine*, by Charles
Singer and E. Ashworth Underwood. Oxford: The Clarendon Press. 1962.)

Up to this point the expanding medium in thermometers was not confined
in a closed system and therefore was subjected to atmospheric pressure as well
as to changes in temperature. This deficiency was eliminated with the invention of the closed bulb thermometer by the Grand Duke Ferdinand II of
Tuscany (1610-1670) around 1640 or possibly later (Middleton 1966; Wolf
1935; Bolton 1900). For his sealed liquid-in-glass thermometers Ferdinand
used spirit of wine (alcohol). At about this time the Accademia del Cimento
of Florence, in the activities of which the Grand Duke was an enthusiastic
participant, had been experimenting with quicksilver as a thermometric liquid.
Records of the Accademia for 1657 show that the members rejected mercury
in favor of spirit of wine for thermometers (Boffito 1929). A year or two
later, Ismael Boulliau (1605-1694) in Paris, after comparing mercury and
spirit, also concluded that the latter was preferable (Middleton 1966). Boulliau

may have been the first to make a sealed mercury thermometer (Maze 1895). The first known drawing of a thermometer is that of Bartolomeo Telioux dated 1611, and this entitles him to some consideration among the contenders for priority (Court 1967). Singer credits Galileo with the invention of the first sealed liquid-in-glass thermometer in 1612.

In England, the recently organized Royal Society took up the study of thermometers in 1661, with Robert Boyle (1627–1691), Christopher Wren (1632–1723) and Robert Hooke (1635–1703) leading the way (Wolf 1935). Although the mercury thermometer had been tried and rejected on the continent, the Fellows of the Royal Society asked Hooke to make two such instruments according to a design of Wren. One of these was incorporated into a "meteorograph" which was to record wind direction, temperature, and hourly rainfall. The Society had planned to entertain Charles II with a demonstration of this thermometer, but the records suggest that the king failed to show up at the appointed time.

Robert Boyle was among the first to question the superiority of spirit of wine over mercury. In this he received partial support from the astronomer Edmond Halley (1656–1742) as a result of experiments carried out in 1693. Halley compared air, water, spirit, and mercury and came to the conclusion that air was best. About mercury he says: "This Fluid being so sensible of gentle Warmth, and withal not subject to evaporate without a good degree of Fire, might most properly be applied to the construction of Thermometers were its Expansion more considerable" (Halley 1693). Hooke preferred rectified spirit of wine "highly ting'd with the lovely colour of Cocheneel, which I deepen the more by pouring some drops of common Spirit of Urine. . . ."
At about the same time, the great Sir Isaac Newton (1642–1727) used linseed oil in his thermometers (cited in Wolf 1935 and Middleton 1964). With these big guns arrayed against it, there is small wonder that mercury did not come into its own until rather late in the history of the thermometer. However, the German Jesuit, Athanasius Kircher (1601?–1680), is said to have described a mercury thermoscope in 1641, the instrument supposedly having been made about 1620 (Middleton 1964).

Many other names have been associated with the early use of mercury in thermometers, particularly in connection with the development of standardized scales. Their contributions have been reviewed by Renou. Most important were the three whose scales have survived, Gabriel Daniel Fahrenheit (1686–1736), René-Antoine Ferchault de Réaumur (1683–1757) and Anders Celsius (1701–1744).

It is one thing to make a thermometer, but quite another to make one that is reliable, accurate and practical. In order to establish accuracy and reliability it is necessary to have a measuring scale by which reproducibility of results can be tested. This in turn necessitates the establishment of fixed points which are constant. Today, the freezing and boiling points of water are used as points of reference, but many others were tried before these were widely accepted. In fact, a major part of the research and most of the controversies in the field of temperature measurement, particularly in the early part of the 18th century, were related to finding suitable points of reference. The survival of the Fahrenheit, Celsius (Centigrade) and Réaumur scales is evidence that no single system has been universally acceptable.

Fahrenheit was a man of science and a skilled technician. Evidence that his talents were recognized by his contemporaries is found in the writings of Hermann Boerhaave who ". . . desired that industrious and incomparable Artist, Daniel Gabriel Fahrenheit, to make me a couple of thermometers, one with the densest of all Fluids, Mercury, the other with the rarest, Alcohol, which should be so nicely adjusted, that the ascents of the included Liquor in the same degree of Heat, should be always exactly equal in both, as might appear by a scale fix'd on the side" (1735 translation). This was certainly earlier than 1732 and probably earlier than 1724. Fahrenheit is generally credited with being the first to make an accurate, reliable mercury thermometer. With it he established the scale which bears his name and which became the standard in the English-speaking world (but not in his native Germany). The date 1714 is commonly given for Fahrenheit's perfection of his mercury thermometer. His scale has 180 degrees, with the melting point of ice at $32°$ and the boiling point of water at $212°$.

The one hundred degree scale (centigrade) was designed in 1742 by Celsius, a native of Sweden, who called the boiling point of water zero and the melting point of ice 100. Celsius used mercury in his thermometers. Shortly after his death, his friend and successor at Uppsala, Carl von Linné (1707-1778), inverted the scale so that zero became the melting point of ice and $100°$ the boiling point of water. A similar inversion had been made by Jean Pierre Christin (1683-1755) of Lyon a year earlier. The result was the centigrade thermometer which became the standard for France and for much of Europe (but not for Celsius's native Sweden).

A third widely-used scale is that developed by Réaumur in 1730. Réaumur used the melting point of ice as his zero and called the boiling point of water $80°$. His system was adopted in Sweden, Russia, and parts of Germany (but not in his native France).

Who invented or made the first mercury thermometers? Was it the Accademia del Cimento whose members tried and rejected mercury? Was it Boulliau? Was it Hooke? Or was it some unknown person? As in many claims for priority, national pride may threaten objective evaluation of available evidence. Perhaps the question will never be settled and, like the syphilis problem, remain as a never-ending source of material for professional as well as amateur historians.

Hypsometers

A precursor to the modern altimeter, the hypsometer consists of a mercury thermometer so designed that the bulb can conveniently be immersed in boiling water. By measuring the temperature at which water boils it is possible to calculate elevation above sea level. According to Middleton (1966) the first hypsometer was designed by Jean André Deluc (1727–1817) around 1770, but it appears that Fahrenheit had made such an instrument some 40 years earlier (Wolf 1935).

Psychrometers

A psychrometer ("cold" + "measure") is used for measuring relative humidity. It is a form of hygrometer employing two mercury thermometers. One of these, called a wet-bulb thermometer, has a sleeve of some absorptive material encasing the bulb. The second thermometer is of the ordinary dry-bulb type. In measuring relative humidity, the sleeve is moistened with distilled water which on evaporating causes a cooling of the mercury. By the use of tables which have been calculated, relative humidity can be determined from the difference in temperature readings between the wet- and dry-bulb thermometers.

Two methods have been used to hasten evaporation around the wet-bulb thermometer: the sling and the aspirating psychrometer. The 20th century industrial hygienist who has come to grief by smashing his sling psychrometer against a machine will perhaps be comforted to know that one of the earliest users of the instrument, H. B. de Saussure, in 1788, lost his sling thermometer because the string which he used for whirling it became frayed and the instrument flew off (Middleton 1966). This kind of accident can be prevented by the use of a swivel, but one must still be careful to have ample room to maneuver. The aspirating psychrometer uses a rubber bulb and tubing to project a stream of air against the wet-bulb thermometer, thus reducing the risk of breakage.

Clinical (medical) Thermometers

Historians generally agree that the first recorded attempt at clinical thermometry was made by Sanctorius in 1611 or 1612, using an instrument designed by Galileo. In 1625 Galileo published a work which explained how the thermometer could be used to study disease. A familiar picture in many histories of medicine is that of an unhappy-looking man who appears to have a snake issuing from his mouth and entering into a small flask nearby. Actually this is a representation of Sanctorius's modification of Galileo's air thermometer, with the air bulb being held in the mouth of the person whose temperature is being measured. In 1632 Jean Rey tested fever patients by placing a bulb filled with water in their hands and noting the expansion by the rise of water in a tube connected with the bulb.

No serious attempts to improve the crude thermometry of Sanctorius and Rey were made for more than a hundred years, when Boerhaave undertook the task. One of his aphorisms (No. 673) states: "External febrile heat is recognized by the thermoscope; internal by the sensation of the patient and by the redness of the urine." One of Boerhaave's pupils, Gerhard L. B. van Swieten (1700-1772), carried on his work and used the term "thermometer" instead of "thermoscope" but it was Anton De Haen (1704-1776), a student of both Boerhaave and Sydenham, who first popularized the use of clinical thermometry.

De Haen's principal contribution was to establish mercury as the fluid of choice for the instrument. Although his thermometers might take as long as ten minutes to reach a maximum reading, he preferred them to other types. With great patience he attempted to measure the range of temperatures in normal subjects of various ages and he also studied the fever curves in a number of diseases. Thus the mercury thermometer became an important tool in diagnosis and prognosis. During the latter half of the 18th century the mercury thermometer was used by Albrecht von Haller (1708-1777), John Hunter (1728-1793) and others (Ebstein 1928) in studying a variety of physiological phenomena. This work continued with increased tempo well into the 19th century and as a result, an extensive literature developed. Wunderlich (1815-1877) undertook the immense task of bringing together all of the earlier important findings and then adding a monumental collection of his own observations. His book, *Das Verhalten der Eigenwärme in Krankheiten*, published in 1868, is one of the classics of medicine.

Slightly less than half of Wunderlich's book is devoted to discussions of the temperature in 33 specific diseases. In his chapter on "The Art of Medical Thermometry" he gives the specifications for a suitable instrument and prescribes that it should be a mercurial thermometer. Other characteristics are described in great detail.

> The reservoirs of metal must be neither too large, or too small. If the bulb is too large, it is wanting in sensitiveness; if too small, it is difficult to retain it in close apposition to the body. A diameter of about 1/3-3/4 of a centimetre seems the most convenient size . . .
> The tube or stem of the instrument must have an even bore throughout, and be of such a diameter that the distance between any two-tenths of a degree C. can be easily divided by the eye into half and quarter parts (so that 1/10° to 1/20° Fahr. can be easily read). The length of the tube must be such that the degrees on the stem are at least 12 centimetres from the bulb, in order that the height of the mercury may be easily read *in situ*.
> For the sake of portability, however, it is well not to have too long a stem; and it will be found sufficient to have a tube a little longer than the probable height to which the mercury will rise when applied to a living human being.

A scale that would allow the eye easily to divide each two-tenths of a degree into quarter parts could scarcely have less than two millimeters between each mark. If it is to record the temperatures from 32.5° to 45°C, as Wunderlich directs, it would have some 60 marks at two millimeter distances, a total of 12 centimeters. When this is added to the 12 centimeters between the bulb and the scale it gives a length of at least 24 centimeters or about 10 inches.

(The translator of the English edition notes that British thermometers in 1871 vary in length from about 5 to 10 inches, or more.) A standard American clinical thermometer of 1970 is about 10 centimeters (four inches) long, has a distance of about two centimeters from bulb to scale (7/8 inch), has a scale from 94° to 108°F or 110°F (34.4°-42.2°C or 43.3°C), and has five marks to each degree for a total of 70 or 80 scale marks. In part, the shortening was made possible by the invention of the registering thermometer in which the mercury remains at its highest point until shaken down, thus obviating the necessity of making readings *in situ*. Sir Thomas Clifford Allbutt (1836–1925) has been credited with bringing medical thermometry to a point where it became a routine procedure (Singer and Underwood 1962).

Wunderlich refers to a Ch. Martin whom he credits with having published in England about 1740, the first accurate observations on temperatures in healthy men and animals. Apparently he means George Martine (1702–1741) whose *De animalium calore* was in fact published in London in 1740. In the same year Martine also published a book entitled *Essays Medical and Philosophical* in which he asks:

> What fluid shall we take for our Thermometers? We have found inconveniences in Air, Oil and Spirits; and water is more exceptionable than any of them. We have, it seems, nothing left but Quicksilver. This is a very moveable and ticklish fluid; it both heats and cools faster than any liquor we know, or have had occasion to try, faster I am sure than water, oil or even spirit of wine; and it bears a great deal of heat before it arrives at a boiling expansion; and, if well purified, does not wet or stick to the inside of the tube.

Additional interesting comments by Martine are that Halley's objections to mercury ". . . may be avoided by making the bulb have a great proportion to the tube." And of mercury thermometers: "It is said that they were first contrived by that curious Mathematician Olaus Roemer. Mr. Fahrenheit in Amsterdam, and other workmen in that country manufacture many of them . . ." but ". . . they are made no where in greater perfection or with greater exactness than by our countryman Wilson at London."

Other Adaptations of the Mercury Thermometer

A number of other adaptations and modifications of the mercury thermometer have played important roles in science and technology. Among these may be mentioned recording thermometers (thermographs), black bulb thermometers and maximum-minimum thermometers. While these are being replaced in many applications by electrical and electronic devices, the mercury thermometer remains as a standard and is often used for calibrating the more elaborate or complicated instruments.

No account of the mercury thermometer would be complete without some mention of the sad saga of Sarah Binks, the Sweet Songstress of Saskatchewan. This story reveals one of the sinister aspects of the thermometer and shows that in fact it may be a lethal instrument. Sarah has been immortalized in a touching biography by Paul Gerhardt Hiebert.

For many years Sarah was the brightest star of the Canadian prairies. She captured the spirit of her environment and distilled it into verse, bringing joy to an otherwise dull and chilly community. Hiebert has selected the words "Burbank, bobolink and snearth" from one of her earlier poems to epitomize how she expressed her feeling for the fields, the sun and the sky. Sarah's more mature phase can be illustrated by:

> This makes me scratch myself and ask,
> When shall my powers fade?
> It puts me severely to the task,
> To face this fact undismayed.

Sarah's genius did not go unrecognized or unrewarded in her own time. Therein, as Hiebert has pointed out, lay the elements of true Greek tragedy, for her success led to her undoing and horrible death. As first prize in a poetry contest sponsored by the local grain and feed merchant, Sarah received a horse thermometer, which, alas, registered six degrees too high. And then came the terrible epidemic of hives which made Sarah and all her neighbors scratch themselves. Her passion for Scotch mints was so strong that she held one in her mouth even when she was taking her own temperature with the inaccurate horse thermometer. Absent-mindedly she began to chew the mint thus breaking the bulb of the thermometer and swallowing the mercury, a full tablespoon according to Hiebert. Death from mercury poisoning followed. The details have been recorded and the case stands as the only one of its kind in all the annals of medical history, a fact that received recognition at an international symposium on mercury in Ottawa in 1971. (See bibliography, "Famed Canadian singer. . . .")

Sphygmometers

The history of sphygmometry is to be found in most textbooks of physiology and in monographs dealing with disturbances in arterial blood pressure (Master, Garfield, and Walters 1952; Singer and Underwood 1962). An excellent monograph on blood pressure measurement is Geddes' published in 1970.

The Reverend Stephen Hales (1677-1761) apparently was the first person to make an accurate measurement of blood pressure. This was in 1733, and was done by introducing a canula directly into the crural artery of a mare and connecting the canula to a glass tube. Hales found that the blood rose to a height of eight feet, three inches.

Nearly a century passed before any attempt was made to refine Hales' technique, for it was not until 1828 that the French physiologist, Jean Léonard Marie Poiseuille (1799-1869) first employed a mercury U-tube manometer for this purpose. The weight of the mercury kept the height of the column within reasonable limits. Poiseuille followed the lead of Hales in making his measurements by direct canulation of the arteries of his experimental animals.

The next significant advance in measuring blood pressure was made by Karl Friedrich Wilhelm Ludwig (1816-1895) in 1847. Using the Poiseuille U-tube mercury manometer, he introduced a small float into the longer limb of the manometer and connected this with a writing arm. This in turn was made to impinge on the revolving drum of a kymograph and thus record changes in blood pressure. This, too, was a direct recording instrument, requiring the introduction of a canula or needle into an artery, and consequently had little practical value in measuring blood pressure in man. (Ludwig, incidentally, invented a mercury blood-pump by means of which gases in the blood could be removed in a partial vacuum and made available for study. This was to become a basic tool in the study of the physiology of respiration.)

The first step toward indirect measurement of blood pressure without canulation of an artery was the "sphygmomètre" of Jules Hérison (1789-?) (Marcy 1881). This device, introduced in 1834, consisted of a metallic half-sphere to which was joined a calibrated glass capillary tube containing mercury. The larger surface of the sphere was covered with a membrane which could be placed over an artery. Changes in the pulse were reflected by a rhythmic rise and fall of the mercury in the capillary tube and the magnitude of the deflection could be read on the graduations of the tube. This, of course gave only the changes in arterial blood pressure during the cardiac cycle and not the pressure itself, but it was a significant step forward.

Other attempts at indirect measurement of the blood pressure were made by Karl Vierordt (1818-1884) in 1855 and by Etienne Jules Marey (1830-1904) in 1860, but these procedures did not entail the use of mercury. They represent only two of the many efforts between 1850 and 1900 to make indirect measurements of blood pressure, some of which used mercury and some of which did not.

The immediate forerunner of the present-day sphygmomanometer was developed by Scipione Riva-Rocci (1863-1936) in 1896. This instrument consisted of a mercury manometer which was connected by means of rubber tubing on one side to a rubber sack and on the other to a rubber bulb. The sack was strapped around the arm in a cuff-like manner and inflated by means of the rubber bulb acting as a pump. The pressure within the system was reflected by a rise in the mercury column of the manometer. The level of the mercury was read at the point where the radial pulse was obliterated by the increasing pressure within the cuff. This represented the systolic blood pressure. Following the introduction of the Riva-Rocci sphygmomanometer various refinements were made, but the present-day instruments using mercury are essentially the Riva-Rocci device.

REFERENCES

Boerhaave, H. 1735. *Elements of chemistry.* Trans. Timothy Dallowe, M. D. Vol. 1, London.

Boffito, G. 1929. *Gli strumenti della scienza e la scienza degli strumenti.* Florence: Libreria Internazionale Seeber.
Bolton, H. C. 1900. *The evolution of the thermometer.* Easton, Pennsylvania: Chemical Publishing Co.
Brombacher, W. G. 1960. In *Mercury barometers and manometers.* By W. G. Brombacher, D. P. Johnson, and J. L. Cross. Washington: U.S. Dept. of Commerce, National Bureau of Standards.
Burckhardt, F. 1902. *Zur Geschichte des Thermometers.* Basel: Reinhardt.
Court, A. 1967. Concerning an important invention. *Science* 156:812–13.
Descartes, R. 1897–1908. *Oeuvres.* Ed. Charles Adam and Paul Tannery. 11 vols. Paris: L. Cerf.
Ebstein, E. 1928. Dreihundert Jahre klinischer Thermometrie. *Klin. Wochschr.* 7:950–53.
Famed Canadian singer dies from mercury poisoning. 1971. In *Proceedings of the Royal Society of Canada special symposium on mercury in man's environment*, 15–16 February 1971, Ottawa, Canada, p. ii. Ottawa: Royal Society of Canada.
Geddes, L. A. 1970. *The direct and indirect measurement of blood pressure.* Chicago: Year Book Medical Publishers.
Hales, S. 1733. *Statical essays: Containing haemostatics, or, An account of some hydraulic and hydrostatical experiments made on the blood and blood vessels of animals.* London.
Halley, E. 1693. An account of several experiments made to examine the nature of the expansion and contraction of fluids by heat and cold etc. *Phil. Trans.* 17:650–56.
Hiebert, P. G. 1947. *Sarah Binks.* London: Oxford University Press.
Hooke, R. 1961. *Micrographia.* Facsimile of the London, 1665, edition. New York: Dover Publications.
Ludwig, C. 1847. Beiträge zur Kenntniss des Einflusses der Respirationsbewegungen auf den Blutlauf im Aortensysteme. *Arch. Anat. Physiol. Wiss. Med.* pp. 240–302.
Marey, E. J. 1881. *La circulation du sang à l'état physiologique.* Paris.
Martine, G. 1740. *Essays medical and philosophical.* London.
Master, A. M.; Garfield, C. I.; and Walters, M. B. 1952. *Normal blood pressure and hypertension.* Philadelphia: Lea and Febiger.
Maze, l'Abbé. 1895. Sur le premier thermomètre à mercure. *Compt. Rend.* 120:722–23.
Middleton, W. E. K. 1964. *The history of the barometer.* Baltimore: Johns Hopkins Press.
Middleton, W. E. K. 1966. *A history of the thermometer and its use in meteorology.* Baltimore: Johns Hopkins Press.
Mitchell, S. W. 1891. The early history of instrumental precision in medicine. *Trans. Congr. Am. Phys. Surg.* 2:159–98.
Poiseuille, J. L. M. 1828. Recherches sur la force du coeur aortique. *J. Physiol. Exp. Pathol.* 9:341–58.
Renou, E. 1876. Histoire du thermomètre. *Ann. Soc. Météorol. France* 24:19–72.
Riva-Rocci, S. 1896. Un nuovo sfigmomanometro. *Gazz. Med. Torino* 47:981–1001.
Singer, C. 1959. *A short history of scientific ideas to 1900.* Oxford: Clarendon Press.
Singer, C., and Underwood, E. A. 1962. *A short history of medicine.* Rev., enl. Oxford: Oxford University Press.
Vierordt, K. 1855. *Die Lehre vom Arterienpuls.* Braunschweig.
Wolf, A. 1935. *A history of science: Technology and philosophy in the 16th and 17th centuries.* London: George Allen and Unwin.
Wunderlich, C. A. 1868. *Das Verhalten der Eigenwärme in Krankheiten.* Leipzig.
Wunderlich, C. A. 1871. *On the temperature in disease: A manual of medical thermometry.* Trans. W. B. Woodman. London.

ADDITIONAL READINGS

Evelyn, J. 1854. *Diary and correspondence.* Vol. 1. London.
Gerland, E. 1896. In Report of the International Meteorological Congress, 21–24 August 1895. Chicago. ed. D. L. Fassig. *U.S. Weather Bur. Bull.* 11(3):690.
Middleton, W. E. K. 1971. *The experimenters: A study of the Accademia del Cimento.* Baltimore: Johns Hopkins Press.
Singer, C.; Holmyard, E. J.; Hall, A. R.; and Williams, T. I. 1954–58. *A history of technology.* 5 vols. Oxford: Clarendon Press.

Science, Technology, and Mercury

Early Uses of Mercury in Science and Technology

Newton's Physics

When it came to writing the *Opticks,* Sir Isaac Newton (1642–1727) abandoned the "Hypotheses non fingo" ("I frame no hypotheses") of the *Principia* and, as I. B. Cohen has put it, "let himself go." (See Newton 1952, p. xxiii.) Quicksilver could not have had a more distinguished sponsor for its introduction into theoretical physics than Sir Isaac, who thus made amends for having rejected the metal in favor of oil for his thermometers, although in the light of present-day knowledge he might have done better to have used mercury thermometers and not to have depended on the elusive element to support some of the ideas put forth in Book Three of the *Opticks.*

By way of illustrating the thesis that ". . . Nature . . . seems delighted with Transmutations," Newton (1952 edition, p. 374) states that "Mercury appears sometimes in the form of a fluid Metal, sometimes in the form of a hard brittle Metal. . . ." This is difficult to reconcile with Boerhaave's statement in 1753 (p. 81) that mercury is not ". . . capable by any known degree of cold, of coalescing into a solid mass." Joseph Black and Henry Cavendish, who experimented with the freezing of quicksilver, were both born after the death of Newton. Some of the language in Question 31, toward the end of the *Opticks,* has a distinct alchemical flavor: ". . . when Mercury sublimate is sublimed from Antimony, or from Regulus of Antimony, the Spirit of Salt lets go to the Mercury, and unites with the antimonial metal which attracts it more strongly, and stays with it till the Heat be great enough to make them both ascend together . . ." (*Opticks,* p. 382). The idea that mercury can be sublimed from antimony suggests that Newton looked upon it as an essence rather than as a metal.

Capillary Action

There may be some question as to whether or not Robert Hooke anticipated Newton in suggesting the laws of universal gravitation, but there can be
no doubt that he offered an explanation for capillary action many years before Sir Isaac had anything to say on the subject. In Observation VI of *Micrographia* (1665), Hooke describes experiments, in which quicksilver was used,
leading to an explanation of ". . . the rising of Liquors in a Filtre, the rising of
Spirit of Wine, Oyl, melted Tallow, and c. in the Week of a Lamp. . . ."

A few years later (1672) there was published in the *Philosophical Transactions* an "Extract of a letter of M. Hugens [sic] . . . attempting to render the
Cause of that odd Phaenomenon of the Quicksilver remaining suspended far
above the usual height in the Torricellian Experiment." Huygens postulated
that there was another pressure stronger than that of air ". . . which without
difficulty penetrates glass, water, quicksilver and all other bodies . . ." and "is
capable to sustain the 75 inches of Mercury, and possibly more. . . ." This
almost certainly refers to capillarity.

Early in the 19th century, that amazing Napoleonic creation, the Society
of Arcueil, became interested in capillarity. The principal investigators were
the astronomer-mathematician Pierre-Simon Laplace (1749–1827), the
chemist Joseph Gay-Lussac (1778–1850), and the mineralogist René-Just
Haüy (1743–1822). Laplace had originally become interested in capillary
action while working with Lavoisier on the standardization of mercury barometers, at which time it had been observed that moisture or grease could
alter the meniscus in the barometer tube. As a result of the studies of this trio,
Laplace was able to derive, test and prove a mathematical equation for capillary forces in tubes of small diameter. Observations were extended from mercury to include water and alcohol as well (Crosland 1967).

Research similar to that conducted by the Arcueil group was going on in
England at the same time, but due to the military-political situation the exchange of scientific information did not enjoy its customary freedom. Laplace
in France and Thomas Young (1773–1829) in England published similar reports in 1805.

Seismoscopes

Instruments for demonstrating tremors in the earth's crust are said to have
been designed first in China, but the date is not clear. According to Boffito it
was 136 B. C., but Gutenberg says it was ". . . aus dem Jahre 136 . . .," not
specifying B. C. or A. D. The Chinese seismoscope depended on the displacement of balls in a manner which would indicate the direction of origin and the
severity of a shock wave. Among the improvements and refinements in seismo-
logical devices made over the years there have been several in which quicksilver
has been used. One of these was described by the Abbé Jean de la Hautefeuille
(1647–1724) in 1703 (Boffito 1929). In principle it was similar to its Chinese
predecessor, depending on the displacement of a body to indicate the occur-

rence of a tremor. This improved seismometer consisted of a mercury-filled basin with a wide brim in which there were eight depressions with channels leading to them from the basin. If an earthquake should occur, the level of the mercury would be disturbed, causing some of it to overflow into the cups. By measuring the amount of mercury which had run into the cups, and observing the side on which the overflow took place, the severity and direction of the shock could be estimated.

A seismoscope similar to that of Hautefeuille was described by Atanasio Cavalli (Fig. IX-1) in 1785 and built by him in 1787 (Boffito 1929). This type of instrument was commonly used by seismologists through the 19th century.

Clocks

Three early uses for quicksilver in instruments for measuring time have been described: the Chinese clepsydra-like clock of the 5th or 8th century

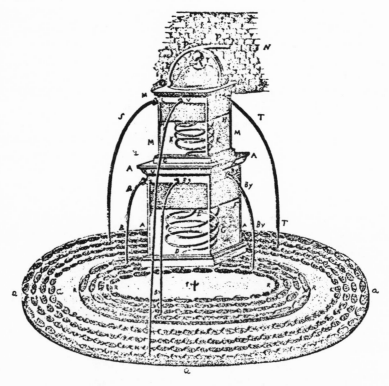

Figure IX-1 Cavalli's Seismoscope. This type of instrument was in use throughout the 19th century. A tremor caused mercury to spill from shallow containers in the tower-like structure into small cups placed in rings below. The spilled mercury was weighed to find the intensity of the shock. (From *Gli Strumenti della Scienza e la Scienza degli Strumenti*, by G. Boffito. 1929.)

A. D., the 13th century "Relogio dell Argen Vivo" of King Alfonso of Spain and Tycho Brahe's measuring device of the 16th century. None of these instruments made any lasting impact on the world of clocks, but in the 18th century a permanent place was found for mercury.

The compensating mercury pendulum was invented by George Graham (1673-1751) in 1721 and has been in use ever since (Bruton 1962). Pendulums of this type carry one or two tubes of mercury which compensate for changes in length of the bob on heating or cooling. If heat causes the bob to lengthen, it also causes the mercury to rise, with converse effects on cooling. This results in the center of oscillation or effective length of the pendulum remaining constant. An improved form was designed by Riefler in 1889. In the Riefler clock the bob of the pendulum is a steel tube nearly filled with mercury; this is said to give better compensation since the adjustment is effective over the entire length of the pendulum.

Latent and Specific Heat

The name of Joseph Black has already been mentioned in connection with his use of mercuric chloride in experiments on alkaline materials. Noted as a great teacher (Benjamin Rush was his student) as well as a great scientist, Black was the first to elucidate the concept of latent heat. In a series of experiments with mercury and water, he showed that mercury took up and liberated heat more quickly than water. "We must, therefore, conclude that different bodies, although they be of the same size, or even of the same weight, when they are reduced to the same temperature or degree of heat, . . . may contain very different quantities of the matter of heat . . ." (Magie 1935). Black's observations were not only of great theoretical importance, but in a practical application they ". . . were of very great service to James Watt (1736-1819) then an instrument maker at Glasgow, in his invention in 1765 of the improved condensing steam-engine" (Partington 1965).

Mercury figured in at least two major studies on heat, in addition to those of Joseph Black. The better known of the two is that of James Prescott Joule (1818-1889), the first to determine the mechanical equivalent of heat. In one set of experiments, Joule measured the heat developed by friction in mercury, this being one step in a series of observations which led him to conclude "1st. That the quantity of heat produced by the friction of bodies, whether solid or liquid, is always proportional to the quantity of force expended. And, 2nd. That the quantity of heat capable of increasing the temperature of water . . . by 1° Fahr. requires for its evolution the expenditure of a mechanical force represented by the fall of 772 lb. through the space of one foot." Less well known than the work of Joule is that of Charles Cagniard de la Tour (1777-1859) on critical temperature (the temperature at which a gas, by pressure, can be reduced to a liquid). By using mercury, de la Tour was able to clarify some otherwise puzzling observations, thus laying the groundwork for subsequent studies on the liquefaction of gases (Magie 1935).

Electricity

Among the more important discoveries in the early history of electricity was that made by George Simon Ohm (1787-1854); "Ohm's law" states that electromotive force (volts) = current (amperes) X resistance. Mercury contacts were used in the apparatus designed by Ohm for the studies which led to the formulation of this principle (cited in Magie 1935). A similar use of mercury was made by Michael Faraday (1791-1867) in his experiments on electromagnetic induction from which stemmed the development of electric motors as a source of power for driving machinery. Joseph Henry (1797-1878) also used mercury contacts in his experiments, which antedated those of Faraday (Magie 1935; Henry 1832).

Toward the end of the 19th century the production of electricity reached a point where it could be made available commercially, creating the need for a device to measure with accuracy and reliability the amount of the commodity that was passing from the producer to the consumer. The invention of a mercury-motor meter by S. Z. Ferranti (1864-1930) in 1883 provided a solution to this problem (Ridding 1964). The original Ferranti meter was designed to measure direct current, but it served as a prototype for more refined meters for both direct and alternating current. Related to the invention of the mercury-motor meter was earlier work by H. C. Oersted (1777-1851) on the action of electric currents on magnets (Magie 1935). Mercury was among the conductors used by Oersted in his experiments.

Electrometers

Mercury's relationships with electricity, of increasing importance in the 20th century, had their beginnings about 1705, when Francis Hawksbee (d. 1713?) observed that flashes of light could be produced when mercury was shaken in a dry container (Hawksbee 1705). Hawksbee probably did not recognize the electrical nature of the phenomenon since he referred to it as "mercurial phosphorus" (Berry 1968, Chap. 3). Newton, too, was aware of this type of phosphorescence.

The difficulties encountered by Davy and Berzelius in producing electrolysis of the alkaline earths were eliminated when a mercury electrode was used (p. 103). Thus began, in 1807, a long series of uses for mercury electrodes, including an application which accounts for the largest single use of the metal at the present time, the mercury cathode in the chlor-alkali process.

The key to three important tools in modern medicine, the electrocardiograph, the electroencephalograph, and the electromyograph, was the capillary electrometer (Geddes and Hoff 1961). Acting on the long-known fact that the curvature of a mercury surface in contact with an aqueous solution will vary depending on the state of electrification, Gabriel Lippmann (1845-1921), a French physicist and Nobel prize winner, began in 1875 an intensive study of relationships between electrical potential and surface tension at mercury/water interfaces. As an aid in his research, Lippmann designed an ultra-sensitive capillary electrometer, adapting for his purposes an instrument in common use in electrochemistry (Berry 1968).

Relationships between muscle contraction and electrical currents were demonstrated by Aloysio Galvani (1737-1798) about 1792, in experiments on the skeletal muscles of frogs (Clendening 1960). Similar phenomena for heart muscle were observed by R. A. von Kölliker (1817-1905) and Johannes Mueller (1801-1858) in 1856, and for brain tissue by Hans Berger (1873-1941) in 1929. Electrophysiology became a field of major interest during the latter part of the 19th century and early part of the 20th, commanding the attention of many of the leading scientists of the day (Berry 1968; Clendening 1960; Singer and Underwood 1962), and leading little by little to applications in clinical medicine. It was E. J. Marey (1830-1904), using the Lippmann capillary electrometer, who was primarily responsible for bridging the gap between experimental and applied electrophysiology (Geddes and Hoff 1961; Marey 1876a and 1876b) through the measurement of electrical currents in heart muscle.

E. D. Adrian (1889-   ) was among the last to use the capillary electrometer. He combined the instrument with a vacuum tube amplifier system which gave greatly increased sensitivity and reduced reaction time. His studies in neurophysiology earned him the Nobel Prize in 1932, which he shared with Sir Charles Sherrington (1857-1952) (Geddes and Hoff 1961).

Mercury Pumps

High on the list of basic scientific instruments in the development of which mercury played an indispensable role is the vacuum pump. The first step, one in which mercury figured, was the demonstration by Berti that a vacuum could exist (p. 108). There followed numerous experiments such as those of Otto von Guericke (1602-1686) with his Magdeburg hemispheres, of Robert Boyle (1627-1691) and of Robert Hooke (1635-1703), directed toward the development of more efficient vacuum pumps (Partington 1965). Various types of pumps with varying degrees of efficiency were designed up until the middle of the 19th century. In 1865, Hermann Sprengel published a description of his mercury pump which is said to be the parent of most of the more complicated instruments devised later for removing air from an enclosed space (Tilden 1926). A diagrammatic representation of the pump is shown in Figure IX-2.

The degree to which pressure can be reduced by the Sprengel pump is limited by the vapor pressure of mercury, which is 0.0001 millimeters at 20°C. The Töpler pump, which uses mercury as a sort of piston, is capable of reducing pressures to as low as 0.00001 millimeters of mercury, and other types of pumps can go much further. High vacuum mercury diffusion pumps are an essential part of many electron microscopes.

Even though the Sprengel mercury pump was not capable of producing a high vacuum, a modified form was used by Sir William Crookes (1832-1919)

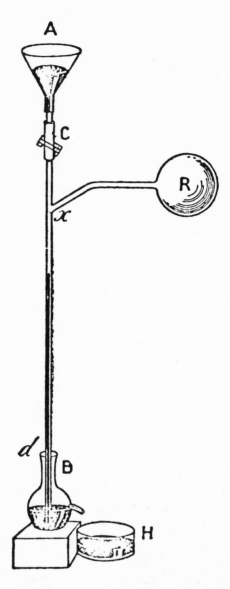

Figure IX-2 Sprengle's Mercury Pump Used to Create a Partial Vacuum.
Mercury falls from funnel (A) through tube to receiving flask (B) which has a
spout through which mercury spills out. The mercury's falling through the
long vertical tube draws air from flask (R). Stopcock (C) keeps funnel from
emptying entirely (which would permit air to enter tube). Mercury collected
in container (H) is poured back into funnel several times and flows down tube
until as much air as possible is drawn from flask (R). Operation is complete
when no air bubbles are in the mercury in vertical tube. The height of the
column of mercury in tube is now equal to that in a barometer. (From H.
Sprengel, *J. Chem. Soc.* 3:9-21 [1865].)

in his early work on vacuum tubes (Tilden 1926). Similarly, the Sprengel pump was an important factor in the production of some of the first incandescent electric lamps. Other important early applications of vacuum tubes were in the discovery by Nobelist Jean Perrin (1870-1942) in 1895 that the cathode discharge carried negative electric charges, the discovery of the x-ray by Roentgen (1845-1923) also in 1895, and the discovery of the electron by J. J. Thompson (1856-1940) in 1897. In a very real sense, therefore, mercury played a key role in the founding of the science of electronics, just as it did, according to Pledge, in the elaboration of the quantum theory.

Perpetual Motion

Nothing could be more natural than the use of quicksilver, with its great versatility and mysterious properties, in attempts to make a "perpetual motion machine." Chinese technicians are said to have designed such a machine in medieval times and an unsuccessful attempt in this direction, using alcohol, was made by von Guericke about 1600 (Pledge 1939). The design and operation of an improved model using quicksilver was published in the Philosophical Transactions in 1685 (Papin).

Modern Uses of Mercury

A comprehensive survey of present uses of mercury and possible new ones was conducted by the Battelle Memorial Institute for the American Quicksilver Institute in 1959 (Cross and Hale). Much of the following material is taken from the Battelle report and gives an idea of the range of scientific and technological fields in which mercury is, or might be, useful.

Electrical Apparatus

At least as far as the United States is concerned, the industries concerned with the manufacture of electrical apparatus have, until 1967, consistently been the largest consumers of mercury. Principal specific uses include mercury vapor lamps of several types (including fluorescent lamps), rectifiers, power cells and batteries, switches and relays. Of special interest in the latter half of the 20th century are the mercury "minibatteries" and similar small power sources which have been indispensable in the exploration of outer space.

Agricultural Chemicals

Agricultural uses of mercury, with origins in folklore (Chap. 2), were reintroduced in the first two decades of the 20th century and greatly expanded in the period starting in the 1930's. Mercurials have been found useful in agriculture as insecticides, fungicides, herbicides, and bactericides. Among the inorganic compounds, the chlorides, oxides, sulfates, sulfides, iodides, nitrates, bromides, cyanides and salicylates are of greatest importance. Three classes of organic mercurials have been used: alkyl, aryl and alkoxy-alkyl. Because of extremely high toxicity, the alkyl compounds such as those of methyl- and ethylmercury, and other short-chain alkyls, are rapidly being replaced by the less toxic aryl and alkoxy compounds. Phenyl mercurials, which have been shown to have relatively low toxicity (p. 173) are widely used not only in agriculture but also as preservatives in paints and adhesives and, in the paper industry, for retarding the growth of slime molds. Several hundred different phenyl mercurials have been prepared in the laboratory; of these the most commonly used are the acetate (PMA), benzoate, nitrate, propionate and oleate. Health aspects of the use of mercurial pesticides are discussed in Chapter 11.

Industrial and Control Instruments

Mercury barometers and thermometers, the oldest applications of the metal to instruments, have already been discussed (Chap. 8). Except for laboratory use, mercury barometers have been largely displaced by the aneroid type, but mercury thermometers remain as one of the most important measuring devices in medicine, science and industry. One prominent manufacturer of scientific instruments lists about 600 different types of mercury thermometers in his catalogue. Thermostats and thermo-regulators represent common applications of mercury thermometers. Hygrometers and psychrometers have already been mentioned. A variety of applications have been found for mercury manometers and flow meters.

Navigational uses for mercury are seen in such diverse applications as gyroscopes and artificial horizons and as low-friction baths to support the revolving units of lighthouses.

Electrolytic Preparation of Chlorine and Caustic Soda

Between 1947 and 1967 the consumption of mercury in the electrolytic preparation of chlorine and caustic soda (NaOH) experienced a striking increase from 700 flasks to 14,300 flasks, thus representing the largest single use and surpassing that of electrical apparatus.

In the electrolytic process, a current is passed through a solution of sodium chloride (brine) in a closed cell with mercury acting as the cathode. Chlorine gas is evolved at the anode and the released sodium amalgamates with the mercury cathode. The amalgam is then decomposed with water; hydrogen gas and caustic soda are formed, and the mercury is freed and re-cycled through the cell in a continuous process.

Antifouling Paint

Antifouling paints are applied to the bottoms of ships to inhibit the growth of barnacles and other forms of aquatic life which grow attached to the surface of the ship, and if left unchecked can interfere with the ship's speed and maneuverability. The red and yellow oxides of mercury have been common ingredients in these paints, frequently in combination with other metals, especially copper, lead, and arsenic. In theory, the mercuric oxide is supposed to react slowly with the salt in sea water to form the toxic bichloride of mercury. Research carried out during World War II in the United States cast serious doubt on the efficacy of mercury in this usage and resulted in its elimination from many antifouling paint formulations. High speed ships, and especially those like the new super-tankers which spend only a few hours in harbor during turn-around, are believed not to require antifouling paint at all, but coatings containing mercury are extensively used on yachts and other small craft to suppress the growth of algae. A related application was that of phenyl mercurials as algicides in swimming pools, but this use was banned in the United States in 1971.

(As part of a campaign in Sweden to reduce the pollution of waters with mercury compounds, a move was initiated in 1969 to eliminate such materials from the antifouling paints used on the bottoms of the ships of the Swedish navy.)

Amalgam Metallurgy (other than dental)

Amalgamation of mercury with gold for purposes of recovery and purification is probably the oldest known technological use of the metal. Although this procedure is now not widely used, amalgam metallurgy is commanding great interest for other reasons, particularly in the preparation of pure forms of lead, copper, zinc, titanium, and zirconium from their chlorides. The availability of large quantities of relatively inexpensive mercury-sodium amalgams from the electrolytic production of chlorine and caustic soda has stimulated interest and research in possible uses of the amalgam in a variety of applications in inorganic and organic chemistry.

Investment Casting

Precision casting in which frozen mercury is used to make the pattern is similar in principle to the ancient "lost wax" process. The pattern is made by pouring mercury at room temperature into the die, freezing the metal in a mixture of solid carbon dioxide (dry ice) and acetone, building a ceramic shell around the frozen pattern and allowing the mercury to thaw (melting point = $-38.87°C$) and run out of the shell which is then dried and fired. This serves as the mold for subsequent casting.

Heat Transfer and Mercury Boilers

A number of applications have been found for mercury as a heat transfer agent, but details are not readily available. Mercury boilers have been developed in which the metal is heated in the tubes of the boiler and the vapor used to drive the generator turbines. This does not constitute a wide-spread use for the metal. Information on the use of mercury in connection with nuclear reactor technology is limited because of security reasons, but there is evidence that it has been or is being used as a heat transfer medium as well as for other purposes. Similar secrecy surrounds the use of mercury compounds in propellants.

In this section, as in the one dealing with chemistry, no attempt has been made to compile a complete catalogue of all uses of mercury. Important examples have been described principally to show the indispensable role which mercury has played in modern science and technology.

REFERENCES

Berger, H. 1929. Über das Elektroenkephalogram der Menschen. *Arch. Psychiat. Nervenkr.* 87:527–70.

Berry, A. J. 1968. *From classical to modern chemistry.* New York: Dover Publications.

Boerhaave, H. 1753. *A new method of chemistry.* Trans. Peter Shaw. 3d ed. 2 vols. London.

Boffito, G. 1929. *Gli strumenti della scienza e la scienza degli strumenti.* Florence: Libreria Internazionale Seeber.

Bruton, E. 1962. *Dictionary of clocks and watches.* London: Arco Publications.

Clendening, L. 1960. *Source book of medical history.* New York: Dover Publications.

Cohen, I. B. 1952. In preface to *Opticks,* by Sir I. Newton. New York: Dover Publications.

Crosland, M. 1967. *The Society of Arcueil.* Cambridge: Harvard University Press.

Cross, J. L., and Hale, R. W. 1959. A technical survey of new and expanded applications for mercury. *Report to the American Quicksilver Institute,* 30 September 1959. Columbus, Ohio: Battelle Memorial Institute.

Descartes, R. 1897–1919. *Oeuvres,* ed. Chas. Adam et Paul Tannery, Paris: L. Cerf.

Faraday, M. 1835. Experimental researches in electricity: Ninth series. *Phil. Trans.* pp. 41–56.

Geddes, L. A., and Hoff, H. E. 1961. The capillary electrometer: The first graphic recorder of bioelectric signals. *Arch. Intern. Hist. Sci.* 14:275–90.

Gutenberg, B. 1929. *Lehrbuch der Geophysik.* Berlin: Borntraeger.

Hawksbee, F. 1705. Several experiments on the mercurial phosphorus. *Phil. Trans.* 24:2129–35.

Henry, J. 1832. On the production of currents and sparks of electricity from magnetism. *Am. J. Sci.* 22:403–8.

Hooke, R. 1961. *Micrographia.* Facsimile of the London, 1665, edition. New York: Dover Publications.

Huygens, M. [Hugens]. 1672. Extract of a letter of M. Hugens. *Phil. Trans.* 7:5027–30.

Joule, J. P. 1850. On the mechanical equivalent of heat. *Phil. Trans.* 140:61–82.

Kölliker, R. A. von, and Mueller, J. 1856. Nachweis der negativen Schwankung des Muskelstroms natürlich sich contrahirenden Muskel. *Verhandl. Physik. Med. Ges. Würzburg* 6:528–33.

Magie, W. F. 1935. *A source book in physics.* New York: McGraw-Hill Book Co.

Marey, E. J. 1876a. Des variations électriques des muscles et du coeur en particulier étudiées au moyen de l'électromètre de M. Lippmann. *Compt. Rend.* 82:975–77.

Marey, E. J. 1876b. Inscription photographique des indications de l'électromètre de Lippmann. *Compt. Rend.* 83:278–80.

Middleton, W. E. K. 1966. *A history of the thermometer and its use in meteorology.* Baltimore: Johns Hopkins Press.

Newton, Sir I. 1952. *Opticks or a Treatise of the reflections, refractions, inflections and colours of light.* Based on the 4th edition, London, 1730. New York: Dover Publications.

Papin, Doctor. 1685. Observations concerning a perpetual motion. *Phil. Trans.* 15:1240–41.

Partington, J. R. 1965. *A short history of chemistry.* 3d ed. New York: Harper and Row, Harper Torchbooks.

Perrin, J. 1895. Nouvelles proprie'te's des rayons cathodiques. *Compt. Rend.* 121:1130–34.

Pledge, H. T. 1939. *Science since 1500.* London: H. M. Stationery Office.

Ridding, A. 1964. *S. Z. de Ferranti: Pioneer of electric power.* A Science Museum Booklet. London: H. M. Stationery Office.

Roentgen, W. C. 1895. Ueber eine neue Art von Strahlen. *Sitzber. Würzburger Physik. Med. Ges.* December, pp. 132–41.

Singer, C., and Underwood, E. A. 1962. *A short history of medicine.* 2d ed. Oxford: Oxford University Press.

Sprengel, H. 1865. Researches on the vacuum. *J. Chem. Soc.* 3:9–21.

Thomson, Sir J. J. 1897. Cathode rays. *Phil. Mag.* 44:293–316.

Tilden, Sir W. A. 1926. *Chemical discovery and invention in the twentieth century.* New York: E. P. Dutton and Co.

Young, T. 1805. An essay on the cohesion of fluids. *Phil. Trans.* pp. 65–87.

ADDITIONAL READINGS

Zittel, C. A. von. 1901. *History of geology and palaeontology.* Trans. M. M. Ogilvie-Gordon. London: Walter Scott.

Part II

10

Normal Mercury in Man

With mercury present in the air man breathes, the water he drinks and the food he eats, it inevitably will be present in the human body. Establishing the magnitude of man's background, or "normal," mercury is a matter of some importance.

Biologically speaking, concepts of "normal" depend on arbitrarily chosen definitions. According to one definition, "normal" is the state of affairs which existed before there was any man-made intervention. This is sometimes called "natural." For mercury in man, "normal" has been defined as the amount found in the tissues and organs, in the urine, blood and other body fluids in the general population, excluding individuals who have had any occupational or therapeutic exposure to the metal or its compounds. This is the definition used here. What is not normal must, by definition, be abnormal, but abnormal does not necessarily mean pathological.

"Normal" Mercury in Man

"Normal" values, or rather ranges, for quantitative factors such as height, weight, blood counts and blood chemistries are sometimes arrived at by performing a suitable number of observations on persons who appear to be in good health, calculating a mean and standard deviation, and defining the normal range as embracing the values which fall within two standard deviations on either side of the mean. This, obviously, is a purely arbitrary method of defining what is normal. It is applicable only when the observations are distributed along the familiar Gaussian curve, in which case a spread of two standard deviations on either side of the mean will embrace about 95 percent of the observations. Properly interpreted, this method implies that if a value

135

falls outside the prescribed range there is a strong probability that some extraneous influence is at work. Determinations of mercury in the urine and blood of persons who have had no unusual exposure to the metal or its compounds, however, are not conveniently distributed along a bell-shaped (Gaussian) curve. This is due, at least in part, to the fact that until recently the analytical methods suitable for large scale surveys were of such limited sensitivity that a majority of the tests were reported as "zero." Measurement in the microgram range is adequate for most clinical purposes but it necessitates a statistical approach different from that which is applicable to "Gaussian" data.

Prior to the middle of the 20th century very little attention had been paid to establishing "normal" values for mercury in man. Analytical methods of sufficient sensitivity and rapidity had not been developed, and furthermore, the accumulation of normal data was not considered to be of great importance. Any detectable amount of mercury in urine or tissues was generally considered abnormal. Analysis of biological materials for metals was largely in the hands of forensic chemists who ordinarily dealt with gross quantities of suspected poisons. Another early application of quantitative analysis of urine was related to the use of mercury in the therapy of syphilis. Here, too, relatively large amounts of the metal were involved, and no attempts were made to establish pre-treatment "normal" values.

Urine

What may appropriately be called the "Second Amalgam War" (see Chap. 20, below) might never have occurred had more attention been paid to "normal" mercury in urine. Stock (1926, 1928), having devised a sensitive analytical method, was able to demonstrate what he considered to be dangerous quantities of the metal in urine samples from persons who had dental cavities filled with mercury amalgams. His conclusion that amalgam fillings constituted a serious threat to health by reason of mercury poisoning was challenged by Borinski, who showed that mercury was present in common foods and that the amount found by Stock in the urine of dental patients was no greater than that in "normal" persons, i.e., up to 10 micrograms (μg) excreted daily. This was certainly one of the first occasions when "normal" values for mercury in urine were applied to the interpretation of allegedly abnormal findings.

Several articles containing data on "normal" urinary mercury in man were published between 1936 and 1963; these were summarized by Jacobs, Ladd, and Goldwater in 1964. Much of the information in these articles is fragmentary, with the omission of such important points as the number of persons studied and the sensitivity of the analytical method employed. Values for "normal" mercury in urine in these studies range from zero to 100 μg per liter or per 24 hours.

An organized effort to obtain a world-wide picture of the "normal" absorption and excretion of mercury by man was undertaken in 1961 as a joint study under the auspices of the World Health Organization and Columbia University. In this study samples of urine and blood were collected under standardized conditions in fifteen countries around the world and shipped to Columbia University for analysis. Because of practical limitations the subjects from whom specimens were obtained did not represent a statistical cross section of the populations in the respective countries. They were, rather, persons who were readily accessible, who would submit to having blood samples drawn, who would provide urine specimens, and who had had no known occupational, medicinal, or other exposure to mercury. Only in this sense were they groups of "normals."

Analyses were done by a method which could measure minimum concentrations of 0.5 μg of mercury per liter of urine or per 100 ml of blood, using aliquots of one or two ml (Jacobs, Goldwater, and Gilbert 1961). Thus, any samples containing concentrations of mercury lower than these limits were recorded as "zero." The findings were considered to be those of normal (unexposed) populations. The results of the analyses of urines in this survey are presented in Tables X-1 and X-2.

Since 95 percent of those surveyed had 20 μg or less of mercury per liter of urine, this is the upper limit of "normal" for the group that was studied. Universal application of this standard is not justified, however, since the populations from which samples were obtained may not have represented a true cross section, and since the study was limited to 15 countries.

Blood

Prior to 1961 no systematic effort had ever been made to determine the amount of mercury in the blood of "normal" humans. Stock in 1936 stated that the normal value for mercury in blood was 0.3 to 0.7 μg per 100 g; and Benning in 1958 gave a range of 0.0 to 0.30 μg per 100 g. Neither of these

TABLE X-1
"Normal" Mercury in Urine:
Data from WHO/Columbia University Study

No. of Countries	15
No. of Samples	1107
No mercury detected (less than 0.5 μg/l)	78%
Less than 5 μg/l	86%
Less than 10 μg/l	89%
Less than 15 μg/l	94%
Less than 20 μg/l	95%
Less than 25 μg/l	96%
No. of cases 25–50 μg/l	22 (=1.9%)
No. of cases more than 50 μg/l	17 (=1.5%)
Highest value	221 μg/l

SOURCE: L. J. Goldwater: Personal observations.

TABLE X-2

"Normal" Values for Mercury in Urine
Data from WHO/Columbia University Study

Locality	No. Samples	Range (µg/l.)	No. with Zero[a]
Argentina	49	0.0– 21.0	41
Chile	35	0.0– 21.0	27
Czechoslovakia	20	0.0– 10.5	17
Egypt	28	0.0– 12.0	22
England	30	0.0– 37.0	26
Finland	46	0.0– 30.0	31
Israel	83	0.0– 94.5	72
Italy	25	0.0– 37.5	19
Japan	40	0.0– 45.0	34
Netherlands	60	0.0– 15.0	52
Peru	64	0.0–107.0	32
Poland	98	0.0–157.0	70
Sweden	30	0.0– 73.5	24
Yugoslavia	65	0.0– 69.0	56
U.S.A.			
California	31	0.0– 15.0	27
Ohio	40	0.0–221.0	26
N.Y.C.	363	0.0– 37.5	291
Totals	1107	0.0–221.0	867 (78.0%)

[a]Zero = less than 0.5 µg/l.

authors stated the number of persons tested nor did they suggest that they were attempting to establish a normal range except on a very limited basis. So meagre was the information on "normal" mercury in blood as late as 1940 that Szép found it worthwhile to report the results on four persons. His values were 0.16 to 0.51 µg of mercury per 100 grams of blood. Thus all three of these early observers obtained results of the same order of magnitude. In addition to his studies on living subjects, Stock analyzed blood taken from 21 cadavers and found mercury in the range of 0.2 to 15.0 µg per 100 g. He does not describe what precautions, if any, were taken to avoid contamination of the specimens in the autopsy room (as, for example, from mercuric chloride in preservative solutions that may be used by pathologists).

As part of the WHO/Columbia University study of "normals," 812 samples of blood from 15 countries were analyzed. (See Tables X-3 and X-4.) Since 95 percent of the samples contained less than 3.0 µg of mercury per 100 ml of blood, this may be taken as the upper limit of "normal" for the group surveyed; but it is of little clinical importance whether this figure or 5.0 µg is accepted. In either case the level is higher than that found by earlier observers, the discrepancy possibly being due to different analytical methods. Interpretation and application of these findings are subject to the same limitations noted in the case of the "normals" for urine.

Several possibilities exist to explain the occurrence of a number of values in excess of the upper limit of "normal" for both blood and urine. Among

TABLE X-3

"Normal" Mercury in Blood
Data from WHO/Columbia University Study

No. of countries	15
No. of samples	812
No mercury detected (less than 0.5 µg/l of whole blood)	77.0%
Less than 1.0 micrograms/100 ml.	85.0%
„ „ 2.0 „	89.0%
„ „ 3.0 „	95.0%
„ „ 4.0 „	97.0%
„ „ 5.0 „	97.2%
No. cases 5–10 micrograms/100 ml.	10 (= 1.2%)
No. cases more than 10 micrograms/100 ml.	10 (= 1.2%)
Highest value	39.6 micrograms/100 ml.

these possibilities may be mentioned: laboratory error, contamination of the specimens in collection or handling, unwitting use of a mercurial preparation (such as a contraceptive jelly or medicinal salve), or environmental exposure of which the subject was unaware.

Saliva

When Johann Keyssler visited the Slovenian quicksilver mines at Idria in 1730 he was given a demonstration of the amalgamation of copper coins by their being held in the mouth or rubbed with the fingers of miners. This

TABLE X-4

"Normal" Values for Mercury in Blood

Locality	No. Samples	Range µg/100 ml.	No. with Zero	Percent with Zero[a]
Argentina	49	0.0–3.0	39	80
Chile	35	0.0–3.0	24	65
Czechoslovakia	20	0.0–2.1	12	60
Egypt	28	0.0–1.0	27	96
England	30	0.0–7.5	28	94
Finland	46	0.0–7.5	32	70
Israel	67	0.0–3.9	60	90
Italy	27	0.0–2.1	21	78
Japan	40	0.0–3.0	32	80
Netherlands	60	0.0–2.1	56	93
Peru	58	0.0–19.5	17	29
Poland	95	0.0–39.6	69	73
Sweden	30	0.0–9.0	27	90
Yugoslavia	67	0.0–27.0	48	72
U.S.A.				
California	33	0.0–5.1	26	79
Ohio	40	0.0–23.8	34	85
New York	87	0.0–39.6	72	83
Total	812		624	77

[a]Zero = less than 0.5 µg/100 ml. whole blood

proved the excretion of mercury in saliva and sweat. The literature on mercury in saliva was reviewed by Joselow et al. in 1968, at which time no quantitative data regarding the amount of mercury that may be present in saliva was found. (The amount and significance of mercury excretion in sweat has not been studied.)

Studies carried out at Columbia University embraced 13 unexposed normals and 40 men occupationally exposed to a variety of mercury compounds. None of the controls had sufficient mercury in their saliva to permit detection by a method (see Jacobs, Goldwater, and Gilbert 1961) with a sensitivity of 0.5 µg per 100 ml. The salivary mercury in the exposed workers had a range of 1.0 to 15.5 µg per 100 ml and a mean value of 5.0 µg per 100 ml. Close correlation between mercury in saliva and in blood was found (Joselow, Ruiz, and Goldwater 1968).

Tissues

Mercury which is found in the urine has already left the tissues and can no longer cause any harm. The same is true of salivary mercury if it is spit out, but the probability is that most of it will be swallowed and at least partially reabsorbed. That in the blood is in a state of mobility and consequently is potentially dangerous until it finds its way to an excretory channel. Mercury which has been taken up by tissues or cells (the so-called body burden) probably has greater toxicological significance than that which is found in urine, blood, or saliva, yet, as far as humans are concerned, it has been studied infrequently. Only through the use of autopsy material is it practicable to perform extensive studies on "normal" mercury levels in human tissues. Data so obtained are of limited value because of the difficulty of obtaining complete information on previous unusual exposures to mercury. In addition, in the ordinary autopsy room there are real dangers of contamination of the specimen, and furthermore, the analytical procedures are complicated.

Only five observations on the mercury content of human tissues were published prior to 1940. These were summarized in 1939 by Bodnár who added three more (Bodnár, Szép, and Weszpremy). Stock contributed six additional "normals" in 1940. A report by Szép, in 1940, has a misleading title referring to the mercury content of the human body while in fact he gives only the results of mercury determinations on blood samples.

One of the most extensive studies of mercury in human tissues is that of Stock, the results of which were published in 1943. Over a period of several years Stock performed analyses for mercury on 419 individual samples of tissues from 63 autopsy cases, 36 males and 27 females, with fairly even distribution among various age groups. His findings are summarized in Table X–5. Stock distinguished between those with and without amalgam dental fillings, implying that this had a significant effect, but a study of his data does not reveal any striking difference between the two groups.

Liver and kidney tissues from 90 routine autopsies were analyzed by Forney and Harger and reported in 1949. A large majority of the kidneys

TABLE X-5

"Normal" Mercury in Tissues

Hg/100g	Kidney No.	Kidney %	Hypophysis No.	Hypophysis %	Olfact. Bulb No.	Olfact. Bulb %	Brain No.	Brain %	Muscle No.	Muscle %
Up to 0.5	0	0	2	3.1	6	8.7	9	13.4	25	38.5
1	0	0	5	7.6	2	2.9	17	25.5	19	29.2
2	4	5.4	5	7.6	2	2.9	19	28.5	13	20.0
3	3	4.1	3	4.6	1	1.5	6	8.9	0	0
5	9	12.1	8	12.0	8	11.6	7	10.4	4	6.3
7	4	5.4	7	10.5	7	10.1	2	3.0	1	1.5
10	6	8.1	5	7.6	10	14.5	1	1.5	3	4.6
15	16	21.6	11	16.8	7	10.1	4	5.9	0	0
20	3	4.1	3	4.6	6	8.7	1	1.5	0	0
30	12	16.2	8	12.0	11	16.0	0	0	0	0
50	3	4.1	1	1.5	7	10.1	0	0	0	0
100	5	6.7	7	10.5	2	2.9	1	1.5	0	0
500	5	6.7	1	1.5	0	0	0	0	0	0
over 500	4	5.4	0	0	0	0	0	0	0	0
Total	74		66		69		67		65	

Additional Data

Liver: 10 samples, 4 less than 10   mg/100g
4 25-99 mg/100g
1 273 mg/100g
1 761 mg/100g

Thyroid: 8 samples, Range 1-32 mg/100g
Blood: 21 samples, Range 0.2-15 mg/100g

All samples fresh weight

Adapted from Stock (1943)

NOTE: Apparently more than one sample was analyzed for some tissues, since only 63 cadavers were studied.

yielded mercury values between 0.10 and 10.0 parts per million (wet weight) but there was one case in which the figure was 127 ppm. Most of the livers contained between 0.10 and 5.0 ppm and the highest value found was 17.2 ppm.

Another study of liver and kidney was reported by Butt and Simonsen in 1950. In 69 "normal" kidneys they found the average value for mercury to be 0.075 mg per 100 g (or 0.75 ppm) wet weight and in the same number of livers the average amount was 0.006 mg per g (0.06 ppm).

Griffith and associates analyzed mercury in the kidneys of 45 cases at autopsy. Of these only 15 had not received mercurial medication. These might have been considered "normals" but for the fact that they had died in congestive heart failure. At least some of the analyses were done by emission spectroscopy on tissues previously dried and ashed and it is not stated what measures, if any, were taken to avoid loss of mercury—which would tend to occur because of mercury's high volatility—during these preparatory steps. The average mercury content, in milligrams per 100 grams of dried tissue, was kidney: 2.05, liver: 0.37 and spleen: 0.12; but obviously these results cannot be compared with those obtained on analyses performed on organs in the wet state.

A more recent study carried out at Columbia University (Joselow, Goldwater, and Weinberg 1967) embraced 236 tissue specimens obtained from the autopsy material of 39 humans who had died either of trauma or of a natural cause. There was no reason to believe that any mercury had been administered in any of these cases, though it was not possible to be absolutely certain of this. There were 27 males and 12 females in the series, with fairly even distribution in age groups from less than one year to over sixty. Kidneys were analyzed in all cases and liver, brain, heart, spleen, lung, and muscle in most. The results, expressed as parts per million of fresh weight, showed that kidneys had the highest levels, with a maximum of 26.3 ppm total mercury, and an average of 2.75 ppm. The average for kidneys was about ten times that of the organ with the next highest mercury content, the liver, which averaged 0.30 ppm of mercury. For brain, the highest value was 0.6 ppm, and the average of 27 cases was 0.10 ppm.

As part of a study of essential and nonessential trace elements in the human body, Liebscher and Smith, using neutron activation analysis, examined 482

TABLE X-6
Mercury in Dry Tissue (parts per million)

Tissue	No. of Samples	Maximum	Minimum	Average	SD[a] + −
Adrenal	18	2.44	0.15	0.80	0.75
Aorta	23	7.30	0.10	1.39	1.61
Blood (whole)	3	0.12	0.06	0.09	
Bone	16	1.04	0.03	0.45	0.39
Brain	21	15.2	0.12	2.94	4.10
Breast	3	2.48	0.68	1.79	
Hair	70	24.4	0.03	5.52	5.21
Heart	22	5.62	0.14	1.76	2.41
Kidney	20	79.3	0.08	9.03	20.2
Liver	22	20.0	0.15	3.66	4.96
Lung	26	10.5	0.30	2.38	2.47
Muscle (pectoral)	15	3.40	0.04	0.71	0.95
Nail	25	33.8	0.80	7.27	8.39
Ovary	10	13.5	0.06	2.14	4.09
Pancreas	26	7.23	0.03	1.14	1.53
Prostate	5	1.53	0.04	0.65	0.58
Skin	18	18.7	0.25	3.34	4.49
Spleen	21	7.41	0.10	1.50	1.93
Stomach	21	13.3	0.06	2.27	3.51
Teeth	59	18.1	0.14	3.15	3.44
Thymus	3	8.90	0.47	4.75	
Thyroid	24	24.6	0.10	3.38	6.23
Uterus	11	4.31	0.13	1.43	1.48

Adapted from Liebscher and Smith (1968).

[a]SD = Standard Deviation

samples of 23 different tissues for mercury content. They do not state the number of cadavers they used, but an inspection of their data indicates that it was between 25 and 30. The results are shown in Table X-6. Since the subjects were previously healthy, had died of violence and were not known to have had any unusual exposure to mercury, they may reasonably be assumed to represent "normals." Although the samples were handled in a way that would avoid contamination during analysis, the fact that they were vacuum dried may have resulted in some loss of mercury. The results cannot be compared with those reported in terms of wet weight, except for the relative concentrations of mercury among the different tissues. The finding of the highest concentrations in kidney is consistent with other studies.

Because of the shortcomings pointed out above, these data on tissue analyses do not warrant any sweeping conclusions. They tend to confirm that mercury absorbed by humans in the course of daily life accumulates in the kidneys to a greater extent than in any other organ, a fact which has been demonstrated before.

Hair and Nails

Analysis of hair for evidence of poisoning has held a special fascination for toxicologists. A theoretical basis for the usefulness of this procedure is found in the known affinity between certain metals, especially arsenic and mercury, and the sulfur-containing amino acids, cystine and cysteine, which are abundantly present in hair and nails. The romantic appeal of "whodunnit" forensic toxicology of the 19th century may also play a role in continued preoccupation with analyses of hair in detecting the cause of mysterious deaths. Considerable publicity has been given to the finding of arsenic in a tuft of hair allegedly removed from Napoleon's head the day after his death (Was Napoleon Poisoned? 1964). In using this finding as evidence that the general had been deliberately poisoned by his British captors the protagonists of the claim conveniently overlook the fact that Bonaparte's diet on St. Helena must have contained a great deal of seafood, a rich source of arsenic. Mercury, too, has been implicated in playing a part in Napoleon's final illness, but only as a therapeutic agent, even though administered in heroic doses.

Mercury poisoning has been suggested as the cause of death of Charles II of England (Wolbarsht and Sax 1961). This proposal finds support in the nature of the king's terminal illness and his known devotion to experiments with mercury during his last years. Neutron activation analysis of hair supposed to have come from the head of Charles II has shown 54.6 ppm, about ten times the mean concentration of mercury found in the hair from 70 human subjects in Glasgow (Lenihan and Smith 1967). This might be interpreted as confirmation of the diagnosis of mercury poisoning, but the possibility of contamination of the specimen (perhaps from embalming fluid) between 1685 and 1967 has not been excluded.

Studies of hair allegedly originating from the heads of famous personages such as Napoleon and Charles II more properly fall into the category of "fun and games" than into the realm of scientific investigation, since accurate histories of the specimens are not available. This does not mean that the determination of mercury in hair cannot be a useful procedure in studying the effects of exposure to and absorption of mercury. Serious investigations along these lines have been undertaken.

As a means of evaluating the absorption of mercury and its potential hazard among dental surgery assistants, Nixon and Smith, using neutron activation analysis, analyzed the head and axillary hair, as well as finger nails and toe nails, of 20 female dental assistants, and from 26 control subjects. The authors state that there was little difference between the mercury content of the finger and toe nails in the controls. The mean value for mercury in the nails of controls was 5.10 ppm; for toe nails in the dental assistants, 9.3 ppm; and for finger nails of dental assistants, 68.76 ppm. Mercury content of both axillary and head hair in the controls had a mean value of 8.8 ppm for both types of hair while the exposed subjects had a mean of 7.88 ppm of mercury in their axillary hair and of 32.25 in the head hair. No mention is made of having washed the samples as a means of distinguishing between mercury that was actually incorporated into the samples and that which was merely present as a contaminant on the surface of the exposed nails and hair. The results of this investigation strongly suggest that the excess mercury in the finger nails and head hair of the dental assistants was due to deposition rather than to absorption. The study of the controls is useful, however, in providing information on "normal" mercury in hair and nails.

A small-scale study similar to that of Nixon and Smith has been reported by Lenihan and Smith.

Results of an extensive investigation of mercury in hair were reported by Yamaguchi in 1966. The study embraced 67 male and 27 female Japanese normals, 14 male Americans living in Japan, and 21 long-term inmates of a mental hospital. As part of the study, analyses were made for mercury content of five popular cosmetic preparations used on the hair and scalp. A unique and extremely valuable feature of this work was the analysis of 15 samples of hair before and after washing with a detergent. The main findings in Yamaguchi's study (which is published in Japanese) are summarized in Table X-7.

In a subsequent investigation it was found that the hair of Nepalese contained considerably less mercury than that of Japanese. Yamaguchi suggests, entirely plausibly, that in part it may be due to the higher mercury content of the Japanese diet and in part to the greater amount of the metal incorporated into the hair from polluted air.*

---

*S. Yamaguchi 1970: personal communication.

TABLE X-7

Mercury in Head Hair (parts per million)

Japanese "Normals"			
	Males	*Females*	*All*
Range	0–11.99	1–7.99	0–11.99
Mean	4.48	3.53	4.21
Median	4.0	4.0	4.0
Mode	3.0– 3.99	4.0–4.99	4.0– 4.99
Number	67	27	94
American Males Living in Japan			
Range	0.69–4.23		
Mean	1.89		
Median	1.80–1.96		
Number	14		
Inmates of Mental Hospitals			
	Males	*Females*	
Range	1. 0–3.19	0.69–3.05	
Mean	2.09	2.02	
Median	1.62–2.30	1.98	
Number	12	21	

Effect of Washing				
	Male		*Female*	
	Before	*After*	*Before*	*After*
Range	4.75–16.1	0.89–3.72	2.36–17.59	1.56–6.44
Mean	11.1	2.71	5.69	4.48
Number	6		9	

From S. Yamaguchi et al. Personal communication (1966).

Mercury as an Essential Trace Metal

Pledge (1947) points out that Quinton in 1897, Macallum in 1926, and Pantin in 1931 all discussed the idea that with respect to its inorganic elements, some of which occur in very small amounts, the "internal environment" of living organisms has the composition of the primeval ocean from which life originated. In the biological sciences the term "trace metal" is one present in foods or tissues in concentrations of the order of one part per million (ppm) or less. Trace metals are sometimes spoken of as micronutrients and are classified as essential or non-essential, though it would be preferable to speak of trace elements as those for which a vital function has been recognized and those for which no such role has as yet been identified.

Traditionally, mercury, along with lead, arsenic and some other metals, has been considered a poison; but a valid concept of a poison must be a quantitative one, as illustrated graphically in Figure X-1. There is no substance known to man which cannot, in overwhelming doses, give rise to adverse physiological

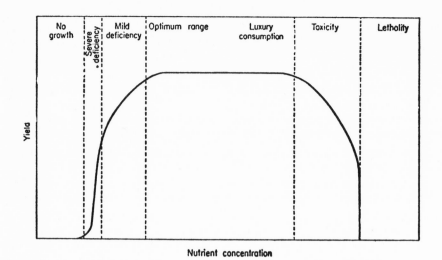

Figure X–1 Diagram Showing Idealized Growth of an Organism as a
Function of the Concentration of an Essential Nutrient. (After P. F. Smith,
Ann. Rev. Plant Physiol. 13:81 [1962]. Courtesy of Annual Reviews, Inc.)

effects. This applies even to such common materials as water, oxygen and
table salt. Conversely, there are numerous substances which are ordinarily
considered to be poisons which, in small doses, are biologically essential or
beneficial.

Before the middle of the 20th century, measurements in the parts per
million range or of microgram quantities were commonly looked upon as advanced micro-analysis. In the 1960's it became common practice to deal with
micrograms (μg) where formerly it was milligrams; nanograms (ng) and pico-
grams (pg) have become familiar terms in the literature of analytical chemistry,
and even single molecules of elements are being detected.

$$
\begin{aligned}
\text{Milligram} &= 1 \times 10^{-3} \text{ gram} \\
\text{Microgram} &= 1 \times 10^{-6} \text{ gram} \\
\text{Nanogram} &= 1 \times 10^{-9} \text{ gram} \\
\text{Picogram} &= 1 \times 10^{-12} \text{ gram}
\end{aligned}
$$

Obviously, then, the concept of "zero" has been revised progressively downward and it has become increasingly rare to speak of elements as being absent
from tissues or from environmental samples (water, soil, etc.).

Limitations in the sensitivity of analytical methods have, in the past, led to
the mistaken belief that many elements were not present in plants and animals
under normal conditions. Mercury falls into this category. As pointed out
above, the concept of "zero" in analytical chemistry is a function of the sensitivity of the procedure. Furthermore, the definition of a poison is a function

of the size of the dose. When it was possible to measure only those concentrations which were in the toxic range, it was quite natural to conclude that any detectable amount of a substance was poisonous.

Bowen has pointed out that the only elements which can be definitely classified as non-essential are actinium, plutonium, polonium, protoactinium, and radium, since these elements are present in living organisms in amounts less than one atom per cell. If this theory is valid, then by definition all other elements are either essential or "neutral." A glance at Table X-8 will show how the number of recognized essential or useful trace elements grew in the 17 years between 1949 and 1966.

A mathematical method for determining the essentiality of trace metals has been proposed by Liebscher and Smith. The theory of this method is more ingenious than convincing, its basis being that the values of essential trace elements in tissues have a "normal" frequency distribution while non-essential elements have a "log-normal" distribution. For essential elements the ratio of the standard deviation to the mean is small, for non-essential elements it is large. Application of this method to a number of human tissues showed that copper, manganese, selenium and zinc are essential while antimony, arsenic, cadmium and mercury are not. This separation happens to coincide fairly closely with current dogma, but it has its weaknesses. For one thing, the classification of copper and zinc as trace metals can be questioned since they are present in amounts greater than one part per million. In Liebscher and Smith's

TABLE X-8
Essential Trace Elements

Humans and Animals[a]	Plants, Animals or both[b]	Man[c]	Plants
Copper	Copper[d]	Copper	Boron
Iron	Iron[d]	Iron	Molybdenum
Zinc	Zinc[d]	Zinc	Silicon
Manganese	Manganese	Manganese	
Iodine	Iodine	Iodine	
Cobalt	Cobalt	Molybdenum (?)	
	Boron	Selenium (?)	
	Molybdenum	Chromium (?)	
	Silicon	Fluoride (?)	
	Nickel		
	Aluminum		
	Arsenic[e]		
	Bromine		
	Selenium		
	Chromium		

[a]From Monier-Williams (1949)
[b]From West et al. (1966)
[c]From Kleiner and Orten (1966)
[d]There is some question as to copper, iron, and zinc properly being designated trace metals.
[e]Arsenic has not been demonstrated as an essential element in animal nutrition, but organic arsenicals have produced improved growth and health in livestock.

published figures the standard deviations for arsenic and antimony for some tissues are smaller than the means, this being one of the criteria they set for essentiality. The possible loss of mercury during the desiccation process raises doubts as to the validity of the analytical results for this element. Furthermore, as shown below, important vital functions have been demonstrated for mercury when it is present in truly trace amounts.

Mercury's known activity as a catalyst in many chemical reactions led to the suggestion by Bodnár and Szép in 1929 that it might serve a similar function in living tissue. This idea was put forth again by Stock in 1934.

While studying the detoxification of diisopropyl fluorophosphate, a cholinesterase inhibitor, Mazur, in 1946, prepared a purified fluorophosphate-hydrolysing enzyme which he found to be totally inhibited by the presence of mercuric ion (Hg^{++}) in concentrations of 2×10^{-5} mole per liter or more. On further investigation, he found that at concentrations of Hg^{++} between 1×10^{-6} and 8×10^{-6} mole per liter enzyme activity was increased, the optimum concentration being 6×10^{-6} which resulted in activity of 240 percent of that of the original untreated enzyme. Mazur referred to this as a "curious phenomenon" probably because it was totally unexpected, and he suggested as a possible explanation that the mercury had caused inactivation of an inhibitor that was present in the enzyme preparation. This observation, of course, points to a possible useful role for trace amounts of mercury in detoxification even if, as Mazur postulated, it were indirectly through the inactivation of an inhibitor. Detoxification of procaine by hydroxy-mercuri-methoxypropyl carbamyl phenoxy acetate (Salyrgan) had been demonstrated by Beutner in 1940.

Studies on physiological adjustment to hypoxia have focused attention on the importance of 2,3-diphosphoglycerate, a substance which is abundantly present in human red blood corpuscles, in mediating the adjustment. It has been shown that Hg^{++} in concentrations between 0.1 and 1.0 millimoles per liter in the substrate enhances the enzymatic splitting of this and several closely related compounds. (A similar effect, incidentally, is produced by silver.) This demonstrates ". . . enzyme activation by mercury . . . [which] appears unique in the field of enzyme chemistry" (Rapoport and Leubering 1951). Further studies have confirmed the activation of the enzyme glycerate-2,3-diphosphatase by mercury and have shown that this activation takes place only in the presence of compounds which contain organic nitrogen (Rapoport, Leubering, and Wagner 1955).

Characterization of mercury activation of enzymes as "a curious phenomenon" and as something "unique in the field of enzyme chemistry" suggests that those who observed this behavior of mercury were, to say the least, surprised to find that the metal might perform a useful function. If this assumption is correct it reflects the widespread belief that, except for therapeutic applications, mercury in almost any amount is poisonous. Yet, as early as 1929 Salant and Brodman had shown experimentally that ". . . mercury exerts a twofold action on the parasympathetic endings in the heart, small doses stimulating and larger amounts depressing their irritability."

On purely theoretical grounds it is not unreasonable to postulate that mercury, and other elements whose role is not completely understood, may have important or even vital biological functions. Man, as well as other forms of animal and plant life, has survived in an environment in which mercury is ubiquitous. It follows, therefore, that these forms of life must have been able to adapt to at least the amount of mercury to which they were exposed. The development of tolerance to mercury (Zador 1949) and a protective action (Haury 1941) of mercury have, in fact, been shown experimentally. Possibly, in the course of aeons, adaptation led to dependence, and those species which developed mechanisms for making optimum use of the elements present in their environment were in a favorable position in the struggle for existence compared with those which could not adapt. Methanogenic bacteria have been found that thrive on mercury. Why not other organisms?

Silver, as well as mercury, can enhance enzyme activity. Arsenic, like mercury generally looked upon as a poison, is ubiquitous and has a well-defined cycle in nature. It is related to phosphorylation and energy transfer in living organisms (Frost 1967) and it stimulates growth in plants, cattle, and poultry (Schroeder and Balassa 1966). Lead, also found everywhere and also viewed as toxic, can enhance the growth of tissue cultures.* Tin and vanadium can be tolerated by mice, with no adverse effects, in concentrations up to 5 ppm in their drinking water, although no essential function for these metals has been identified (Schroeder and Balassa 1966). Finally, deficiency states have been produced in rats maintained in an artificially created environment designed to exclude all trace elements (Smith and Schwarz 1967).

REFERENCES

Benning, D. 1958. Outbreak of mercury poisoning in Ohio. *Ind. Med. Surg.* 27:354–63.

Beutner, R.; Landay, J.; and Lieberman, A. 1940. Evidence for the local effect of mercurial diuretics. *Proc. Soc. Exp. Biol. Med.* 44:120–22.

Bodnár, J., and Szép, O. 1929. Ultramikromethode zur Bestimmung des Quecksilbers. *Biochem. Z.* 205:219–29.

Bodnár, J.; Szép, O.; and Weszpremy, B. 1939. Über den natürlichen Quecksilbergehalt des menschlichen Organismus. *Biochem. Z.* 302:384–92.

Borinski, P. 1931. Sind kleinste Quecksilber-mengen gesundheits-schädlich? *Deut. Med. Wochschr.* 57:1060–61.

Bowen, H. J. M. 1966. *Trace elements in biochemistry.* New York: Academic Press.

Butt, E. M., and Simonsen, D. G. 1950. Mercury and lead storage in human tissues. *Am. J. Clin. Pathol.* 20:716–23.

Forney, R. B., and Harger, R. N. 1949. Mercury content of human tissues from routine autopsy material. *Federation Proc.* 8:292.

Frost, D. V. 1967. Arsenicals in biology. *Federation Proc.* 26:194.

Griffith, G. D.; Butt, E. M.; and Walker, J. 1954. The inorganic element content of certain human tissues. *Ann. Internal Med.* 41:501–9.

Haury, V. G. 1941. Protective action of mercury and lead salts against procaine convulsions. *Proc. Soc. Exp. Biol. Med.* 46:309–10.

---

*H. Sobkowicz 1967: personal communication.

Jacobs, M. B.; Goldwater, L. J.; and Gilbert, H. 1961. Ultramicrodetermination of mercury in blood. *Am. Ind. Hyg. Assoc. J.* 21:276–79.

Jacobs, M. B.; Ladd, A. C.; and Goldwater, L. J. 1964. Absorption and excretion of mercury in man: VI. Significance of mercury in urine. *Arch. Environ. Health* 9:454–63.

Joselow, M. M.; Goldwater, L. J.; and Weinberg, S. B. 1967. Absorption and excretion of mercury in man: XI. Mercury content of "normal" human tissues. *Arch. Environ. Health* 15:64–66.

Joselow, M. M.; Ruiz, R.; and Goldwater, L. J. 1968. Absorption and excretion of mercury in man: XIV. Salivary excretion of mercury and its relationship to blood and urine mercury. *Arch. Environ. Health* 17:35–38.

Keyssler, J. G. 1751. *Neuste Reisen durch Deutschland, Boehmen, Ungarn, die Schweiz, Italien und Lothringen.* Hanover.

Kleiner, I. S. and Orten, J. M. 1966. *Biochemistry.* 7th ed. St. Louis: C. V. Mosby Co.

Lenihan, J. M. A. and Smith, H. 1967. Activation analysis and public health. In *International Atomic Energy Agency Symposium on nuclear activation techniques in the life sciences,* 8–13 May 1967. Amsterdam.

Liebscher, K., and Smith, H. 1968. Essential and nonessential trace elements: A method of determining whether an element is essential or nonessential in human tissue. *Arch. Environ. Health* 17:881–90.

Mazur, A. 1946. An enzyme in animal tissues capable of hydrolyzing the phosphorus-fluorine bond of alkyl fluorophosphates. *J. Biol. Chem.* 164:271–89.

Monier-Williams, G. W. 1949. *Trace elements in foods.* New York: John Wiley and Sons.

Nixon, G. S., and Smith, H. 1965. Hazard of mercury poisoning in the dental surgery. *J. Oral. Therap. Pharmacol.* 1:512–14.

Pledge, H. T. 1947. *Science since 1500.* New York: Philosophical Library.

Rapoport, S., and Leubering, J. 1951. Glycerate-2, 3-diphosphatase. *J. Biol. Chem.* 189:683–94.

Rapoport, S.; Leubering, J.; and Wagner, R. H. 1955. Ueber die Quecksilber-Aktivierung der Glycerinsaure-2, 3-diphosphatase. *Hoppe-Seylers Z. Physiol. Chem.* 302:105–10.

Salant, W., and Brodman, K. 1929. The effect of mercury on cardiac inhibition. *J. Pharmacol. Exp. Ther.* 36:195–202.

Schroeder, H. A., and Balassa, J. J. 1966. Abnormal trace metals in man: Arsenic. *J. Chronic Diseases* 19:85.

———. 1967. Arsenic, germanium, tin and vanadium in mice: Effects on growth, survival and tissue levels. *J. Nutr.* 92:245–52.

Smith, J. C., and Schwarz, K. 1967. A controlled environment system for new trace element deficiencies. *J. Nutr.* 93:182–88.

Stock, A. 1926. Die Gefährlichkeit des Quecksilberdämpfes und der Amalgame. *Z. Angew. Chem.* 39:984–89.

———. 1928. Die Gefährlichkeit des Quecksilbers und der Amalgam-Zahnfüllungen. *Z. Angew Chem.* 41:663–72.

———. 1936. Die chronische Quecksilber und Amalgamvergiftung. *Arch. Gewerbepathol. Gewerbehyg.* 7:388–413.

———. 1940. Der Quecksilbergehalt des menschlichen Organismus. *Biochem. Z.* 304:73–80.

———. 1943. Der Quecksilbergehalt des menschlichen Organismus: II. Uber Wirkung und Verbreitung des Quecksilbers. *Biochem. Z.* 316:108–22.

Stock, A., and Cucuel, F. 1934. Die Verbreitung des Quecksilbers. *Naturwissenschaften* 22/24:390–93.

Szép, O. 1940. Weitere Beiträge zur Kenntnis des Quecksilbergehalt des menschlichen Körpers. *Biochem. Z.* 307:79–81.

Was Napoleon poisoned? 1964. *Pfizer Spectrum* 12:12–14.

West, E. S.; Todd, W. R.; Mason, H. S.; and Bruggen, J. T. 1966. *Textbook of biochemistry.* 4th ed. New York: Macmillan Co.

Wolbarsht, M. L., and Sax, D. S. 1961. Charles II, a royal martyr. *Roy. Soc. London Notes Records* 16:154–57.

Yamaguchi, S.; Matsumoto, H.; Oomura, K.; and Akitake, K. 1966. On the amount of mercury in human hair. *J. Sci. Labour* (Japan. ed.) 21(11): 33–37.

Zador, L. 1949. Tolerance of rats to mercuric chloride poisoning. *Orvosi Hetilap* 89:126–28.

11

Toxicology

Mercury found in man as a result of its absorption from activities of daily living, the so-called "normal" mercury, gains entry to the body via the lungs and gastrointestinal tract. Abnormal amounts of mercury may enter through these portals and also through the intact or broken skin and by mean of injections; entry may be accidental or deliberate.

Metallic Mercury

Ingestion (swallowing)

Dioscorides was in error when he said that quicksilver ". . . hath a pernitious faculty being drank, eating through ye inward parts by its weight." Rhazes was correct when he wrote that it is ". . . passed out unchanged" (cited in Iskandar 1959). His conclusion is supported by innumerable accounts of the oral administration of mercury by physicians of the 18th and 19th centuries in doses of from a few ounces up to a pound or more. Taylor, a prominent toxicologist, wrote in 1875 that "Metallic Mercury is not commonly regarded as a poison. It has been prescribed and taken in large doses by patients suffering from obstruction of the bowels, without injury to health or causing any uneasiness, except that which might arise from its great weight." He mentions the case of a woman who swallowed two pounds of mercury which ". . . remained nine days in her body, and was perceptible to the feel through the abdomen." The mercury had been fully expelled by the 14th day at which time a slight salivation appeared, but ". . . this aftereffect was speedily subdued."

Modern confirmation of the harmlessness of ingested mercury is found in a report issued by the Poison Control Center of the New York City Department of Health in May 1957: "Since the establishment of the Poison Control

Center, March, 1955, 18 incidents involving metallic mercury ingestion were reported: 12 in females and six in male individuals, the age ranging from eight months to 44 years; 12 anxious queries were associated with the ingestion of mercury from a thermometer. Most of the incidents were inconsequential and required no special therapy." One of the cases involved loss of mercury from a Miller-Abbott tube (see below), the patient suffering no ill effects. The Bulletin of the National Clearinghouse for Poison Control Centers, September–October 1966 carries an account of an accident in which a two-year-old child ingested more than 200 grams of mercury and had no adverse reaction. The report concludes with a statement that "The ingestion of such a large amount of metallic mercury without any definite signs or symptoms lends evidence to the innocuousness of ingestion of thermometer mercury or that which is found in small toys."

An exception to the general rule is found in a case reported by Gibb in 1873 (cited in Taylor 1875). The patient was a girl who swallowed four and a half ounces (by weight) of mercury in order to produce abortion. There was no effect on the uterus ". . . but in a few days she suffered from a trembling and shaking of the body (mercurial tremors) and loss of muscular power. These symptoms continued for two months, but there was no salivation and no blue marks on the gums."

When Wilkins described the mercury-weighted stomach tube in 1928 he pointed out its many advantages over previous types and did not overlook the possibility that the mercury might escape into the gastro-intestinal tract. He minimized the danger which might result by pointing out that mercury in the intestine is inert and harmless. His tube, illustrated in Fig. XI-1, was 46 inches in length and made of soft, flexible rubber. The compartment holding the mercury measured 3/16 by 4½ inches and was sealed by a rubber diaphragm.

The Miller-Abbott intestinal tube was introduced in 1934 and differed from that of Wilkins in that it had a double lumen and a rubber bag at the distal end (Miller and Abbott 1934). While this tube was superior in many respects to the single lumen models, it was often difficult to pass the tip with its balloon through the pylorus and into the small intestine. A solution to this problem was suggested by Harris in 1944 when he combined the mercury of the Wilkins tube with the double lumen feature of the Miller-Abbott tube. He believed that the danger of escape of the mercury from the balloon was re-

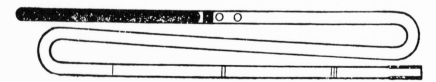

Figure XI-1 Wilkin's Mercury-Weighted Stomach Tube. (From James A. Wilkins, *J. Amer. Med. Assoc.* 91:395–396 [1928].)

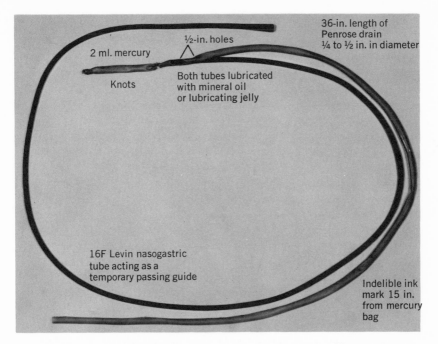

Figure XI-2 A Modified Penrose Nasogastric Feeding Tube. Mercury in one end facilitates passage of the tube. (From L. E. Davis and W. Hofmann, *J. Amer. Med. Assoc.* 209:685-686 [1969].)

mote and that there would be little harm in the event of leakage of the metal. Figures are not available on the number of times the Harris modification of the Miller-Abbott tube has been used successfully, but the possibility of adverse effects following rupture of the mercury-containing balloon is amply documented. The proponents of intestinal intubation have minimized the danger from mercury that might escape into the gut, but they did not anticipate another hazard, namely, aspiration into the lungs. In spite of the dangers, the use of mercury to facilitate the passage of gastric tubes remains in vogue (Davis and Hofmann 1969). (See Figure XI-2.)

Mercury escaping from a ruptured balloon of a Miller-Abbott tube does not usually constitute a serious threat but it does not always stay in the gut. Occasionally, as pointed out by Lindsey and Becker (1968), ". . . rupture of the mercury-filled balloon fixed at the end of a long intestinal tube may lead to serious complications." These physicians describe a case of intestinal obstruction due to mercury granulomas resulting from an intubation accident, and cite a similar report by Crikelair (1953). The latter, in turn, refers to additional cases of the same nature. Other complications described in articles listed in their bibliography by Lindsey and Becker include appendicitis (Birnbaum 1947), intestinal ulceration (Kristofferson 1959), rectal fistula

(Bradford, Hugo, and Quinn 1957; Lindenmuth 1949), and a reaction simulating osteomyelitis (Vare 1953). Several additional articles call attention to
dangers resulting from the accidental escape of mercury during intestinal intubation (Hoffman 1948; Hanselman and Meyer 1962; Farrell and Reams
1957).

The report by Cantor in 1951, describing a single case in which fecal
fistulas resulted from mercury which had escaped from an intestinal tube, is
of considerable historical importance. As pointed out by Cantor, the use of a
pound or two of quicksilver to relieve intestinal obstruction by sheer weight
had been practiced for several hundred years. It had been shown by Dujardin-
Beaumetz in 1886 that the mercury did not remain in a single mass but was
broken up into small aggregates in the intestines. This pretty much put an end
to this use of mercury. The significance of Cantor's case is that it was the first
in which the break-up of the mass of mercury was demonstrated by x-ray examination. Of secondary importance was the outcome of the case, the fecal
fistulas healing when a temporary colostomy was performed.

In an attempted murder by means of oral administration of metallic mercury, the intended victim was a six month old girl, born out of wedlock. The
father hoped to put her out of the way by pouring quicksilver into her
mouth. The attempt was discovered when globules of the metal were noticed
in the child's stools. The quantity given amounted to no more than a few
grams and the infant showed no adverse effects (Kockel 1928).

Aspiration

Aspiration of metallic mercury into the lungs other than as a vapor, is a
rare occurrence. A case was reported by Schultze in 1958 in which the patient had swallowed quicksilver in a suicidal attempt and aspirated some of it
while vomiting. Mercury globules in the lungs were demonstrated roentgeo-
graphically immediately after the episode and were still present on follow-up
five years later, the patient having become asymptomatic. A pneumonitis was
also found in the initial x-ray examination but this could have been caused by
the vomitus.

Accidental rupture of a Miller-Abbott tube resulted in the aspiration of
mercury by a patient described by Zimmerman in 1969. Of the 10 ml of
metal present in the bag, about four ml were immediately expelled by coughing and gagging, an additional amount was found to have entered the
stomach, and 0.5 ml was recovered after postural drainage. X-ray films several hours after the aspiration showed mercury distributed in the bronchioles
and alveoli of both lungs. The patient died on the fifth day following the accident after developing severe cough, bloody sputum, tachypneia, and cyanosis. In the meantime, 8.2 ml of mercury had been recovered. This case is complicated by the fact that the patient had Hodgkin's disease, had just received
an intensive course of radiation therapy, and had had intestinal obstruction.
The role of mercury in contributing to the fatal outcome must remain in
doubt.

Aural Administration

The possibility of administering mercury through the ear has not gone unnoticed. In summarizing the *Compendium Medicinae* of Gilbertus Anglicus, written about 1240, Handerson calls attention to what he calls a novel form of poisoning described by Gilbert. This consists of pouring metallic mercury into the ear, which is said to cause pain, delirium, convulsions, epilepsy, apoplexy, and if the metal penetrates to the brain, ultimate death (Handerson 1918). The mercury can be removed by inserting gold foil into the ear, producing amalgamation of the mercury with the gold, a method suggested earlier by Rhazes (cited in Iskandar 1959).

Edward II of England (1284–1327) was probably not murdered by means of mercury in the ear, but the assassin hired to do the job was familiar with the aural route of administering poisons and of the lethal properties of quicksilver. In Marlowe's play "The Troublesome Raigne and Lamentable Death of Edward, the Second, King of England,"Act V, Scene IV, Lighthorne says:

> You shall not need to give instructions:
> 'Tis not the first time I have killed a man.
> I learned in Naples how to poison flowers,
> To strangle with a lawn thrust through the throat,
> To pierce the windpipe with a needle's point,
> Or whilst one is asleep, to take a quill
> And blow a little powder in his ear,
> Or open his mouth and pour quicksilver down.

Could the players in Hamlet have used mercury in the ear?

Injection

Various preparations which contain mixtures of mercury either as a salt or as the pure metal have been given by injection for their therapeutic effects but for the most part parenteral entry of plain metallic mercury into the human body is accidental or suicidal. Injection may be into an artery, a vein, beneath the skin, or, during catheterization, directly into one of the chambers of the heart.

A case of attempted suicide by means of intravenous injection of quicksilver was reported by Umber in 1923, the subject being a young woman who injected two cc of the metal into her right cubital vein. There was no immediate effect, but after three days, diarrhea and stomatitis occurred, followed by gingivitis. At the end of three or four months all signs of mercurialism had disappeared, although a chest x-ray showed a miliary distribution of radio-opaque bodies (presumably mercury) throughout both lungs. The patient died ten years later of an intercurrent disease not related to mercury poisoning.

Greater success was met with by another young woman reported by Johnson and Koumides in 1967. The victim of this suicidal act injected between one and two ml of mercury into her left forearm. Some of the metal apparently remained at the injection site, causing necrosis, and some must have entered a vein, as multiple pulmonary emboli were found on x-ray and at post-mortem examination. Evidence of kidney injury was present both ante-mortem and at autopsy.

These two cases show that mercury administered intravenously can produce systemic poisoning as well as embolization, and that the outcome may or may not be fatal. A third type of response is illustrated by a case reported by Conrad et al. in 1957. The subject of this suicidal attempt was a young man who injected himself intravenously with an undetermined amount of mercury. Although pulmonary embolization resulted, there were no signs of systemic poisoning such as gingivitis, diarrhea or kidney injury, and recovery followed.

In a case reported by Hill (1967) the subject unsuccessfully attempted suicide by injecting 20–40 grams of quicksilver into his thigh. A patient seen by the author (Goldwater:unpublished) received two cc of mercury intramuscularly as the result of a physician's order being misunderstood. Surgical excision of the injected areas was performed in this case and in Hill's, although all of the metal was not thus removed. In neither case were there any untoward effects other than local ones.

Intra-arterial injection of mercury, when it has occurred, has always been accidental (Lathem et al. 1954; Schulz and Beskind 1960; Buxton et al. 1965; Devlin and Sudlow 1967). In drawing blood for purposes of gas analysis it is important to maintain anaerobic conditions and this is accomplished by using as a seal in the syringe a liquid which will not react with the blood gases. Prior to about 1945 it was customary to employ mineral oil but this has been replaced by mercury, thus applying the principle first discovered by Cavendish in 1766 when he designed the pneumatic trough.

An article on mercury embolization published in 1965 by Buxton and his associates notes that "While the advances of cardiovascular surgery have increased the demand for blood-gas analysis, it is certain that parallel caution with this procedure has not yet been obtained." The authors report nine cases of mercury embolization in a review of 1,063 patients who had had blood drawn for gas analysis, samples having been taken, in various cases, from arteries, veins and all chambers of the heart. There were three fatalities in this series and a variety of complications, but the case histories given do not provide sufficient information to justify any firm conclusions as to the role of mercury either in embolization or as a poison. The article contains a description of an apparatus designed to eliminate the accidental entry of mercury into the circulation when blood samples are drawn.

Two cases of alleged mercury embolization have been reported by Schulz and Beskind (1960). Their case histories are fragmentary and do not give conclusive evidence that they were in fact examples of mercury embolism. The article is valuable, however, in that it contains a review of the literature on the subject.

In spite of all precautions, accidental mercury embolization continues to occur. Guirgis (1971, in press) observed a premature female infant who received about 700 mg of metallic mercury by injection through an umbilical artery into the abdominal aorta, with subsequent distribution of the metal in the intercostal and spinal arteries. When last examined at the age of six months, the infant was growing and developing in a normal manner, while carrying sustained blood mercury levels of between 50 and 200 μg per 100 ml of whole blood.

Absorption through the Skin

Ramazzini, in describing the hazards to those who administer mercury inunctions, noted that "Though they wear a glove when so engaged, it is impossible for them to prevent the mercurial atoms from penetrating the leather and so reaching the hand of him who applies the ointment . . ." It might be implied that if it can pass through the leather of the glove it could also pass through the skin of the person doing the rubbing. That this can actually happen has been demonstrated. The French translation of Ramazzini by Fourcroy (Paris: 1777, Chap. 3, pp. 48–49) says that the mercury ". . . s'insinue facilement par leurs pores." The original text is "Licet autem chirotheca in hac re uti soleant, non satis tamen munire se possunt, quin mercuriales atomi corium, per quod Mercurius alioquin exprimi solet et perpurgari, pervadant & Unctoris manum pertingant . . ." The word "corium" can mean skin, hide, or leather and in this case means the last of these since pressing through leather was a usual means of purifying quicksilver. The metal reached the skin of the "Unctor" but the thought that it entered the pores is insinuated by Fourcroy.

A case of direct passage of mercury through the skin during inunction was reported by Rixford in 1895. The patient had sustained multiple fine puncture wounds of the extremities upon falling into a prickly pear tree. To relieve the pain, a companion rubbed mercury into the affected parts. Multiple small tumors later developed at the puncture sites and when these were excised they were found to contain globules of mercury, some as large as small peas. In this case it cannot be said that the mercury entered through intact skin.

Passage of quicksilver through unbroken skin is described in Stammel's case reported in 1929. The patient in this instance had rubbed blue ointment into his wife's scalp about three weeks prior to the appearance of a painful swelling of his right thumb. When this was incised mercury exuded. Subsequently, several additional "tumors" developed and these, too, were found to contain mercury. A roentgenogram showed multiple opaque deposits in the area affected (Fig. XI-3). Up to six months later, the patient had shown no evidence of systemic mercurialism.

Although the title of an article published in 1929 (Drügg) implies that mercury poisoning resulted from subcutaneous implantation of mercury from a broken thermometer, a reading of the report leaves room for doubt. The patient was a student nurse who broke a thermometer and in so doing caused

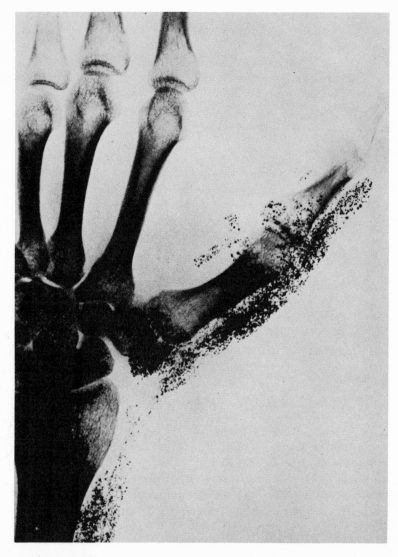

Figure XI-3 X-Ray Showing Metallic Mercury in Subcutaneous Tissue following Mercury Inunction. (From C. A. Stammel in *Military Surgeon* 65: 528–530 [1929].)

some of the liberated mercury to penetrate the skin of a finger. This part of the account is not open to question. Three weeks after the accident she complained of pain in the region of the liver, in the back and in the abdomen, and later of dryness of the lips, but this is hardly the classical picture of mercurialism. Four years later this case was used as an excuse for a general article on the subject of traumatic mercury poisoning (Borchard 1933). Thus does the literature become enriched.

Inhalation of Mercury Vapor

Acute respiratory effects from the inhalation of mercury vapor were described first by Dioscorides in his account of what happens when cinnabar is heated. He says that "In ye furnace it changeth into a very lively and flaming colour, but it hath amongst metals a choaking smell, and therefore ye workmen there put bladders about their faces, that they may see, but not draw in the vapours" (1934 translation). The roasting of cinnabar has always been the standard procedure for driving mercury from the ore (Chap. 3) and the "choaking smell" certainly implies that the vapors caused a feeling of tightness in the chest, probably accompanied by coughing. Recognition of the necessity of protecting the lungs is shown by the use of protective masks. Pliny describes a similar use of respirators by workmen exposed to dust in polishing cinnabar, these two references being the earliest to describe the employment of protective masks to reduce the hazard from vapors and dusts.

Toxic effects as the result of mercury fumigations are described by Serapion (fl. ninth cent.), with paralyses and other neurological manifestations being most prominent (cited in Mettler 1947). The occurrence of paralysis is confirmed by Avicenna who states also that mercury vapors can cause blindness, muscular spasms, foul breath and tremors (Chap. 14). This appears to be the first mention of what later became known as "the hatters' shakes." It is clear that the cardinal signs and symptoms of both acute and chronic mercury poisoning were known to at least some of the Arab physicians of the 10th century.

Acute pneumonitis due to mercury vapor, while less common than chronic systemic poisoning, is by no means rare. Young children appear to be particularly vulnerable.

When Campbell, in 1948, reported a case of acute mercury pneumonitis in a child of four months, he commented that he had found no similar cases in the *Index Medicus* in the 20th century. Had he extended his search of the literature he probably would not have found any in earlier centuries. In Campbell's case, exposure had occurred when the child's father had accidentally dropped a half teaspoonful of mercury on the top of a kitchen range and was so amazed at the rapid volatilization that he repeated the procedure for the edification of his wife. The infant, present in the kitchen at the time, first showed symptoms about two hours after the exposure. Four days later the child died in acute pulmonary edema, a condition which was confirmed on autopsy. In addition, there was generalized edema of the body, dilatation of the right ventricle of the heart, a necrotic appearance of the gastric mucosa and degenerative changes in the tubules of the kidneys.

In a case reported by King in 1954 the degree of exposure was similar to that in the case described above. The victim was an amateur prospector who was attempting to recover gold from an ounce of "gold sand" by means of amalgamation, using about a teaspoonful of mercury. An hour after heating the mixture on a kitchen stove he becam acutely ill with severe coughing accompanied by vomiting, fever, dyspnea and cyanosis. X-ray examination confirmed the diagnosis of acute pneumonitis. The patient recovered after treatment with oxygen, dimercaprol and antibiotics.

A household accident in which mercury vapor caused fatal pneumonitis in three children was reported by Matthes et al. in 1958. A space heater had been painted with a mixture of aluminum, mercury and turpentine and then lighted to assist in the setting of the coating. The children, ages four, 20 and 30 months respectively, and their mother all slept in the room in which the heater was located. All developed pneumonitis from which the infants did not recover. On autopsy they were found to have exudative pulmonary edema and necrotizing bronchiolitis (Teng and Brennan 1959).

Mercury pneumonitis in a young man described as an amateur alchemist was reported by Haddad in 1963. This latter-day adept had heated some mercury in the course of his experiments and subsequently developed symptoms suggestive of metal fume fever: chills, fever, constriction in the chest, cough and rales. The onset occurred about four hours after the exposure to mercury vapor. The alchemist's wife, whose contact with the vapors was less intimate, experienced similar, but less severe, symptoms. Both recovered.

In an episode reported by Hallee (1969), five members of a family showed evidence of pulmonary irritation when the father accidentally dropped 30 ml of mercury into a red hot pan, causing almost instantaneous vaporization of the metal. The wife and three children exhibited mild responses, having been in less intimate contact with the vapors than the father. The latter developed a severe interstitial pneumonia and still showed impaired pulmonary function a year after the exposure.

Most of the reported cases of acute mercury pneumonitis have occurred as a result of household accidents involving not more than a few ounces of the metal. In striking contrast is an incident described by Tennant et al. (1961) in which "several tons" were involved. Rupture of a tube in a mercury vapor boiler in a power generating plant released this huge amount of heated mercury into the premises at a temperature which is described as ". . . warm but not hot enough to burn the skin." Eight men, without any respiratory protection, were involved in attempting to recover the spilled metal, and they spent about five hours at the task. Symptoms suggestive of metal fume fever appeared in five of the men several hours after the exposure and all five subsequently developed pneumonitis. There was one fatality, the autopsy showing pulmonary edema, congestion and fibrinous changes. The kidneys were not affected.

Milder degrees of mercury pneumonitis due to occupational exposure were described by Hopmann in 1927, four workers being involved. Two had symptoms suggestive of metal fume fever and all showed signs of pulmonary irritation: cough, rales and fever, along with tightness of the chest.

The outbreak of mercury poisoning on board H.M.S. Triumph, and the lesser outbreak on the Phipps are favorite topics among historians of quicksilver. The exposures to mercury vapor (resulting from ruptured shipping containers) were severe and manifestations devastating, but the picture was that of systemic poisoning rather than of pulmonary irritation. One of the best descriptions of the episode is that given by Earles (1964).

Reports on positive findings are generally more interesting to writers and readers than those describing negative results. This creates a biased picture since the literature will contain innumerable articles on adverse effects and relatively few describing exposures which have caused no harm. The effect of inhaling mercury vapor is a case in point, there being but few reports on negative results and many on intoxication.

Therapeutic Fumigations

Mercury inhalations or fumigations were first used in the treatment of syphilis in the first decade of the 16th century and were still being used at least as late as 1928. (See Chap. 15, below.) As a result of this practice, thousands of humans were exposed to mercury vapor; and in the later years the exposures were made under controlled conditions. The treatments frequently extended over a period of months or years, long enough for toxic manifestations to appear if they were to occur. With human material like this available it is amazing how much effort was put into experiments on lower animals during the first decades of the 20th century.

In an article published in 1922, Cole reviewed the literature on the use of the fumigation treatment from the earliest times and reported on inhalation experiments on six men. Using a device for controlled inhalation of heated mercury vapor, Cole exposed his subjects to amounts of mercury vapor varying from 225 mg in two weeks to 750 mg in three weeks. He found no appreciable pulmonary irritation nor other adverse effects, either local or systemic.

A much more ambitious study was undertaken by Engelbreth who published his findings in 1928. He designed an inhalation apparatus by means of which different concentrations of mercury vapor could be generated by varying the temperature. A diagram of the apparatus is shown in Fig XI-4, and the vapor concentrations at different temperatures in Table XI-1. Engelbreth estimated the degree of absorption by measuring urinary mercury excretion, as shown in Table XI-2.

Over a period of 15 years, Engelbreth subjected about 2,000 individuals to a total of more than 100,000 exposures to mercury vapor, an average of 50 exposures per person. His standard "dose" was an exposure of 20 minutes at 100°C, but this was varied in some cases so that the amount of mercury absorbed, according to his calculations, was 10-20 centigrams at each exposure. It is not clear how he arrived at the dosage figures unless it represented the amount of mercury placed in the reservoir of the apparatus rather than the amount actually absorbed. Using Engelbreth's schedule of 20 minutes at 100°C and his stated concentration of 105.0 mg of mercury vapor per 100 liters of air, and assuming a tidal air of as little as 20 liters per minute, or 400

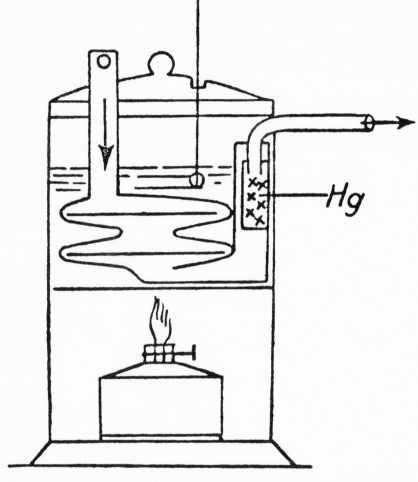

Figure XI-4 Diagram of Mercury Vaporizer for Engelbreth's Inhalation Apparatus. (From C. Engelbreth in *Dermatol. Wochschr.* 86: 572–578 [1928].)

liters, the total amount of mercury entering the respiratory tract would be
420 mg or 42 centigrams. Possibly the full concentration of 105.0 mg per
100 liters was not maintained throughout the 20 minutes or perhaps the
treatment was not continued until all of the mercury in the reservoir was volatilized. The frequency of exposure sessions is not clearly given but it appears
to have been anything from daily to weekly.

In commenting on possible adverse effects, Engelbreth states that his
subjects were not aware of the slightest discomfort during the inhalations and
that there was no irritation either of the upper air passages or of the lungs. As
to systemic effects, he notes that they occurred in only a few cases and that
they were mild and transitory: two cases of albuminuria, one of gingivitis and

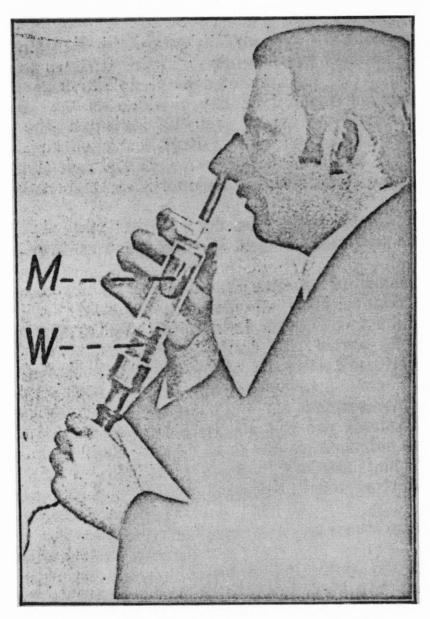

Figure XI-5 Engelbreth's Inhalation Apparatus in Use with Nose Cone. (From C. Engelbreth in *Dermatol. Wochschr.* 86: 572–578 [1928].)

TABLE XI-1

Mercury vapor concentrations from mercury surrounded
by water bath at various temperatures

Temperature of Water Bath	Mercury Vapor per 100 Liters of Air
40°C	4.4 mg
60°C	21.4 mg
80°C	53.0 mg
100°C	105.0 mg

SOURCE: Engelbreth 1928

TABLE XI-2

Mercury output in urine following
inhalation of mercury vapor

Inhalation Dose	Mercury per Liter of Urine
20 minutes at 60°C	1.4 mg
15 minutes at 80°C	2.3 mg
15 minutes at 100°C	4.9 mg
20 minutes at 100°C	5.6 mg

NOTE: Temperatures are the temperatures of the waterbath surrounding the
mercury being vaporized.
SOURCE: Engelbreth 1928

four of tremor of the hands. Since his subjects were all under treatment for
syphilis they tended to remain under observation for from two to four years,
thus providing ample opportunity for any complications to have been detected. If the values for excretion of mercury of 5.6 mg per liter of urine are
accepted it is obvious that there was substantial absorption of the metal. The
fact that absorption sufficient to result in excretion at this high level should
have caused negligible toxic manifestations is truly remarkable. Unfortunately for sceptics, conditions conducive to a repetition and validation of
Engelbreth's findings are not likely to be encountered.

One of the questions raised by Engelbreth's works is that of the degree of
retention or absorption of inspired mercury vapor, this being one of the factors which influence the possibility of toxic effects. Kudsk has published several important papers on this subject in which he has reviewed the literature
and recorded his own observations (1965a; 1965b; 1968). He points out that
previous investigators had reported 25–100 percent retention of inhaled mercury vapor, some of the variation being accounted for by differences in experimental methods. In his own experiments, using four human volunteers,
Kudsk showed that at exposure levels from 50 to 350 μg per cubic meter, and
when allowance is made for the dead space in the pulmonary tract, absorption
is close to 100 percent. His calculations were made by comparing mercury
vapor concentrations in the inspired and expired air. This high degree of ab-

sorption adds to the impressiveness of Engelbreth's finding of negligible toxic effects.

Inorganic Compounds

Toxic actions due to mercury were recognized in the first century A. D. by Dioscorides who described the effects of swallowing the metal and of inhaling its vapors (Chap. 14 below). Pliny, also in the first century, believed that mercury sulfide (minium) was poisonous, but he does not describe the toxic manifestations (Chap. 14). Celsus included minium in several of his prescriptions, showing that he did not consider it a poison in the prescribed doses, but Galen took a firm stand on the opposite side of the question without giving any description of the nature of the ill effects. This is where matters stood for nearly a thousand years.

Experimental toxicology in ancient times was confined principally to the development of theriacs (pastes used as antidotes to poisons) and mithridates (universal antidotes) and to the use of slaves or servants to taste food and drink to be sure that they contained no poisons. What may be the first recorded use of a non-human experimental animal was that carried out by Rhazes when he fed quicksilver to an ape to see what would happen (Chap. 14, below). There may have been some question in the mind of Rhazes and of his fellow physicians about the safety of administering mercury by mouth but after his experiment he concluded that it might cause some temporary abdominal discomfort and that otherwise no great harm would result. Haly Abbas (d. 994) was of the same opinion, but pointed out that when mercury was oxidized or sublimed it became highly toxic, causing severe abdominal cramps, bloody diarrhea, and suppression of urine (cited in Neuberger 1910–25, Vol. 2, pt. 1). This is an accurate description of the clinical picture commonly seen in an individual who has been poisoned by oral administration of mercuric chloride.

Mercury Sublimate

Best known of the soluble salts of mercury is its bichloride $HgCl_2$), also known as corrosive sublimate and sometimes simply as sublimate. The latter designation is based on the nature of the reaction between mercury, vitriol (probably copper sulfate), and common salt. When these substances are heated together, hydrochloric acid is generated and this reacts with the mercury to form mercuric chloride which sublimes, thus yielding corrosive sublimate. This process is first described in the 10th century in Rhazes, *De Aluminibus et Salibus* (see Chap. 6). In the opinion of Multhauf (1966), ". . . this preparation marks the beginning of the most significant period in the history of the science of matter between ancient times and the organization of the science of chemistry in the 18th century . . ." since "It marks the beginning of the systematic pursuit of synthetic chemistry." It also initiates a series of important developments in the history of medicine and toxicology. Rhazes was familiar with the toxic effects of mercury sublimate in causing colic and bloody diarrhea.

The name "corrosive sublimate" indicates that the corrosive properties of the compound must have been known from the earliest times after its synthesis. Peter of Abano (1250–1316) in his *De Venenis* states that argentum vivum extinctum (mixed with fat) or sublimatum causes corrosion of the tongue, stomach and intestines (Abano 1949). Minium and cinnabar cause similar injury, he wrote, and call for treatment with butter and rape seed, enemas of oil and honey, and a bezoar mixed with urine. The use of oily demulcents is certainly reasonable; the bezoar-urine combination might be related to the doctrine of *similis similibus curantur* since the evacuation of both feces and urine is affected by sublimate.

The poisonous nature of bichloride of mercury must have been known in 14th century Italy since, according to Thompson (1923), as early as 1365 "... a statute was passed in Siena rendering it illegal to sell red arsenic or corrosive sublimate to any slave, freed or otherwise, or to any servant or person under 20 years of age." The sale was supposed to be restricted to adults and only to those who were well known to the apothecary. Thompson states further that Perugia in 1378 and Genoa in 1488 enacted similar laws designed to curb the traffic in poisonous materials. The Augsburg interdiction of 1582 against the sale of mercury preparations (Crosland 1962) is discussed in Chap. 16, below.

Mild restrictions such as those cited above could hardly have been effective in keeping sublimate out of the hands of anyone bent on using it for criminal or suicidal purposes. The reason for its infrequent use by amateur or professional poisoners is pharmacological rather than legal. The effects of ingested mercuric chloride being immediate and dramatic while those of arsenic are slow and insidious, it is obvious why the latter was preferred. Furthermore, the effects of corrosive sublimate were well known in Renaissance times to at least the more sophisticated layman. Benvenuto Cellini relates that within less than a half hour after eating some suspicious dishes "... I felt as though my stomach was on fire. ... During the night I got no sleep, and was constantly disturbed by motions of my bowels. When the day broke, feeling an intense heat in the rectum, I looked eagerly to see what this might mean, and found the cloth covered with blood. ... I made my mind up that they must have administered a dose of sublimate in the sauce ... inasmuch as sublimate produces all the symptoms I was suffering from." This diagnosis was confirmed by Benvenuto's attending physician (Cellini 1924 translation).

Less fortunate than Cellini was Sir Thomas Overbury who, in 1613, became involved in a sordid intrigue at the court of James I and was marked for elimination. When he stubbornly refused to succumb to mercuric chloride and white arsenic given him in his food, he was finished off with an enema of corrosive sublimate (Parry 1928).

Details of the clinical course of Sir Thomas Overbury following the homicidal corrosive sublimate enema which was administered to him have not been recorded but it may very well have been similar to that of a 10 year old girl reported by Schwittay in 1938. The child's mother, in accordance with instructions from a physician, gave an enema containing 0.5 grams of mercuric chloride. Further instructions to follow the enema with a soda and salt flushing were not observed. The result was a bloody diarrhea, vomiting and anuria. Fortunately the child recovered, but only after a very stormy course.

Bichloride of mercury was well known as a suicidal agent during the 19th and early part of the 20th centuries but perhaps its use was not as common as generally believed. In London, according to Taylor, "This compound is not often taken as a poison. In the coroner's report for 1837-8, there are about 15 fatal cases of mercurial poisoning, in 12 of which corrosive sublimate was the poison taken . . . . It is freely retailed to the public at the rate of twopence for one or two drams . . . ."

Suicides with sublimate continued to occur, but with decreasing frequency as suicides from barbiturates increased. Among 1,000 attempted suicides reported in Lyon, France, in 1967, bichloride is not mentioned as one of the poisons used nor did it appear as the cause of death in 500 successful suicides (Vedrinne, David, and Vitoni 1967). The chief medical examiner of Suffolk County, New York, stated in November 1969 that he had not seen a case of bichloride suicide in the last 10 years.*

Following Rhazes' experiment with the ape there seems to have been a long hiatus in the history of experimental mercury poisoning. Among the colorful stories related by Thompson (1923) there is one which describes an experiment with an antidote against corrosive sublimate: terra sigillata, a form of clay. The experiment was conducted at the court of Prince William, Landgrave of Hesse, in 1580 by one Bertold of Oschatz and under the supervision of the court physicians, Maurice Thauer and Lawrence Hyper. The Prince commanded them

> . . . to make a perfect tryall of the said earthe, whereupon the saide Doctors in Physicke to satisfy their Prince, did make a double proffe of the deadliest poysons that might be, which were, Mercurie Sublimate, Aconitum, Nereum Apocinum, and of some one of these they gave half a dramme apeece to eight dogges, to four of them they gave the earth, after the poyson, and to the other foure the poyson alone: of these foure that tooke it alone, the first that tooke Apocynum: dyed within halfe an houre, the second that has taken Nereum died within foure houres: the third that swallowed Mercurie, died within nine houres after . . . The other foure dogges to whom the poysons before named with the like quantities of this Terra Sigillata was given, for three houres after the receiving of it, were

---

*S. Weinberg. 1969: personal communication.

very sicke and feeble, . . . the next day they were all well and did eate their meate greedily, so as there appeared scarse any token of poyson.

Thus did mercury participate in what must have been one of the first instances of a controlled pharmacological or toxicological experiment. That terra sigillata, a form of clay, should have had antidotal efficacy is not entirely fanciful since clay is highly absorptive.

The classical effects of ingested mercury sublimate are all too well known. The only major manifestation not mentioned by early writers (except Haly Abbas) is kidney injury, primarily to the tubules, resulting in severe albuminuria followed by anuria and death in uremia. So firmly has this picture of mercury poisoning become embedded in generations of physicians that some find it difficult to believe that mercurialism can occur without any of these effects being present. Prompt administration of dimercaprol and use of the artificial kidney (hemodialysis) may be life-saving in acute sublimate poisoning.

Calomel

Corrosive sublimate, designating mercuric chloride ($HgCl_2$), suggests the harsh and harmful character of this compound. In contrast, the names which have been given to mercurous chloride ($HgCl$) suggest gentleness and even beauty: Celestial Eagle more white than Snow, mercurius dulcis, sweet sublimate, and calomel, to name a few.

Speculation on the origin of the term calomel has given rise to some colorful explanations. A popular story, and one that is repeated in the Oxford English Dictionary, is that a "whimsical chymist" who employed a black in his laboratory, used calomel in alluding to the complexion of the servant (melas—black) and the beautiful (kalo-s) white of the powder. It has even been suggested that the "whimsical Chymist" was Turquet de Mayerne (1573–1655) who is known to have experimented with mercury salts. Synthesizing words of this type from two adjectives is contrary to the general rule. A more reasonable explanation would have the "mel" derived from the Greek root meaning honey. This is in consonance with "mercuris dulcis" and analogous to the well-established hydromel and oxymel of the old formularies.

Because it is relatively insoluble, calomel is less drastic in its action than sublimate and has been more widely used in medicine, particularly as a cathartic, anti-syphilitic and diuretic. Administered orally to young children as a teething powder it has been responsible for numerous cases of poisoning, and has been implicated as the cause of acrodynia or "pink disease" of children (Warkany and Hubbard 1948).

Other Inorganic Compounds

Compared with the two chlorides of mercury, sublimate and calomel, the other inorganic compounds are of relatively little importance from the toxicological point of view. Mercury nitrate is of historical interest as the cause of poisoning in the felt-hat industry (Chap. 19) but its use in the felting process has been almost completely abandoned. The red oxide of mercury, used in anti-fouling paints for ships' bottoms, has been responsible for a few cases of poisoning during its application (Goldwater and Jeffers 1942; Schrager 1964) as well as in its manufacture.* Adverse reactions to the yellow oxide (Greuer 1940), bromide (Proksch 1891), and protiodide (Warren 1965) are occasionally reported following therapeutic uses. Poisoning due to the sulfate (turpeth mineral) has been known to occur (Taylor 1875).

Tattooing

Mercuric sulfide (cinnabar), because of its low order of solubility, is relatively non-toxic if swallowed but can cause serious disturbances when implanted in the skin for decorative purposes, as in tattooing. This practice is of ancient origin and is said to have been used by the Pharaoh Ikhnaton and his queen, Nefertiti, in the 14th century B. C. (Tattooing 1966). The Mosaic prohibition that "Ye shall not make any cuttings in your flesh for the dead, nor print any marks upon you" (Leviticus) almost certainly refers to this practice. There is no information available on the nature of the pigments used during those early times nor in the tattooing that was in vogue among the early Greeks or Romans. It is reasonable to assume that when and where it was done, use would be made of whatever coloring materials were available. Red has long been a favorite color in tattoos, and cinnabar was known in antiquity (Chap. 5). There is some evidence that this pigment was used in tattoos by the pre-Inca peoples of Peru (Chap. 5), and it is certainly being used at the present time. Numerous reports in the literature attest to the frequency with which the use of cinnabar in tattoos results in significant skin reactions (Lane, Burman, and Mescon 1954; Goldstein 1967; Biro and Klein 1967).

Virchow made use of cinnabar tattoos to illustrate the effect of foreign bodies on the lymphatic system (1860 translation):

> In proof of the inevitable obstruction to the passage of solid particles through the lymphatic glands, a very pretty experiment is afforded by a custom prevalent among the lower classes . . . the well-known practice of tattooing the arms and occasionally other parts. When a workman or a soldier has a number of punctures made upon his arm . . . nearly always, in consequence of the great number of punctures, some of the superficial lymphatic vessels are injured. . . . Afterwards a substance is rubbed in which is insoluble in the fluids of the body, such as cinnabar . . . or the like, and which . . . causes permanent coloration. . . . a certain number of the particles find their way into the lymphatic vessels . . . are carried along in spite of their heaviness . . . and reach the nearest lymphatic glands, where they are separated by filtration. We never find that any particles are conveyed beyond the lymphatic glands. . . .

A specimen taken from the arm of a soldier who had been tattooed fifty years earlier was shown by Virchow to illustrate this point.

---

*A. W. Hoover 1969: personal communication.

Organic Compounds

Toxicologically, the organic compounds of mercury can be considered in broad, general categories. One classification distinguishes between those which, on entering the human body, are stable, and those which are not. Mercurial diuretics have been shown to be unstable (Shoemaker 1957; Baer and Beyer 1966) and the behavior of phenyl (aryl) compounds indicates that they, too, are readily broken down, forming mercuric ion (Hg^{++}) and the organic radical. There is evidence that some members of the alkoxy alkyl group behave similarly to the aryl mercurials in being unstable. Alkyl compounds, of which those containing methyl and ethyl groups are best known, are, on the other hand, of greater stability than the aryls and alkoxy alkyls.*

Alkyl Mercurials

Differences in stability result in differences in metabolism and differences in toxicological action, the latter being both qualitative and quantitative. The unstable compounds behave very much like inorganic mercurials, one feature being the rapidity with which they are excreted. For some of the mercurial diuretics it has been estimated that as much as 100 percent of the mercury can be excreted within 24 hours (Shoemaker 1957). By way of contrast, methyl mercury compounds are excreted so slowly that only one half may leave the body in ten weeks (cited in Karolinska Institute 1968). While in the body, alkyls are strongly attached to erythrocytes and have a marked affinity for lipids, most significantly for those of the brain. This is the basis for the severe damage to the central nervous system which is the most prominent feature of alkyl mercurial poisoning in humans. Mental retardation and cerebral palsy have been found even in infants born to mothers who had been exposed to methyl mercury (Englesson and Herner 1952; Matsumoto, Koya, and Takeuchi 1965).

Several important reviews of the toxicity of organic mercurials have been published, the first being that of Swensson in 1952. The author summarizes 32 cases of human poisoning, all except three of which were attributed to alkyl compounds of mercury. He properly emphasizes the importance of damage to the central nervous system and states that the case reports in the literature do not point to any conclusions in respect to the relative toxicity of the various organic compounds. This opinion is based in part on the incorrect assumption that three cases reported by Koelsch in 1937 were due to phenyl mercurials.

---

*Ingested methyl mercury compounds, ordinarily quite toxic to mammals, may be rendered less toxic by the presence of selenium in the diet. Like sulfur, selenium tends to combine readily with mercury and mercury compounds. Selenium in seafoods may lessen the danger to man (and other mammals) of mercury in fish and other seafoods. (See H. E. Ganth et al. 1972. *Science* 175:1122–1124.)

A comprehensive review of the early history of occupational poisoning due to organomercurials is given by Teleky in his book on industrial diseases published in 1955. Teleky is one of several authors (Bidstrup 1964; Hunter 1955) who describe the clinical picture presented by two young laboratory workers who succumbed to the toxic effects of dimethyl mercury in 1865 and 1866 respectively. The manifestations of the poisoning were primarily related to the central nervous system: weakness, ataxia, slurred speech, impairment of vision and hearing. Mindful of this tragedy, Hepp, in 1885, sounded a warning against the use of the related compound, ethyl mercury, by enterprising syphilologists who at this time were injecting every known mercury compound into victims of "the pox."

In an article on Minamata disease published in 1960 Kurland summarizes 39 cases of poisoning due to organic mercury compounds. He includes all of the cases reported by Swensson and adds seven that appeared in the literature in 1954. His interpretation of Koelsch's cases is the same as that followed by Swensson.

Bidstrup in 1964 compiled 45 cases, again embracing those first collected by Swensson as well as those added by Kurland, but adding six to the total. She, too, accepts the same evaluation of Koelsch's cases as was followed by Swensson and Kurland. Four cases of alkyl mercury poisoning published by Katsunuma in 1963 and 18 cases with eight fatalities recorded by Koelsch in 1950 are not included in any of the reviews mentioned above.

The reason for repeated reference to Koelsch is that his article published in 1937 dealing with organic mercury compounds is frequently quoted, but not always correctly. On careful reading it becomes apparent why this article should be confusing. In speaking of the compound known as Ceresan (the name is a trademark), Koelsch shows by means of a question mark that he is uncertain as to its composition. This should serve as a warning to anyone who is tempted to use the cases of exposure to Ceresan as examples of any particular type of poisoning other than that caused by an unspecified organic mercurial. Another case, in which there allegedly was exposure to Germisan (the trademark for cyanmerkurikresolnatrium), is described in terms which strongly suggest that the patient was suffering from acute rheumatic fever with heart failure or possibly from an acute exacerbation of chronic glomerular nephritis. The exposure to the organomercurial Germisan could only have been incidental.

Koelsch gives short histories of three workers in a chemical research laboratory who had been exposed to several alkyl mercurials. Most of the abnormalities found in these three individuals were in the form of irritation of the upper respiratory tract, although one had a slight tremor, and all had some non-specific subjective complaints. No information is presented on the levels of exposure. While there is no doubt that Koelsch's patients were exposed to various organomercurials and that they all suffered from some form of illness, it is not possible, from the descriptions given, to say much more than that.

Specifically, these cases cannot be offered as evidence that phenylmercury compounds are in any way similar to the alkyls in respect to toxicological properties. In addition to the differences in chemical behavior pointed out above, there is extensive clinical experience to show their dissimilarity.

Outbreaks of alkyl-mercury poisoning in humans and animals in Japan and in birds in Sweden has stimulated research into the possibility of biotransformation of inorganic mercurials into alkyl compounds (Yamaguchi and Matsumoto 1969; Kondo 1964; Ui 1968; Jernelöv 1969). Some of the early work was reviewed by Wood in 1968. Using methanogenic bacteria isolated from mud, Wood and his associates showed that methyl groups could be transferred from methyl-cobalamine to mercury by non-enzymatic as well as by enzymatic mechanisms. An incidental finding was the increased rapidity with which the conversion took place in the presence of increased concentrations of mercury, this being one of several mechanisms so far recognized in which the metal plays a vital role in biotransformations and related metabolic functions. The authors note that methyl mercury was banned as a pesticide in Sweden in 1966 and that restrictive legislation was enacted in Japan following the Minamata episode, and they suggest that these examples be followed by other countries. Steps in this direction began in the United States in 1970 (Chap. 1).

Human poisoning due to the ingestion of grain treated with mercurial fungicides has been reported. Outbreaks involving hundreds of persons occurred in Iraq in 1956 and 1960 (Jalili and Abbasi 1961), with ethyl mercury p-toluene sulphonanilide (Granosan M-Dupont) as the offending agent. The dressed seed had been distributed for planting, and the farmers had been adequately warned that it should not be eaten, but apparently hunger got the better of prudence. The exact number of cases of poisoning is not known, but in the 1956 outbreak more than 100 cases, with 14 deaths, were admitted to Mosul Hospital and in 1960, 221 patients were treated in one hospital in Baghdad (Damluji 1962). The clinical picture showed injury to kidneys, gastro-intestinal tract, skin, heart and muscles, but the most common abnormalities were those of the central nervous system. A similar outbreak occurred in Guatemala in 1965, involving 45 cases with 20 deaths (Mercury Poisoning in Guatemala 1966). The predominating manifestations were referable to the central nervous system, with weakness progressing to paralysis, loss of vision and, in the severe cases, coma and death. A similar illness involving three members of a New Mexico family occurred during the latter part of 1969 (Storrs et al.). In this episode the victims had eaten the meat of hogs fed on grain that had been treated with methyl mercury dicyandiamide. Shortly thereafter the U.S. Department of Agriculture banned the use of alkyl mercurials as agricultural fungicides. A third outbreak in Iraq occurred in 1972.

Aryl (phenyl) Mercurials

Extensive, continuing studies of humans exposed to phenylmercurials were initiated at Columbia University in 1961, with reports being published from time to time (Goldwater, Jacobs, and Ladd 1962, 1963; Goldwater and Joselow 1967; Goldwater, Ladd, and Jacobs 1964; Jacobs and Goldwater 1965; Jacobs, Ladd, and Goldwater 1963, 1964; Ladd, Goldwater, and Jacobs 1963, 1964; Joselow and Goldwater 1967; Joselow, Ruiz, and Goldwater 1968). Most of the subjects under study were in fact handling mixtures of inorganic and phenyl compounds but the predominant exposure was to the latter. Limited observations on humans exposed only to several phenylmercurials revealed responses similar to those having the mixed exposures (Ladd, Goldwater, and Jacobs 1964; Goldwater 1964). Noteworthy among the findings has been the almost complete absence of any abnormalities of clinical significance that could be attributed to the absorption of mercury, in spite of exposures greatly in excess of the accepted Threshold Limit Value of 0.1 mg per cubic meter of air (Goldwater 1966). This, however, is in accord with the theoretical concept that aryl mercurials are relatively unstable, readily liberating Hg^{++} ions which are rapidly excreted and in accord with observations made on experimental animals (Gage and Swan 1961).

In spite of the evidence of relatively low toxicity of phenylmercurials, steps to eliminate their use as biocides were initiated by governmental agencies in the United States in 1970 and 1971.

Alkoxy Alkyl Mercurials

Relatively few articles dealing with the toxicity of alkoxy alkyl mercury compounds have appeared in the literature. One such report, published in 1952 (Wilkening and Litzner) has a misleading title, implying that exposure had been to an alkyl mercurial while in fact the substance was methoxy-ethylmercury silicate known as Abavit (CH_3-CH_2-CH_2-Hg-silicate). The picture is further complicated by the authors' description of the workplace in which the exposure took place, as they relate that there was gross contamination of the floor, with the presence of free metallic mercury. The abnormalities found in six of twelve workers were gingivitis, erethism and retention of urea. This suggests that metallic mercury rather than an organic form was responsible for the trouble. Three of the cases were very mild, and three slightly more severe; all made complete recoveries when exposure was terminated.

REFERENCES

Abano, Pietro d'. 1949. *Il trattato "De venenis."* Ed. Prof. A. Benedicenti. Florence: Leo S. Olschki.

Baer, J. E., and Beyer, K. H. 1966. Renal pharmacology. *Ann. Rev. Pharmacol.* 6: 261–92.

Bidstrup, P. L. 1964. *Toxicity of mercury and its compounds.* Amsterdam: Elsevier Publishing Co.

Birnbaum, W. 1947. Inflammation of the vermiform appendix by metallic mercury. *Am. J. Surg.* 74:494–96.

Biro, L., and Klein, W. P. 1967. Unusual complications of mercurial (cinnabar) tattoo. *Arch. Dermatol.* 96:165–67.

Borchard, A. 1933. Über traumatische Quecksilbervergiftung. *Zentr. Chir.* 60:2930–33.

Bradford, F. E.; Hugo, G. J.; and Quinn, W. F. 1957. Persistent rectal fistula due to metallic mercury. *Am. J. Surg.* 93:74–76.

Buxton, J. T.; Hewitt, J. C.; Gadsen, R. H.; and Bradham, G. B. 1965. Metallic mercury embolism: Report of cases. *J. Am. Med. Assoc.* 193:573-75.

Campbell, J. A. 1948. Acute mercuric poisoning by inhalation of metallic vapour in an infant. *Can. Med. Assoc. J.* 58:72-75.

Cantor, M. O. 1951. Mercury lost in the gastrointestinal tract. *J. Am. Med. Assoc.* 164: 560-61.

Cellini, B. 1924. *The life of Benevenuto Cellini.* Trans. J. A. Symonds. New York: Charles Scribner's Sons.

Cole, H. N.; Gericke, A. J.; and Sollmann, T. 1922. The treatment of syphilis by mercury inhalations: History, methods and results. *Arch. Dermatol. Syphilol.* 5:18-33.

Conrad, M. E.; Sanford, J. P.; and Preston, J. A. 1957. Metallic mercury embolism: clinical and experimental. *Arch. Internal Med.* 100:59-65.

Crikelair, G. F., and Hiratzka, T. 1953. Intraperitoneal mercury granuloma. *Ann. Surg.* 137:272-75.

Crosland, M. P. 1962. *Historical studies in the language of chemistry.* London: Heinemann Educational Books.

Damluji, S. 1962. Mercurial poisoning with the fungicide Granosan M. *J. Fac. Med. Baghdad Iraq* 4:83-103.

Davis, L. E., and Hofmann, W. 1969. A long-term nasogastric feeding tube made from modified Penrose tubing. *J. Am. Med. Assoc.* 209:685-86.

Devlin, H. B., and Sudlow, M. 1967. Peripheral mercury embolization occurring during arterial blood sampling. *Brit. Med. J.* 1:347-48.

Dioscorides, P. 1934. *The Greek herbal.* Englished by John Goodyer, A. D. 1655. Edited and first printed, A. D. 1933, by Robert T. Gunther. Oxford: Oxford University Press.

Drügg. 1929. Thermometerverletzung mit Quecksilbervergiftung. *Deut. Med. Wochschr.* 55:1637-38.

Earles, M. P. 1964. A case of mass poisoning with mercury vapour on board H. M. S. Triumph at Cadiz, 1810. *Med. Hist.* 8:281-86.

Engelbreth, C. 1928. Die Quecksilberinhalation durch einen neuen Inhalationsapparat. *Dermatol. Wochschr.* 86:572-78.

Englesson, J., and Herner, A. B. 1952. Alkyl mercury poisoning. *Acta Paediat.* 41: 289-94.

Farrell, J. J., and Reams, G. B. 1957. Abuse of intestinal intubation. *Am. Surg.* 23: 401-8.

Gage, J. C., and Swan, A. A. B. 1961. The toxicity of alkyl and aryl mercury salts. *Biochem. Pharmacol.* 8:77.

Goldstein, N. 1967. Mercury-cadmium sensitivity in tattoos: a photoallergic reaction in red pigment. *Ann. Internal Med.* 67:984-89.

Goldwater, L. J. 1964. Occupational exposure to mercury. The Harben Lectures, 1964. *J. Roy. Inst. Public Health Hyg.* 27:279-301.

Goldwater, L. J. 1966. Occupational exposure to mercury. In *Proceedings of 15th International Congress on Occupational Health,* 19-24 September 1966, Vienna, Paper B III-17.

Goldwater, L. J.; Jacobs, M. B.; Ladd, A. C. 1962. Absorption and excretion of mercury in man: I. Relationship of mercury in blood and urine. *Arch. Environ. Health* 5:537-41.

Goldwater, L. J.; Jacobs, M. B.; Ladd, A. C. 1963. Absorption and excretion of mercury in man: IV. Tolerance to mercury. *Arch. Environ. Health* 7:568-73.

Goldwater, L. J., and Jeffers, C. P. 1942. Mercury poisoning from the use of antifouling plastic paint. *J. Ind. Hyg. Toxicol.* 24:21-24.

Goldwater, L. J., and Joselow, M. M. 1967. Absorption and excretion of mercury in man: XIII. Effects of mercury exposure on urinary excretion of coproporphyrin and delta-aminolevulinic acid. *Arch. Environ. Health* 15:327-31.

Goldwater, L. J.; Ladd, A. C.; Jacobs, M. B. 1964. Absorption and excretion of mercury in man: VII. Significance of mercury in blood. *Arch. Environ. Health* 9:735-41.

Greuer, W. 1940. Psoriasisbehandlung mit Hg praecipitat. flav. unter tödlichem Ausgang. *Dermatol. Wochschr.* 111:939-40.

Haddad, J. K., and Stenberg, E. 1963. Bronchitis due to acute mercury inhalation. *Am. Rev. Respirat. Diseases* 88:543-45.

Hallee, J. T. 1969. Diffuse lung disease caused by inhalation of mercury vapor. *Am. Rev. Respirat. Diseases* 99:430-36.

Handerson, H. E. 1918. *Gilbertus Anglicus: Medicine of the 13th century.* Cleveland: Cleveland Medical Library Association.

Hanselman, R. C., and Meyer, R. H. 1962. Complications of gastrointestinal intubation. *Intern. Abstr. Surg.* 114:207-8.

Harris, F. I. 1944. A new rapid method of intubation with the Miller-Abbott tube. *J. Am. Med. Assoc.* 125:784-85.

Hepp, P. 1885. Ueber Quecksilberäthyl. *Z. Klin. Med.* 6:665-66.

Hill, D. M. 1967. Self-administration of mercury by subcutaneous injection. *Brit. Med. J.* 1:342-43.

Hoffman, I. L. 1948. Spontaneous evacuation of metallic mercury from the vermiform appendix. *Bull. U. S. Army Med. Dept.* 8:802-3.

Hopmann, A. 1927. Acute Quecksilberdampfvergiftungen. *Zentr. Gewerbehyg.* 4: 422-23.

Hunter, D. 1955. *The diseases of occupations.* Boston: Little, Brown and Co.

Iskandar, A. Z. 1959. A study of AR-Razi's medical writings with selected texts and English translations. Unpublished Ph. D. dissertation, University of Oxford.

Jacobs, M. B., and Goldwater, L. J. 1965. Absorption and excretion of mercury in man: VIII. Mercury exposure from house paint; A controlled study. *Arch. Environ. Health* 11:582-87.

Jacobs, M. B.; Ladd, A. C.; and Goldwater, L. J. 1963. Absorption and excretion of mercury in man: III. Blood mercury in relation to duration of exposure. *Arch. Environ. Health* 6:634-37.

———. 1964. Absorption and excretion of mercury in man: VI. Significance of mercury in urine. *Arch. Environ. Health* 9:454-63.

Jalili, M. A., and Abbasi, A. H. 1961. Poisoning by ethyl mercury toluene sulphonanilide. *Brit. J. Ind. Med.* 18:303-8.

Jernelöv, A. 1969. conversion of mercury compounds. In *Chemical Fallout,* ed. M. W. Miller and G. G. Berg, pp. 68-74. Springfield, Ill.: C. C. Thomas.

Johnson, H. R. M., and Koumides, O. 1967. Unusual case of mercury poisoning. *Brit. Med. J.* 1:340-41.

Joselow, M. M., and Goldwater, L. J. 1967. Absorption and excretion of mercury in man: XII. Relationship between urinary mercury and proteinuria. *Arch. environ. Health* 15:155-59.

Joselow, M. M.; Ruiz, R.; and Goldwater, L. J. 1968. Absorption and excretion of mercury in man: XIV. Salivary excretion of mercury and its relationship to blood and urine mercury. *Arch. Environ. Health* 17:35-38.

Karolinska Institute. 1968. Recommendations on maximum allowable concentrations of mercury and its compounds. *Report of an International Committee,* Karolinska Institute, 4-7 November 1968, Stockholm.

Katsunuma, H.; Suzuki, T.; Nishi, S.; and Kashima, T. 1963. Four cases of occupational organic mercury poisoning. *Rep. Inst. Sci. Labour (Japan)* 61:33-40.

King, G. W. 1954. Acute pneumonitis due to accidental exposure to mercury vapor. *Ariz. Med.* 11:335.

Kockel, H. 1928. Mordversuch mit metallischem Quecksilber. *Arch. Kriminol.* 83: 309-10.

Koelsch, F. 1937. Gesundheitsschädigungen durch organische Quecksilberverbindungen. *Arch. Gewerbepathol. Gewerbehyg.* 8:113-16.

Koelsch, F. 1950. Vergiftungen durch organische Quecksilberverbindungen. In *Proceedings of the International Congress of Occupational Medicine,* 1950, Milan, pp. 103-6.

Kondo, T. 1964. Studies on the origin of the causative agent of Minamata disease: IV. Synthesis of methyl (methyl-thio) mercury. *J. Pharm. Soc. Japan* 84:137-41.

Kristoffersen, K. 1959. Metallic mercury as the cause of intestinal ulceration. *Nord. Med.* 62:1388-89.

Kudsk, F. N. 1965a. Absorption of mercury vapor from the respiratory tract in man. *Acta. Pharmacol. Toxicol.* 23:250-62.

———. 1965b. The influence of ethyl alcohol on the absorption of mercury vapor from the lungs in man. *Acta. Pharmacol. Toxicol.* 23:263-74.

_____. 1968. Absorption of mercury vapor from the respiratory tract in man. Working paper, Symposium on MAC-values, November 1968, Stockholm.

Kurland, L. T.; Faro, S. N.; and Siedler, H. 1960. Minamata disease. *World Neurol.* 1: 370-95.

Ladd, A. C.; Goldwater, L. J.; and Jacobs, M. B. 1963. Absorption and excretion of mercury in man: II. Urinary mercury in relation to duration of exposure. *Arch. Environ. Health* 6:480-83.

_____. 1964. Absorption and excretion of mercury in man: V. Toxicity of phenylmercurials. *Arch. Environ. Health* 9:43-52.

Lane, R. A. G.; Burman, H.; and Mescon, H. 1954. Mercurial granuloma occurring in tattoo. *Can. Med. Assoc. J.* 70:546-48.

Lathem, W.; Lesser, G. T.; Messinger, W. J.; and Galdston, M. 1954. Peripheral embolism by metallic mercury during arterial blood sampling: Report of two cases. *Arch. Internal Med.* 93:550-55.

Leviticus, 19:28.

Lindenmuth, W. W. 1949. Fecal fistula due to metallic mercury from a Miller-Abbott tube. *J. Am. Med. Assoc.* 141:986-87.

Lindsey, E. S., and Becker, W. F. 1968. Intestinal obstruction due to mercury granuloma. *Arch. Surg.* 97:568-69.

Matsumoto, H. G.; Koya, G.; and Takeuchi, T. 1965. Fetal Minamata disease: A neuropathological study of two cases of intrauterine intoxication by a methyl mercury compound. *J. Neuropathol. Exp. Neurol.* 24:563-74.

Matthes, F. T.; Kirschner, R.; Yow, M. D.; and Brennan, J. C. 1958. Acute poisoning associated with inhalation of mercury vapor. *Pediatrics* 22:675-88.

Mercury Poisoning in Guatemala. 1966. *U. S. Morbid. Mortal. Weekly Rep.* 15:34-35.

Mettler, C. C. 1947. *History of medicine.* Philadelphia: Blakiston Co.

Miller, T. G., and Abbott, W. O. 1934. Intestinal intubation: A practical technique. *Am. J. Med. Sci.* 187:595-99.

Multhauf, R. P. 1966. *The origins of chemistry.* London: Oldbourne.

Neuberger, M. 1910-1925. *History of medicine.* Trans. Ernest Playfair. 2 vols. London: Oxford University Press.

Parry, L. A. 1928. *Some famous medical trials.* New York: Charles Scribner's Sons.

Plinius Secundus, C. 1938-63. *Natural history.* Trans. H. Rackham, W. H. S. Jones and D. E. Eichholz. 10 vols. London: W. Heinemann.

Proksch, J. K. 1891. *Die Litteratur ueber die venerischen Krankheiten.* Bonn.

Ramazzini, B. 1964. *Diseases of workers.* Trans. Wilmer Cave Wright. New York: Hafner Publishing Co.

Rixford, E. 1895. Lesions resembling cutaneous gummata caused by metallic mercury, and accompanied by pus formation. *Occidental Med. Times* 9:533-36.

Schrager, G. O. 1964. Acute mercury poisoning in a child following contact with marine antifouling paint. *J. Pediat.* 65:780-82.

Schultze, W. 1958. Roentgenologische studien nach aspiration von metallischen Quecksilber, *Fortschr. Roentgenstr.* 89:24-30.

Schulz, E., and Beskind, H. 1960. Systemic deposition of metallic mercury. *J. Pediat.* 57:733-37.

Schwittay, A. M. 1938. Mercurial poisoning from bichloride of mercury enema. *Wisconsin Med. J.* 37:558-59.

Shoemaker, H. A. 1957. The pharmacology of mercury and its compounds. *Ann. N. Y. Acad. Sci.* 65:504-9.

Stammel, C. A. 1929. Metallic mercury in subcutaneous tissue: Case report. *Military Surg.* 65:528-30.

Storrs, B.; Thompson, J.; Fair, G.; Dickerson, M. S.; Nickey, L.; Barthell, W.; and Spalding, J. E. 1970. Organic mercury poisoning: Alamogordo, New Mexico. *U. S. Morbid. Mortal. Weekly Rep.* 19:25-26.

Swensson, A. 1952. Investigations on the toxicity of some organic mercury compounds which are used as seed disinfectants. *Acta Med. Scand.* 143:365-84.

Tattooing. 1966. *Pfizer Spectrum.* Fall:84–87.

Taylor, A. S. 1875. *On poisons.* 3d American ed. Philadelphia: Henry C. Lea.

Teleky, L. 1955. *Gewerbliche Vergiftungen.* Berlin: Springer-Verlag.

Teng, C. T., and Brennan, J. C. 1959. Acute mercury vapor poisoning. *Radiology* 73: 354–61.

Tennant, R.; Johnston, J. H.; and Wells, J. B. 1961. Acute bilateral pneumonitis associated with the inhalation of mercury vapor. *Conn. Med.* 25:106–9.

Thompson, C. J. S. 1923. *Poison mysteries in history, romance and crime.* London: Scientific Press.

Ui, J. 1968. A short history of Minamata disease's research and present situation of mercury pollution in Japan. In *Nordisk Kvicksilversymposium,* 10–11 Oktober 1968, Lidingö.

Umber, F. 1923. Quecksilber-Embolien der lebenden Lunge durch intravenose Injektion von metallischen Quecksilber. *Med. Klin.* 19:35.

Vare, V. B., Jr. 1953. Extravasated metallic mercury simulating osteomyelitis. *U. S. Armed Forces Med. J.* 4:773–75.

Vedrinne, J.; David, J. J.; and Vitani, Ch. 1967. Comparative study of suicide in the hospital and in the morgue. *Bull. Med. Legale Toxicol. med.* 4:223–24.

Virchow, R. 1860. *Cellular pathology.* Trans. from the 2d ed. New York. R. W. De Witt.

Warkany, J., and Hubbard, D. M. 1948. Mercury in the urine of children with acrodynia. *Lancet* 254:829–30.

Warren, D. E. 1965. nephrotic syndrome caused by protiodide of mercury. *Arch. Dermatol.* 91:240–42.

Wilkening, H., and Litzner, S. 1952. Über Erkrankungen insbesondere der Niere durch alkylquecksilberverbindungen. *Deut. Med. Wochschr.* 77:432–34.

Wilkins, J. A. 1928. Mercury-weighted stomach tube. *J. Am. Med. Assoc.* 91:395–96.

Wood, J. M.; Kennedy, F. S.; and Rosen, C. G. 1968. Synthesis of methyl-mercury compounds by extracts of a methanogenic bacterium. *Nature* 220:173–74.

Yamaguchi, S., and Matsumoto, H. 1969. Occurrence of methyl-mercury compound in an environment. In *Proceedings of the 16th International Congress on Occupational Health,* 22–27 September 1969, Tokyo.

Zimmerman, J. E. 1969. Fatality following metallic mercury aspiration during removal of a long intestinal tube. *J. Am. Med. Assoc.* 208:2158–60.

ADDITIONAL READINGS

Zangger, H. 1930. Erfahrungen über Quecksilbervergiftungen. *Arch. Gewerbepathol. Gewerbehyg.* 1:539–60.

12

Pharmacology

When the new science of organic chemistry began to take shape, it was not long before mercury and sulfur became involved. In the 1830's, W. C. Zeise (1834) was studying reactions between sulfides and salts of ethyl-sulfuric acid, and was looking for new bindings between these organic sulfides and metallic oxides and chlorides. In the course of this work he noted one particular compound ($C_4H_{10}S_2$ + H_2) which had an unusually strong affinity for mercury. He decided to call this compound *mercaptan* from *corpus mercurium captans* (Goldwater 1965). (The term "mercapto" is now used synonymously for thiol and sulfhydryl [-SH] groups.) Within a year of this event, the term "mercaptide," for a metallic salt of a mercaptan, was created. The reaction of mercury with thiols, with the formation of mercury mercaptides, is at the heart of the pharmacological activity of mercury.

This affinity between mercury and sulfur has been known since the 11th century A. D. when Theophilus described a simple process for making cinnabar from quicksilver and sulfur, and it may have been recognized even earlier by alchemists. The contrived obscurity of language in their writings makes it difficult to evaluate the extent of the alchemists' knowledge, but their preoccupation with mercury and sulfur lends credence to the assumption that they must have had considerable experience with reactions between these two elements. Textbooks of chemistry, from those of Beguin and Lemery in the 17th century up to the present time, have taken note of the combining affinity between mercury and sulfur.

Interest in the mechanisms whereby toxic agents produce injury is by no means of recent origin. Dioscorides, who was the first to record the fact that quicksilver could cause harmful effects, explained this on the basis that it ". . . hath a pernitious faculty being drank, eating through ye inward parts by its weight." The idea that weight was a factor in the toxicity of mercury was still alive in the 18th century. Both Galen and Pliny, contemporaries of

Dioscorides, were aware of the poisonous properties of minium (mercuric sulfide), but neither offered any suggestion as to how it acted.

Older Views of the Actions of Mercury

During medieval and early modern times theories of pharmacological and toxicological action were based on properties of, and interactions among, the four "elements": earth, air, fire, water; the four properties: wet, dry, hot, cold; the four humors: blood, phlegm, yellow bile, black bile; and the four temperaments: sanguine, phlegmatic, choleric, melancholic. Proper balance and functioning of the three Spirits (animal, vital, and natural) were also of major concern. All foods and drugs were classified according to their properties, quicksilver being predominantly cold and moist.

Nothing of real significance was contributed to an understanding of the mechanisms of mercury toxicity during the 14th, 15th, and 16th centuries. An example of 17th century pharmacology can be found in Lemery's *Cours de Chimie*. "Quicksilver . . . is . . . dangerous, when it happens into the hands of Quacks, who use it upon all occasions . . . and give it to all sorts of persons without any respect to the Temperament . . . ." This (temperament) is an obvious reference to what today would be called the "host factor" as an important consideration in influencing the response to a potentially harmful agent. By way of further explanation of the mechanism of action, Lemery states that: "Those who . . . work much with it, do often fall into the Palsie, by reason of Sulphurs that continually steam from it; for these Sulphurs, consisting of gross parts do enter through the Pores of the Body, and fixing themselves rather in the Nerves by reason of their coldness, than in other Vessels, do stop the Passage of the Spirits, and hinder their course."

A contrary view was to appear in the 18th century, as will presently be shown. When mercury is given by mouth for "Miserere" (intestinal obstruction) Lemery states that ". . . it is better to take a great deal of it than a little, because a small quantity might be apt to stop in the circumvolutions of the Guts, and if some Acid humours should happen to joyn with it, a *Sublimate Corrosive* would be there made; but when a large quantity of it is taken, there's no need fearing this Accident, because it passes quickly through by its own weight." By a large quantity he means two or three pounds. The harmful effect of corrosive sublimate, he says, is due to the fact ". . . the Sublimate sticks to some one or more of the vessels, and coroding their membrane, causes grievous Hemorrhagies, as I have seen happen several times, and among others to a man in Languedock, who voided in half an hours time twelve pints of blood by mouth, without dying of it notwithstanding, because he was a very stout lusty man." Stout indeed!

The theory that mercury exerts its toxic action through the nerves was accepted by Ramazzini who states that ". . . when those vapors have gained entrance to the dwelling-house of life and are mixed with the blood, they pervert and pollute the natural composition of the nervous fluid, and the result is palsy, torpor, and the maladies above-mentioned"—i.e. dyspnoea, phthisis, apoplexy, paralysis, cachexy, swollen feet, loss of teeth, ulcerated gums, pains in the joints and palsy.

Another theory is suggested by Ramazzini to explain the extreme danger from heated mercury: "Are we to suppose that this happens because mercury, when its structure is broken up by the strong action of fiery heat, is reduced to infinitesimal and highly penetrating particles, and hence, when taken in by the mouth and nose, it makes its way to the lungs, heart and brain? By this means it may easily cloud the animal spirits and induce narcosis in the whole mass of bodily fluids. . . ."

Recognition of "host factors" is shown by Ramazzini in his discussion of treatment for occupational poisoning among potters in that ". . . they suffer from yet another drawback, I mean that they are very poor, so that we must resort to the medicine of the poor. . . ." (1964 translation).

An example of toxicological theory of the early part of the 18th century is found in the words of John Quincy M. D. (1730) who wishes his readers

> To understand . . . the Manner of Operation, and particularly, how a Metal of no remarkable Efficacy is changed into a violent Poison, in making it into the common Sublimate. . . . The Salt (of Aqua Fortis) being drove into the mercurial Globules, gives them Points which they had not before; and the mercurial Globules add to the Saline Particles a Gravity and Force, which they had not without them: that is, Crude Mercury by its Weight, when in Circulation in the Juices, would strike hard upon whatsoever it met with, but, for want of Angles, or Points, could not vellicate the Part; and the saline Particles, though they had Points, have not Force enough to drive them into the Membranes, so as to do much Harm. But when by this Process, they are join'd together, the Weight of the Mercury drives the saline Spiculae like Wedges, and makes them cut and tear to Pieces whatsoever comes in their Way. So that these Crystals, or arm'd Balls, as so many Knives and Daggers, wound and stab the tender Coats of the Stomach and Guts, and all Parts they pass through, whereby they abrade their natural Mucus, tear off the Extremities of the Vessels, and draw Blood itself.

Still another mechanism of toxic action is put forward by "The Learned Belloste" as part of his justification for the therapeutic use of quicksilver. In disputing Lemery's theory, and in deprecating the relevance of occupational poisoning, Belloste (who is quoted by Dover [1733]) writes:

> There is no Absurdity in believing that these Workmen, being continually surrounded with the volatile Vapours of Mercury, draw it in with the Air at their Nostrils; presently after which they are elevated 'till they meet with Opposition from the internal convex Part of the Cranium, whose pores they cannot penetrate, and then fall down again, like Water boiling in a Pot, when stopp'd by the Lid. These Vapours being at length united, form small Globules, which drop down again by their own proper Weight, toward the Basis of the Cranium, and compress the Nerves in their Origine, and thus occasion a Palsy. The same thing happens to Gilders for the same Reason,

but more frequently; for they make Use of Mercury over a Fire, which elevates it more easily, so that they breathe in the very Substance of it.

This, in effect, is Agricola's distillation *per descensum* applied to toxicology.

During the 19th century, particularly the latter part, mercury was spoken of as a "protoplasmic" poison, a term which was not clearly defined. Sollmann, whose studies of mercury have been outstanding, wrote in 1948 that "Mercury, particularly the mercuric salts, has a strong toxic action on protoplasm." By way of amplification he says that in acute poisoning "the immediate effects are due to coagulation, irritation and superficial corrosion, to which mucous membranes are highly susceptible." Ten years later, a new edition of Sollmann's textbook contains the same wording, but in addition there is a discussion of the inhibitory effect of mercury on enzymes, particularly those containing thiol (-SH) groups (Sollmann 1957).

Recent Views of Mercury Pharmacology

Modern concepts of the mechanisms of mercury toxicity have much in common with those expressed by Lemery and Ramazzini with a new technical vocabulary replacing the old. In 20th century terminology,

> Damage to the central nervous system following exposure to elemental mercury vapor appears to be due to a high uptake of mercury by the brain. Two steps are involved in this uptake process, first the diffusion of metallic mercury from the blood stream into the brain and second, the oxidation in the brain of rapidly diffusible metallic mercury to mercuric mercury. Since mercuric mercury binds strongly to tissue proteins, it will diffuse slowly out of the brain (Clarkson 1968).

The mechanism described here involves, in fact, three steps: diffusion, oxidation and binding.

The concept of "binding" plays an important role in present-day theories of the nature of toxic effects due to various heavy metals, including mercury. In simplest terms this means that mercury is capable of entering into reactions with certain chemical compounds, known as ligands, which are present in living organisms. Among the many ligands in biological materials there are varying degrees of affinity for different heavy metals and all ligands do not have the same degree of affinity for individual metals. In other words, many ligands can bind many metals with differences in the relative affinity of each metal for each ligand. The concentration ratios of different ligands in living systems is not known, but it *is* known that "in living systems the large numbers of reactive substances (ligands) compete for traces of the heavy metal" (Passow, Rothstein, and Clarkson 1961).

One factor in determining the degree of harm done by a heavy metal is the extent of binding that takes place in sites which are sensitive to small changes and which are concerned with highly important functions. A small amount of metal can cause a large amount of injury if there is a particularly strong affinity between the metal and a sensitive, vital site. Competition for and diversion of the metal on the part of less sensitive sites may afford some degree of protection. Cell membranes are particularly susceptible to the action of metals and, furthermore, metals are known to interfere with a number of important enzymes and other important proteins. Enzymes which contain sulfhydryl (-SH) groups (thiols) are particularly attractive to a number of metals, and especially to mercury, their combinations being designated mercury mercaptides.

Applying these general considerations to the mechanism of mercury toxicity, several significant points can be made:

1. The principal reaction of mercury in biological systems is with thiols (Hughes 1957; Clarkson 1968a).
2. Cell membranes are particularly rich in thiols and consequently are important binding sites for mercury.
3. The binding of mercury on cell membranes may result in changes in ion distribution, altered movement of fluid across the membrane, changes in electrical potential on the surfaces of the membrane, alterations in permeability and perhaps other disturbances (Passow, Rothstein, and Clarkson 1961).

As a result of these alterations or perhaps concomitantly but independently, there may be loss of functional integrity not only of cell membranes but also of subcellular structures such as mitochondria (Rodin and Crowson 1962) and lysosomes (Norseth 1968). In addition, or sometimes underlying these alterations, there may be loss of function in enzymes, particularly in those which are rich in sulfhydryl groups. An extensive review of the subject was published by Brown and Kulkarni in 1967.

Mechanisms of toxicity can operate directly by damaging a histological or biochemical unit, or indirectly by inhibiting a normal detoxication action. In the latter case, injury results from the accumulation of compounds which would otherwise be rendered harmless or relatively so. Detoxication processes may be interrupted before they are complete, thus causing the liberation of intermediates which may be more toxic than either the initial material or the end products of complete detoxication. Experimentally it has been shown that the organic compounds methylmercurihydroxide, methylmercuridicyandiamide, phenylmercurihydroxide and p-chloromercuriphenylsulfonic acid can interfere with oxidative demethylation, a detoxication process normally taking place in the liver. This interference in turn permits a build-up of amines which are able to react with deoxyribonucleic acid (DNA) and ribonucleic acid (RNA), constituting a potential threat to genetic integrity (Arrhenius 1967). It remains to be seen whether or not the in vitro experiments have any significance in relation to living cells in man or in other animals.

In addition to the enzyme systems already mentioned several others have been suggested as playing a part in the mechanisms of mercury toxicity.

Bidstrup (1964) states that "There is evidence that mercury also specifically inhibits other enzyme systems in vitro including phenolsulphate conjugation, citrulline phosphorolysis, oxidative mitochondrial phosphorylation and serine biosynthesis." Inhibition of cytochrome C oxidase has been described (Seibert, Kreke and Cook 1950) and depression of 2,3-diphosphoglycerate has been suspected. Inhibitory effects are not necessarily limited to enzymes which contain thiol groups (Cook and Perisutti 1947).

A number of studies on the effects of mercury compounds on cells were summarized by A. J. Clark (1933). The investigations discussed by Clark were concerned principally with bacteria, spores, yeasts and protozoa, but one of the studies had to do with the fixation of mercuric chloride on the surface of red blood corpuscles. In the preface to his book, "The author apologizes for the fact that so much of the argument in the volume is destructive criticism of hypotheses that are attractive in their simplicity." He concludes that ". . . the action of drugs on cells must be far more complex than has been assumed." Perhaps the unpleasant clean-up job helped to open the way for subsequent studies which indeed showed the complexities involved in this type of basic toxicology and pharmacology.

One of the hypotheses which Clark found untenable was the so-called Arndt-Schulz Law, first proposed in 1885, which held that "Weak stimuli excite, medium stimuli partially inhibit and strong stimuli produce complete inhibition" (Clark 1933). He accepted the validity of that part of the law which said that strong stimuli could cause paralysis and death if this meant that huge doses could be lethal, but he rejected the other parts of the hypothesis. Of interest in this connection is the demonstration by Goldstein and Doherty in 1951 that low concentrations of mercuric chloride cause irreversible inhibition of cholinesterase while the inhibition caused by higher concentrations is reversible. This is an example of a small dose having higher toxicity than a large one.

Mercury, in common with many pharmacologically active chemical agents, has been found to exhibit great variability in its effects on different human subjects. (Clinical and occupational observations on this are discussed in Chapter 19). This phenomenon is commonly called individual susceptibility and by Roger J. Williams biochemical individuality.

Mercuric iodide and mercuric chloride were among the compounds mentioned by Clark to illustrate individual variations in responses to chemical agents. When the chloride was applied to the skin of 35 human subjects, it was found that in one case skin irritation was produced by a solution of 1:100,000, in another by a strength of 3:100,000; five responded to 10: 100,000, eleven to 30:100,000, thirteen to 100:100,000, while four showed no response to the highest of these concentrations. Results with mercuric iodide were similar. Comparable observations had been reported in 1904 by Brouardel when he found that in 37 cases of fatal poisoning resulting from obstetrical uses of bichloride of mercury the concentrations of solutions had varied from 1:1,000 to 1:5,000. (Additional material on mechanisms of action is presented in Chapter 13 in the section, "Biochemical Tests.")

REFERENCES

Arrhenius, E. 1967. Effects of organic mercury compounds on the detoxification mechanism of liver cells in vitro. *Oikos. Suppl. 9*:32–35.

Bidstrup, P. L. 1964. *Toxicity of mercury and its compounds.* Amsterdam: Elsevier Publishing Co.

Brouardel, P. 1904. Intoxications accidentelles par le mercure. *Ann. Hyg. Publ. Méd. Leg.* 1:5–26.

Brown, J. R., and Kulkarni, M. V. 1967. A review of the toxicity and metabolism of mercury and its compounds. *Med. Serv. J.* 23:786–808.

Clark, A. J. 1933. *The mode of action of drugs on cells.* Baltimore: Williams and Wilkins Co.

Clarkson, T. W. 1968a. Biochemical aspects of mercury poisoning. *J. Occupational Med.* 10:351–55.

———. 1968b. The metabolism and mode of action of mercury compounds. Working paper for a conference on MAC values, November 1968. Stockholm.

Cook, E. S., and Perisutti, G. 1947. The action of phenylmercuric nitrate: III. Inability of sulfhydryl compounds to reverse the depression of cytochrome oxidase and yeast respiration caused by basic phenylmercuric nitrate. *J. Biol. Chem.* 167:827–32.

Dover, T. 1733. The ancient physician's legacy to his country. 2nd ed. London.

Goldstein, A., and Doherty, M. E. 1951. Properties and behavior of purified human plasma cholinesterase: Inactivation by mercuric chloride. *Arch. Biochem.* 33:35–49.

Goldwater, L. J. 1965. The birth of mercaptan. *Arch. Environ. Health* 11:597.

Hughes, W. L. 1957. A physicochemical rationale for the biological activity of mercury and its compounds. *Ann. N. Y. Acad. Sci.* 65:454–60.

Lemery, N. 1686. *A course of chymistry.* 2d ed., enl. and trans. from the 5th edition in the French, by Walter Harris, M. D. London.

Norseth, T. 1968. The intracellular distribution of mercury in rat liver after a single injection of mercuric chloride. *Biochem. Pharmacol.* 17:581–93.

Passow, H.; Rothstein, A.; and Clarkson, T. W. 1961. The general pharmacology of the heavy metals. *Pharmacol. Rev.* 13:185–224.

Quincy, J. 1730. *Lexicon physico-medicum or A new medicinal dictionary.* London.

Ramazzini, B. 1964. *Diseases of workers.* Trans. Wilmer Cave Wright. New York: Hafner Publishing Co.

Rodin, A. E., and Crowson, C. N. 1962. Mercury nephrotoxicity in the rat: Investigation of the intracellular site of mercury nephrotoxicity by correlated serial time histologic and histoenzymatic studies. *Am. J. Pathol.* 41:485–99.

Seibert, M. A.; Kreke, C. W.; and Cook, E. S. 1950. The mechanism of action of organic mercury compounds on cytochrome oxidase. *Science* 112:649–51.

Sollmann, T. 1948. *A manual of pharmacology.* 7th ed. Philadelphia: Saunders.

———. 1957. *A manual of pharmacology.* 8th ed. Philadelphia: Saunders.

Williams, R. J. 1963. *Biochemical individuality.* New York: John Wiley and Sons.

Zeise, W. C. 1834. Das Mercaptan, nebst Bermerkungen über einige neue Producte aus der Einwirkung der Sulfurete auf weinschwefelsaure Salze und auf das Weinöl. *Ann. Physik. Chem.* 31:369–431.

ADDITIONAL READINGS

Crosland, M. P. 1962. *Historical studies in the language of chemistry.* London: Heinemann Educational Books.

Joannes, de Sancto Amando. 1893. *Die Areolae des Johannes de Sancto Amando (13. Jahrhundert).* Ed. J. L. Pagel. Berlin.

13

Analytical Methods

When the severity of occupational exposure to mercury (or anything else) is such that it regularly produces classical manifestations of intoxication, diagnosis becomes obvious and presents no difficulties. In modern industrial hygiene the objective is to eliminate all adverse effects, not only severe ones. This creates real problems, first, in recognizing even the slightest degree of poisoning; second, in establishing acceptable limits of exposure; and third, in applying control measures that will eliminate unacceptable exposures. Two of these problems are primarily medical, the third, engineering. All require the assistance of the laboratory, or of its projection in the form of instruments used in the field. Current concepts of toxicology make it necessary to measure mercury in the range of parts per billion, and modern technology makes it possible.

Early Attempts at Quantitative Analysis

Laboratory procedures for diagnosing mercury poisoning came into common use among forensic toxicologists toward the middle of the 19th century. By present-day standards these older methods would seem rather crude, their sensitivities being in the range of 1:8000 to 1:15000. When gross quantities of corrosive sublimate were involved they were identified by the formation of crystals; smaller amounts were detected by means of amalgamation with copper and subsequent condensation in the form of mercury globules. Commonly the tests were performed on the contents of the stomach or intestines in cases where suicide or murder was suspected. Little was known and little was done about detecting mercury in urine. Accounts of the kind of chemical evidence accepted in the mid-19th century as bases for conviction or acquittal of suspected poisoners can lead to no other conclusion than that justice had a somewhat shaky foundation. It was believed, for example, that a person might be fatally poisoned by corrosive sublimate and no mercury be found in the organs if a period of four days or more had elapsed between the time of administration and the time when the analyses were performed (Taylor 1875).

Laboratory analysis of saliva for the detection of mercury was practiced at least as early as 1875, amalgamation with copper being the accepted method. According to Taylor, "In a case of inunction with mercury the metal was thus detected in the saliva on the third day. . . . This analysis of the saliva may not only furnish evidence that the patient was under the influence of mercurial poisoning, but it will prove, in a case otherwise doubtful, whether the salivation from which a person is suffering is owing to mercury or some other cause."

Industrial hygienists of the 20th century have the syphilitics of the 19th century to thank for some of their most useful analytical procedures. Interest in the absorption and excretion of mercury used in the inunction treatments stimulated research into precise, sensitive methods of measuring mercury in urine and other biological materials. An appropriate starting point from which a direct line can be traced may be chosen, somewhat arbitrarily, perhaps, in the work of Neubauer, published in 1878. It is true that Neubauer refers to earlier methods of Ludwig (1877), Fresenius (1875) and Merget (1871), but the work of these and other chemists was directed primarily toward analytical methods for mercury in general, with no particular application to biological materials.

Neubauer's procedure uses zinc or copper to amalgamate the mercury in the urine with subsequent driving off of the mercury with heat and reacting the vapor with iodine, thus forming crystals of mercuric iodide. While essentially qualitative, the method can be used quantitatively by liberating the mercury and measuring the size of the droplet under a microscope. Refinements of this so-called microscopic-volumetric technique were to follow, reaching their widest application in the hands of Stock a half century later (vide infra).

Early efforts to develop a colorimetric method for the quantitative determination of mercury in biological materials are represented by the work of Klein, published in 1889, using the Nessler reaction. Nessler's reagent, which contains mercury, is ordinarily used to determine nitrogenous bodies. Klein used the action in reverse, with Nessler's reagent providing the mercury and the "unknown" sample providing the nitrogen, on the assumption that mercury in the tissue would intensify the yellow or yellow-brown color characteristic of the reaction. This application of the Nessler reaction is more ingenious than useful, as it is not a sensitive method for measuring mercury quantitatively. It is of interest historically, as mentioned above, because it introduces the concept of colorimetric procedures into the laboratory study of the absorption and excretion of mercury in man.

A method depending on isolation of urinary mercury by amalgamation and a final gravimetric determination was described by Hofmeister in 1889. This procedure requires a minimum of 1,500 milliliters of urine, the sample being made to trickle over copper foil at the rate of 50 drops per minute. The apparatus for this is illustrated in Figure XIII-1. After primary amalgamation with copper, the mercury is driven off by heat and the vapor passed over previously weighed gold foil. The added weight after this step gives the amount of mercury. The sensitivity of the method is not stated, but Hofmeister notes that the highest value for mercury found in medical specimens was 0.0028 grams per liter.

Another early, and somewhat crude, attempt to devise a colorimetric procedure was devised by Vignon in 1893. This method depends on the formation of a brown color when mercury reacts with hydrogen sulfide. The author claims that the sensitivity is in the parts per 10,000 range, but the method is neither precise nor specific. It illustrates, however, the striving of analysts toward the application of colorimetry to the quantitative determination of mercury in urine.

Interested in measuring the amount of mercury excreted and retained during the mercurial treatment of syphilis, Jolles, in 1895, presented a method which he describes as simple (*einfach*) and sensitive (*empfindlich*). He refers to older methods which depend on the reduction of mercury by tin chloride as being unsatisfactory and expresses preference for amalgamation. For this he employed gold, his procedure being a simplification of that of Hofmeister. Jolles claimed that he could measure as little as 0.0002 grams of mercury in 100 milliliters of urine and is one of the first analysts in this field to report recovery experiments to validate his method. Returning the compliment, Hofmeister in 1896 published a modification of the Jolles method.

More directly related to modern analytical techniques than any so far described is a method published by Caseneuve in 1900. This is an early application of diphenylcarbazide to the quantitative colorimetric determination of heavy metals, including mercury. The author specifically and significantly mentions the transformation of diphenylcarbazide into diphenylcarbazone on contact with salts of mercury and copper. He notes that mercuric diphenylcarbazone has a strong blue color and that the complex is soluble in benzine and chloroform. A sensitivity in the parts per million range is claimed.

Concern over the practicality of older methods is expressed by Schumacher and Jung (1900) who point out that as much as three or four days may be required for the completion of an analysis. They describe a procedure for determining mercury in urine which can be completed in 24 hours, using amalgamation on gold, with pre- and post-amalgamation weighing of the gold. The applicability of this method to biological substances other than urine is suggested. Authors label their technique not only as simple (*einfach*) but also as reliable (*zuverlässig*), perhaps indirectly casting doubt on earlier work. Their claim to reliability is supported by recovery experiments which by modern standards would not be impressive.

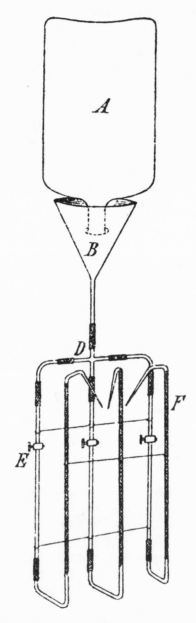

Figure XIII-1 Hofmeister's Apparatus. Urine is fed through a manifold (D) into tubes (F) containing copper foil, and mercury is extracted by amalgamation. (From F. Hofmeister in *Z. Anal. Chem.* 28: 753–756 [1889].)

The apparatus used by Schumacher and Jung is shown diagrammatically in Figure XIII-2. The purpose of reproducing this diagram is for comparison with one portraying a method developed about 60 years later (Fig. XIII-3).

A combination of amalgamation and electrolysis forms the basis of a method described by Jänecke in 1904. He collected all of the urine passed by his test subjects for fifty days, amounting to an average of 75 liters per case! (The problem of storage must have been formidable.) Analyses were performed on aliquots of 250 milliliters each containing from 0.01 to 1.0 milligrams of mercury, these amounts being commonly found in patients under mercury therapy. Primary separation is performed by amalgamation with a spiral of copper wire, the mercury next being driven off by heat and the vapors trapped in a solvent mixture. The dissolved mercury is then electrolyzed, with the use of a gold wire as the cathode. The amount of mercury is calculated on the

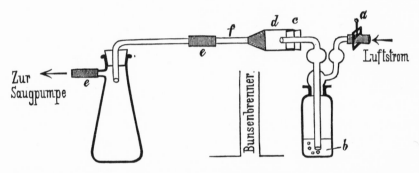

Figure XIII-2 Mercury Analysis Apparatus of Schumacher and Jung (1900). Compare with Fig. XIII-3. (From Schumacher [II] and W. L. Jung in Z. Anal. Chem. 39: 12–17 [1900].)

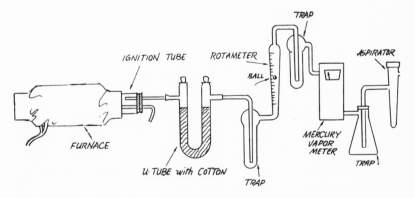

Figure XIII-3 Apparatus for analysis of Mercury in Apples (1961). (From Morris B. Jacobs and Leonard J. Goldwater in Food Technol. 15[8]: 357–360 [1961]. Copyright © 1961 by Institute of Food Technologists.)

basis of the weight of the gold wire before and after electrolysis. Sensitivity is given in the range of around 0.01 milligrams of mercury in two liters of urine.

After reviewing the early amalgamation methods, Raaschou in 1910 expressed the opinion that they were not usually completely quantitative and that for precise results either accurate weighing or an improved colorimetric technique was required. He implies that as yet neither of these had been perfected. His article contains a useful summary of many of the early analytical methods: gravimetric, colorimetric, and microscopic-volumetric, as well as those using amalgamation. In his own modification of the microscopic-volumetric approach, the final quantitative step is the measurement of the size of a mercury droplet. The sensitivity of Raaschou's method is not clearly stated but it appears to be less than 0.2 milligrams of mercury per liter of urine.

Of the analytical procedures developed during the latter part of the 19th century and which found wide application, the microscopic-volumetric and the colorimetric (diphenylcarbazone) methods are most significant. The microscopic-volumetric technique was refined by Stock and used by him in his extensive studies (Chap. 10). Some industrial hygienists in the United States still rely on this method, to which the name of Stock has become attached.

Colorimetric methods received great impetus from the work of H. Fischer when he demonstrated the grouping of metals in their reactions with diphenylthiocarbazone (Dithizone). The formation of metal-dithizone complexes having characteristic colors is the basis of this method (Fischer 1929). Various modifications of the original procedure have been made from time to time, some of these being given official status by the American Conference of Governmental Industrial Hygienists (n. d.) and the Association of Official Agricultural Chemists (Horwitz 1960). At present (1971) the dithizone method for determining small amounts of mercury in biological materials is still used but is rapidly being supplanted by flameless atomic absorption spectrophotometry.

Modern Techniques

Writing in 1963, Stokinger expressed the opinion that "A spectrographic method suitable for the quantitative determination of microgram quantities of Hg in types of specimens of industrial hygiene interest has not been reported." This does not mean that such methods had not been used before nor that they have not been used since that time. One of the most extensive studies of industrial mercurialism, that conducted by the U.S. Public Health Service in the fur-felt and felt-hat industries (Chap. 19, below), depended on spectrographic analysis in determining the amount of mercury in the urine of exposed workers. The anomalous results found in the reports of this study may have been part of the evidence used by Stokinger to support his opinion. At least some of the shortcomings of direct spectrographic analysis of biological materials are eliminated in a method described by Dal Cortivo and associates in 1964, utilizing the time-honored technique of amalgamating the mercury with copper before the sample is placed in the arc. This represents a distinct improvement over the older methods but because of its complexity it is not likely to find large-scale application in medicine, toxicology, or industrial hygiene.

Recent trends in quantitative analysis have been away from purely chemical methods and toward physical methods. Spectrographic analysis illustrates this trend, but because of technical difficulties it does not lend itself to routine use, particularly when simplicity and rapidity are desired. A physical method which depends on the well-known phenomenon of absorption by mercury vapor of ultraviolet light of 2536.5Å wave length has found a useful place in industrial hygiene and toxicology. This is commonly spoken of as photometric detection and measurement but it could just as well be designated atomic absorption since it involves the absorption of ultraviolet light by atoms of mercury (Yamaguchi and Matsumoto 1968). Separation of the mercury from its organic matrix is generally necessary, but a method which eliminates this troublesome procedure has been described (Magos and Cernik 1969).

An instrument for the direct measurement of mercury vapor in air, based on ultraviolet absorption, was described by Woodson in 1939 and patented in the following year. A similar device, designed to measure the concentration of certain chlorinated hydrocarbons in air, was developed in 1941 (Hanson) but in practice this instrument was found to be more sensitive to mercury vapor than to the compounds for which it was intended. Further modifications and improvements have led to the design and construction of mercury vapor meters of sufficient simplicity to warrant their commercial development by instrument manufacturers. Small, portable instruments weighing but a few pounds have been made available and are in wide use, being calibrated so that direct readings can be made in the field. Difficulties due to absence of a central power supply and to setting the zero point on these instruments in contaminated areas have been overcome (Jacobs and Jacobs 1965), and adaptations have been devised which permit their use in the analysis of biological materials in the laboratory (Jacobs et al. 1960; Jacobs, Goldwater, and Gilbert 1961; Kudsk 1965; Smith, Vorwald and Patil 1969).

Detailed descriptions of analytical methods applicable to biological materials are found in standard reference works on industrial hygiene laboratory methods, notably in the comprehensive book by Jacobs (1967).

Advances in electronics and nuclear physics have resulted in the development of analytical methods of great sensitivity and specificity which are applicable to biological as well as to inorganic materials. A great advantage of some of these techniques is that the specimen being analysed need not be damaged or destroyed, as it is in chemical analysis. A disadvantage is that expensive instruments and highly skilled operators are needed.

One of the newer methods is electron microanalysis. The essential apparatus consists of an electron source, a means of focusing the electron beam, an x-ray spectrometer and scanning mechanisms. In performing an analysis, the electron beam is focused on the sample, thereby producing excitation of atoms. The excited atoms emit x-rays having patterns which are unique or specific for each element, as shown by measurement in the x-ray spectrometer. This permits positive identification with a sensitivity down to 10^{-12} grams (Ogilivie 1967). To date there have been no reported applications of electron microanalysis in toxicology or industrial hygiene or to the determination of mercury in biological materials, but it is not unlikely that such uses will be developed.

Neutron activation analysis is another non-destructive technique of high sensitivity and specificity but it requires the use of an atomic reactor and complicated auxiliary instruments. With this method of quantitative analysis the sample is irradiated with a neutron beam which imparts radioactivity to the elements so treated. Every element has a characteristic decay curve or half-life which can be measured by means of gamma-ray spectroscopy. Neutron activation analysis has been applied to the determination of mercury in biological materials (Sjöstrand 1964; Christell et al. 1965; Erwall 1967; Brune 1967; Lambert and Goldwater [unpublished] 1969) and may be expected to be developed further. Erwall (1967) has found it possible under routine conditions to measure as little as 0.1 nanograms in samples weighing a few hundred milligrams, signifying a sensitivity of 0.5 ng/g.

Tracer techniques using radioactive mercury, ^{197}Hg and ^{203}Hg, have become standard methods used by neurosurgeons for localizing tumors of the brain. A radioactive isotope of mercury (^{203}Hg) has also been administered to human volunteers for purposes of studying the metabolism of methyl mercury compounds (Aberg et al. 1969). There can be little doubt that the nuclear physicist will play an increasingly important role in laboratory studies of the behavior of mercury in man.

At this time (1971) flameless atomic absorption spectrophotometry appears to be coming to the fore as the method of choice in the quantitative analysis of mercury in biological samples. As is true for most other methods, total mercury is measured. For the identification of organic mercury compounds, gas chromatography and mass spectrometry are being employed, not without difficulties. Technical refinements have not yet reached a point where one laboratory can regularly duplicate the findings of another; in fact, many laboratories have difficulty in reproducing their own results.

Biochemical Tests

More than a century ago Claude Bernard wrote:

When we reach the limits of vivisection we have other means of going deeper and dealing with the elementary parts of organisms where the elementary properties of vital phenomena have their seat. We may introduce poisons into the circulation, which carry their specific action to one or another histological unit . . . poisons are veritable reagents of life, extremely delicate instruments which dissect vital units . . . I am of the opinion that studious attention to agents which alter histological units should form the common foundation of general physiology, pathology and therapeutics (Bernard 1857).

These concepts have direct application in the study of overt and suspected cases of mercury poisoning, by directing attention to the effects of an agent rather than to the agent itself.

In modern industrial hygiene practice severe occupational poisoning is unacceptable; cases of intoxication exhibiting classical manifestations have become increasingly rare, although they have not disappeared entirely. This applies to mercury as well as to all other materials presenting occupational health hazards. Greater attention is being paid to possible subtle effects on vital histological and biochemical units.

Many of the studies of the effects of mercury at the cellular or subcellular level have been performed on experimental animals. These investigations will not be given detailed consideration here, not because they are unimportant, but simply because predictions for man based on animal experimentation are often unreliable (Brodie, Cosmides, and Rall 1965). On the other hand, animal experiments serve a useful purpose in suggesting where in man to look for the effects of functional or structural derangement.

Affinity between mercury and sulfur, known to the alchemists and reflected in the word *mercaptan*, serves as the basis for much of the work that has been done in searching for evidence of subtle effects on histological and biochemical units. This is due primarily to the importance of sulfur, and in particular to the vital role of thiols, in many essential enzymes (Hughes 1957). Of the many enzyme systems in man which may be affected by mercury, those involved in the syntheses of heme have been most extensively studied: d-aminolevulinic acid dehydrase, porphobilinogen deaminase, heme synthetase, uroporphyrinogen decarboxylase and coenzyme A. Effects of mercury on these enzymes individually have not been studied in man but rather the possible disturbance of the synthesis of heme, as reflected by urinary excretion of d-aminolevulinic acid (ALA) and of coproporphyrin. Reports published to date have not shown any significant abnormalities in the urinary excretion of ALA in factory workers exposed to relatively high concentrations of mercury (Goldwater and Joselow 1967) but increased coproporphyrin excretion has been found (Goldwater and Joselow 1967; Suzuki 1962; Tsuchiya 1964).

Enzyme studies of a different type have been performed on a group of seed dressers who had been exposed primarily to phenylmercurials (Taylor, Guirgis, and Stewart 1969). Glutathione reductase in blood serum showed no diminution, but phosphoglucose isomerase activity was definitely reduced. The activity of serum cholinesterase and the level of reduced glutathione in erythrocytes have also been studied in man as possibly indicators of sub-clinical mercury intoxication (Wada et al. 1969).

Before the advent and ready availability of barbiturates, bichloride of mercury was a popular suicidal agent. Taken by mouth in doses of more than half gram it usually produced the desired result. Local corrosive effects would occur in the mouth, esophagus, stomach and even in the intestines, but the cause of death was renal failure. The kidney lesion was predominantly necrosis of the renal tubules resulting in pronounced albuminuria followed by total anuria and death in azotemia. This dramatic and horrible form of death was responsible for the formation in the minds of physicians of a strong association between mercury and albuminuria (proteinuria).

Occupational poisoning resulting from the inhalation of mercury and some of its compounds over a period of months or years produces manifestations quite different from those caused by ingestion of large doses of mercuric chloride. One of the important differences is the reaction of the kidneys in that albuminuria is not a prominent feature. Occupational mercurialism with marked albuminuria and a nephrotic picture has been known to occur (Friberg, Hammarstrom, and Nystrom 1953; Goldwater 1953; Bidstrup 1964; Kazantzis 1965; Smith and Wells 1960), but it is relatively uncommon. Lesser degrees of albumin in the urine also are found, but for the most part persons having occupational exposure to mercury do not exhibit sufficient urinary albumin to permit its detection by routine procedures. When more sensitive quantitative analytical methods are used, protein can be found in the urine of practically all human beings, and when groups having occupational exposure to mercury are compared with similar persons who have had no exposure, slightly, but significantly, greater quantities of albumin are found in the former. This justifies the assumption that mercury has had an effect on the kidneys but it does not tell whether the effect is a result of a physiological disturbance or of loss of structural integrity, that is, a pathological change in kidney cells. Whether the abnormality is transitory or permanent also is not revealed. The fact that cases of severe nephrosis caused by occupational exposure to mercury have made apparently complete recoveries suggests that at least in some instances the kidney lesion responsible for slight increases in urinary protein is of a temporary nature, but additional long-term observations are necessary before definite conclusions can be drawn.

Realization of the fact that different compounds of mercury may have totally different toxicological properties has stimulated research to distinguish between the different forms of mercury that may be present in biological materials. Recognition of the possibility of biochemical synthesis of mono- and di-methyl mercury by aquatic microorganisms has given added impetus to this line of investigation since the alkyl compounds are so much more toxic than the inorganic forms which undergo this biotransformation. Gas chromatography, thin layer chromatography, and mass spectrometry are the laboratory procedures used in distinguishing between organic (alkyl) and inorganic mercury compounds (Yamaguchi and Matsumoto 1969; Yamaguchi et al. 1969; Jensen and Jernelöv 1968; Widmark 1967; Merten and Wortley 1967). When combined with an electron capture detector, the sensitivity of gas chromatography is in the nanogram range. Thin layer chromatography is sensitive only in the microgram range.

Quantitative analysis of urine for mercury, introduced in the 19th century, became and has remained a standard industrial hygiene procedure in studying problems of exposure and intoxication. Probably because of the technical difficulties in removing interfering substances and in isolating the mercury, blood analyses did not come into common use until the middle of the 20th century (Goldwater, Ladd, and Jacobs 1964). The first studies in which analyses of blood for mercury were performed in significant numbers were those of Beani in 1955 and of Turrian in 1956. Following the development of an analytical method which gives reliable results on blood samples of one milliliter or less (Jacobs et al. 1960; Jacobs, Goldwater, and Gilbert 1961), it became practicable to analyze mercury in blood as a routine procedure. This is an important adjunct to studies of mercury in urine, particularly when there has been exposure to alkyl mercurials (Goldwater, Ladd, and Jacobs 1964).

Mercury in Air

Harmful effects from the inhalation of mercury were recognized in the 15th and 16th centuries but methods for demonstrating the presence of the vapor were not described until the middle of the 19th century. A textbook published in 1875 states that:

> Noxious effects may be apprehended when any operations with metallic mercury are carried on in small and ill-ventilated apartments, heated to a temperature above $70°$. The best test for the detection of this vapor is suspension of a slip of pure gold-leaf in the apartment. If mercury be present, this will become slowly whitened by amalgamation. It is easy to prove by this experiment in a closed vessel that the mercury is volatilized at all temperatures (Taylor 1875).

The principle of allowing mercury vapor to impinge on a reactive surface as a means of measuring atmospheric concentrations was employed by Nordlander in a device he described in 1927. Instead of gold-leaf he used paper coated with selenium sulfide, which turns black when in contact with mercury vapor. The degree of blackening of the selenium sulfide paper is dependent on the rate of flow of air striking it, the concentration of mercury vapor and the time of exposure of the paper. The time and rate of air flow can be controlled so that the degree of blackening can be compared with known standards. The instrument is simple, and when properly operated gives reasonably good accuracy and sensitivity. It measures vapor only; additional tests must be performed to measure mercury which may be present in the form of dust.

For measuring total mercury in air a commonly used method is that in which a measured sample of air is passed through a solution of iodine-potassium iodide in an absorption tube. The resulting compound is treated with a mixture of copper sulfate and sodium sulfide and compared colorimetrically with standard solutions containing known amounts of mercury. Another colorimetric method for determining total mercury in air uses dithizone, the air having been passed through water or a water-alcohol mixture as the medium for trapping the mercury (Jacobs 1967).

When the presence of the classical manifestations of mercurialism was the sole criterion in establishing a diagnosis of mercury poisoning, the task was simple. Subclinical effects were not considered and in fact were not sought for. Now that even the most minute deviations from normal can be detected through the laboratory it has become increasingly difficult to interpret the findings. Unless the most scrupulous care is exercised in the collection, handling and analysis of samples, and in watching the elusive decimal point in the calculations, serious errors can creep in.

Analytical methods developed originally for use in industrial hygiene and toxicology found wide applicability in broader environmental fields starting in 1970.

REFERENCES

Aberg, B.; Ekman, L.; Falk, R.; Greitz, U.; Persson, G.; and Snihs, J. O. 1969. Metabolism of methyl mercury (^{203}Hg) compounds in man. *Arch. Environ. Health* 19:478-84.

American Conference of Governmental Industrial Hygienists, Committee on Analytical Methods. N. d. *Analytical methods manual.* Cincinnati.

Beani, L. 1955. Contributo allo studio des mercurialismo cronico. *Med. lavoro* 46:633-45.

Becker, C. G.; Becker, E. L.; Maher, J. F.; and Schreiner, G. E. 1963. Nephrotic syndrome after contact with mercury. *Arch. Internal Med.* 110:178-86.

Bernard, C. 1857. *Leçons sur les effets des substances toxiques et médicamenteuses.* Paris.

Bidstrup, P. L. 1964. *Toxicity of mercury and its compounds.* Amsterdam: Elsevier Publishing Co.

Brodie, B. B.; Cosmides, G. J.; and Rall, D. P. 1965. Toxicology and the biomedical sciences. *Science* 148:1547-54.

Brune, D. 1967. Low temperature irradiation applied to neutron activation analysis of mercury in human whole blood. *Oikos Suppl.* 9:12-13.

Cazeneuve, P. 1900. Sur la diphénylcarbazide, réactif très sensible de quelques composés métalliques: cuivre, mercure, fer au maximum, acide chromique. *Bull. Soc. Chim. Paris* 23:701-6.

Christell, R.; Erwall, L. G.; Ljundgren, K.; Sjöstrand, B.; and Westermark, T. 1965. Methods of activation analysis for mercury in the biosphere and in foods. In *Proceedings of the 2d International Conference on Modern Trends in Activation Analysis,* 1965, A. and M. College of Texas, College Station, Texas, pp. 380-82.

Dal Cortivo, L. A.; Weinberg, S. B.; Giaquinta, P.; and Jacobs, M. B. 1964. Mercury levels in normal human tissue: I. Spectrographic determination of mercury in tissue. *J. Forensic Sci.* 9:501-10.

Erwall, L. G. 1967. Determination of mercury by means of neutron activation analysis. *Oikos Suppl.* 9:11.

Fischer, H. 1929. Über den Nachweis von Schwermetallen mit Hilfe von "Dithizon." *Z. Angew. Chem.* 42:1025-27.

Fresenius, C. R. 1875. Zur Analyse der Schwefelwasser. *Z. Anal. Chem.* 14:321–24.

Friberg, L.; Hammarstrom, S.; and Nystrom, A. 1953. Kidney injury after chronic exposure to inorganic mercury. *AMA Arch. Ind. Hyg.* 8:149–53.

Goldwater, L. J. 1953. Kidney injury after chronic exposure to inorganic mercury. *AMA Arch. Ind. Hyg.* 8:588.

Goldwater, L. J., and Joselow, M. M. 1967. Absorption and excretion of mercury in man: XIII. Effects of mercury exposure on urinary excretion of coproporphyrin and delta-aminolevulinic acid. *Arch. Environ. Health* 15:327–31.

Goldwater, L. J.; Ladd, A. C.; and Jacobs, M. B. 1964. Absorption and excretion of mercury in man: VII. Significance of mercury in blood. *Arch. Environ. Health* 9:735–41.

Hanson, V. E. 1941. Ultraviolet photometer: Quantitative measurement of small traces of solvent vapors in air. *Ind. Eng. Chem. Anal. Edition* 13:119–23.

Hofmeister, F. 1889. Zur quantitativen Bestimmung von Quecksilber im Harn. *Z. Anal. Chem.* 28:753–56.

————. 1896. Nachweis und Bestimmung von Quecksilber im Harn. *Z. Anal. Chem.* 35:635–37.

Horwitz, W., ed. 1960. *Official methods of analysis.* 9th ed. Washington: Association of Official Agricultural Chemists.

Hughes, W. L. 1957. A physiocochemical rationale for the biological activity of mercury and its compounds. *Ann. N.Y. Acad. Sci.* 65:454–60.

Jacobs, M. B. 1967. *The analytical toxicology of industrial inorganic poisons.* New York: Interscience Publishers.

Jacobs, M. B., and Jacobs, R. 1965. Photometric determination of mercury vapor in air of mines and plants. *Am. Ind. Hyg. Assoc. J.* 26:261–65.

Jacobs, M. B.; Goldwater, L. J.; and Gilbert, H. 1961. Ultramicrodetermination of mercury in blood. *Am. Ind. Hyg. Assoc. J.* 22:276–79.

Jacobs, M. B.; Yamaguchi, S.; Goldwater, L. J.; and Gilbert, H. 1960. Microdetermination of mercury in blood. *Am. Ind. Hyg. Assoc. J.* 21:475–80.

Janecke, E. 1904. Über eine Methode zur quantitativen Bestimmung und zum Nachweis zehr geringer Quecksilbermengen im Harn unter Zuhilfenahme der Nernst'wage. *Z. Anal. Chem.* 43:547–52.

Jensen, S., and Jernelöv, A. 1968. Biosynthesis of mono- and dimethyl mercury. In *Symposium on mercury*, 4–7 November 1968, Stockholm.

Jolles, A. 1895. Über eine einfache und empfindliche Methode zum qualitativen und quantitativen Nachweis von Quecksilber im Harn. *Monatsch. Chem.* 16:684–92.

Kazantzis, G. 1965. Chronic mercury poisoning: Clinical aspects. *Ann. Occupational Hyg.* 8:65–71.

Klein, J. 1889. Über den Nachweis des Quecksilbers als Oxydimercuriammoniumjodid. *Arch. Pharm.* 227:73–77.

Kudsk, F. N. 1965. Determination of mercury in dithizone extracts by ultraviolet photometry. *Scand. J. Clin. Lab. Invest.* 17:171–77.

Ludwig, E. 1877. Eine neue Methode zum Nachweis des Quecksilbers in thierischen Substanzen. *Medizin. Jahrbüch.* pp. 143–151.

Magos, L., and Cernik, A. A. 1969. A rapid method for estimating mercury in undigested biological samples. *Brit. J. Ind. Med.* 26:144–49.

Merget, A. 1871. Sur la diffusion des vapeurs mercurielles. *Compt. Rend.* 73:1356–61.

Merten, D., and Wortley, G. 1967. *Report on an "Expert Meeting on Mercury Contamination in Man and His Environment,"* 13 May 1967, Amsterdam.

Neubauer, C. 1878. Eine neue Methode zum Nachweis von Quecksilber in thierschen Substanzen. *Z. Anal. Chem.* 17:395–408.

Nordlander, B. W. 1927. Selenium sulfide: A new detector for mercury vapor. *Ind. Eng. Chem.* 19:518–21.

Ogilvie, R. E. 1967. Electron microanalysis and the history of art. *Technol. Rev.* Dec.: 21–23.

Raaschou, P. E. 1910. Eine mikrochemische Quecksilberstimmungmethode. *Z. Anal. Chem.* 49:172–204.

Schumacher (II), und Jung, W. L. 1900. Eine einfache und zuverlässige Methode zur quantitativen Bestimmung des Quecksilbers im Harn. *Z. Anal. Chem.* 39:12–17.

Sjöstrand, B. 1964. Simultaneous determination of mercury and arsenic in biological and organic materials by activation analysis. *Anal. Chem.* 36:814-19.

Smith, J. C., and Wells, A. R. 1960. A bio-chemical study of the urinary protein of men exposed to metallic mercury. *Brit. J. Ind. Med.* 17:205-8.

Smith, R. G.; Vorwald, A. J.; and Patil, L. S. 1969. A study of the effects of exposure to mercury in the manufacture of chlorine. Paper presented at a meeting of the American Industrial Hygiene Association, 15 May 1969, Denver, Colorado.

Stokinger, H. E. 1963. Mercury. In *Industrial hygiene and toxicology*, ed. F. A. Patty, D. W. Fassett, and D. D. Irish, 2d rev. ed., vol. 2. New York: Interscience Publishers.

Suzuki, T. 1962. Exposure to inorganic mercury and urinary excretion of coproporphyrin. *Japan J. Exp. Med.* 32:45-53.

Taylor, A. S. 1875. *On poisons in relation to medical jurisprudence and medicine.* 3d American ed. Philadelphia: Henry C. Lea.

Taylor, W.; Guirgis, H. A.; and Stewart, W. K. 1969. Investigation of a population exposed to organomercurial seed dressings. *Arch. Environ. Health* 19:505-9.

Tsuchiya, K. 1964. Coproporphyrins in lead and mercury workers. *Ind. Health* 2:162-71.

Turrian, H.; Grandjean, E.; and Turrian, V. 1956. Industriehygienische und medizinische Untersuchungen in Quecksilbertrieben. *Schweiz. Med. Wochschr.* 86:1091-96.

Vignon, L. 1893. Dosage de mercure dans les solutions étendu de sublimé. *Compt. Rend.* 116:584-86.

Wada, O.; Toyokawa, K.; Suzuki, T.; Suzuki, S.; Yano, Y.; and Nakao, K. 1969. Response to a low concentration of mercury vapor. *Arch. Environ. Health* 19:485-88.

Widmark, G. 1967. Analytical problems: Mercury in the general environment: Introduction. *Oikos Suppl.* 9:9-10.

Woodson, T. T. 1939. A new mercury vapor detector. *Rev. Sci. Instr.* 10:308-11.

Yamaguchi, S., and Matsumoto, H. 1968. Ultra-microdetermination of mercury in biological materials by atomic absorption photometry. *Japan. J. Ind. Health* 10:29-37.

——. 1969. Ultra-microdetermination of alkylmercury compounds by gas chromatography. *Kurume Med. J.* 16:33-42.

Yamaguchi, S.; Matsumoto, H.; Hoshide, M.; and Akitake, K. 1969. Microdetermination of organic mercurials by thin-layer chromatography. *Kurume Med. J.* 16:53-56.

ADDITIONAL READINGS

Joselow, M. M. and Goldwater, L. J. 1967. Absorption and excretion of mercury in man: XII. Relationship between urinary mercury and proteinuria. *Arch. Environ. Health* 15:155-59.

14

Early Medical Uses

Pre-Galenic Period

Conclusions, even opinions, on the early medicinal use of mercury compounds depend heavily on such considerations as understanding of terminology, reliability of dates, and our knowledge of the availability of materials through trade and commerce in the past. In China and India, for example, terminology may not be a serious problem, but separation of fact from legend is. In Assyria and Egypt dates can be established with reasonably good accuracy, but terminology presents great difficulties; this is true also for Greece and Rome. Some of the problems in terminology carry over into medieval times and even later, due largely to the persistent influence of Greek and Latin works such as those of Dioscorides and Pliny.

Although one may eschew "preoccupation with priorities," an historical work has legitimate concern with "who did what and when." At the same time, it is desirable to distinguish several degrees of reliability: conjecture, opinion, reasonable conclusion and established fact, realizing that as new information becomes available, some things may move from one category to another. Published works dealing with the history of mercury in medicine have not always followed these simple principles.

Egypt

Except for the "Mercury of Kurna," the ancient dating of which is open to most serious question (Chap. 5), there is no evidence that quicksilver was used in medicine, or was even known, in Egypt in dynastic times. The case of cinnabar (mercury sulfide) is less clear, hinging on terminology.

Relatively speaking, a rich supply of Egyptian medical writings from the period of about 1900 B. C. to about 1200 B. C. is available for study. There is much duplication and overlapping among the eight principal medical papyri, and some of them are merely small fragments. The most important sources of information on the Egyptian materia medica are the Ebers and the Hearst papyri, particularly the former (Leake 1952).

The Egyptian terms prš and mnšt are of possible relevance in the present discussion. Ebers (Ebers Papyrus 1852, 1875 translations) repeatedly uses the German "Mennige" as his translation of mnšt and gives its chemical composition as Bleioxyd or red oxide of lead. In another late 19th century interpretation of the Ebers Papyrus (1890), Joachim lists more than thirty prescriptions which contain "Mennige" and notes that these formulations were used both internally and externally for a variety of ailments. Commenting on this, Partington (1935) says that "Minium is sometimes used internally and thus could not have been red lead: it is perhaps MILTOS, i.e., red ochre; Berthelot thinks, improbably, that it is sometimes litharge [PbO]." Partington's theory that minium (presumably the Egyptian pršt and the German Mennige) could not have been red lead because it was taken internally would presuppose that the Egyptians were familiar with the toxicology of lead and mercury compounds, something which cannot be assumed. Dawson (1964) seems to have followed the same reasoning as Partington.

Bryan's commentary on the Ebers papyrus (Ebers Papyrus 1931 translation) is based on the Joachim version and does nothing to help clarify the meaning of pršt. The German Mennige is translated as red lead, with no consideration of other possible meanings, and the mention of "haematite from Elefantine" introduces another element of confusion. Following Bryan by a few years, Ebbell published a full translation of the Ebers papyrus (1937 translation) which Clendening (1960) calls excellent and which Sigerist (1967) chose to follow in his discussion of empirico-rational medicine in Egypt. Ebbell translates mnšt as red ochre and notes its use in a number of prescriptions. He also mentions minium as though it were not the same as mnšt, but he does not give its Egyptian equivalent. Both minium and hematite are on Dawson's list of mineral drugs used in ancient Egypt, with no indication of the original Egyptian terms for these two compounds (Dawson 1964).

The lexicon of Egyptian drug names prepared by von Deines and Grapow, and published in 1959, contains a comprehensive review of terminology, interpretations and uses of ancient Egyptian medicaments, including pršt and mnšt. This book, although useful as a compendium of previously published work, contributes nothing to a clarification of the meaning of the two terms. No help is found in the earlier work by Stern nor in that by Alfred Schmidt. Von Meyer does not include mercury in his list of Egyptian drugs.

At least two students of ancient Egyptian materia medica have suggested the possibility that cinnabar was used. In a detailed discussion of the Hearst papyrus, Leake (1952), after noting the many similarities between it and the Ebers papyrus, states that "Some of the ingredients in the Hearst Papyrus are not satisfactorily classified. For example, 'dragon's blood' (pršt), which occurs in prescriptions 104, 124, and 229, may be either cinnabar, or red mercuric sulfide, or, on the other hand, Resina draconis from an East Indian palm tree. This confusion regarding 'dragon's blood' has persisted in European formularies through medieval to relatively modern times."

In a book that is more popular than scholarly in style, Ghalioungui (1963) gives a list of drugs used by the ancient Egyptians which includes ". . . salts of lead, copper, mercury and antimony. . . ." He does not give the Egyptian equivalent of any of these compounds.

One striking similarity is found among all of the authors cited above: they all fail to mention any attempt to establish the nature of mnšt or pršt by means of chemical analysis. Unless and until this is done there can be nothing but conjecture on the use of mercury compounds in the medical practice of ancient Egypt.

Assyria

Assyrian physicians at the time of Ashur-bani-pal (7th century, B. C.) and possibly two or three centuries earlier used in their prescriptions materials known as IM.KAL. GUG. and kalgukku (Thompson 1936; and A. L. Oppenheim 1969 [personal communication]). Thompson holds that "IM.KAL. GUG corresponds to the use of mercury satisfactorily . . ." and that it was used in treating diseases of the eyes and ears. He also states that it was used to strengthen uterine contractions in difficult labor, as well as for intestinal and pulmonary disorders. Some of the Assyrian prescriptions called for its administration in beer or in oil and beer. Oppenheim, of the Oriental Institute of the University of Chicago, on the other hand, maintains that the meaning of IM.KAL. GUG and the related kalgukku is by no means clear and almost certainly is not mercury. Medicinal uses similar to those described by Thompson are recognized by the Chicago group, so the question of Assyrian use of mercurials in medicine hinges on the meaning of these terms.

China

Early Chinese medical history is a morass of uncertainty. According to Haeser, definite information on Chinese medicine goes back only to 240 B. C. Earlier data can be typified largely as hearsay and myth and much of what has been published is sheer nonsense. One monograph connects Taoist influence with writings allegedly originating twenty-seven centuries B. C. and follows this with the assertion that "Avicenna in India used the five clays in a strikingly similar manner to the Chinese and that may indicate a diffusion of culture from China to India" (Morse 1938).

Wang Chi Min, another historian of Chinese medicine, thinks that reliable history dates from the middle of the Dhow (Chou) Dynasty, or about 722 B. C. which he says, incredibly, was the age of Lao-tze, Confucius, and Mencius. More acceptable is his statement that Chinese medicine begins properly in the

Han Dynasty, circa 170 B. C. He lists Tsang Kung, Chang Chung-King and Hua To as the three great figures of the period, but makes no mention of the use of mercury or cinnabar as a drug, even though the latter was known in China at least as early as the 18th Century B. C.

The Yellow Emperor's Classic of Internal Medicine (Huang Ti Nei Ching Su Wen) which is believed by some sinologists to date from the 26th century B. C. and by others between the 5th and 3rd centuries (Veith 1967), does not mention any form of mercury. Statements in this work to the effect that people of the West are most successfully cured with poison medicines and that poisons and medicines attack the evil influences might be interpreted as meaning that mercurials were used, but this would be sheer conjecture.

No useful purpose would be served in belaboring the fact of uncertainty in the early history of Chinese medicine. There is no credible evidence to support any claim that the Chinese were the first to employ mercury or cinnabar as a medicine, even though actually they may have been. The difficulties are due not only to uncertainty of dates but also to the virtual impossibility of separating fact from legend.

India

Attempts to establish the early history of the use of mercury in Hindu medicine lead quickly to confusion and ultimately to frustration. This was recognized by Sigerist (1961, p. 182) when he said: "The great Indian medical collections of Charaka, Susruta and Vaghbata . . . like so many other Indian books, present great chronological problems . . . There can be no doubt that they were written relatively late, not before the beginning of our era, but it is equally certain that their content must be much older, the result of a long, for the most part oral, tradition." A noted Indian historian, P. Ray, states the case tersely and even more pessimistically: ". . . the chronology relating to the compilation of many ancient Indian treatises—literary, scientific or religious— is in a state of hopeless confusion."

On the other hand, Neuberger and Pagel, in their monumental three volume work express a less skeptical point of view. In speaking of Indian medicine they say: "Mercury (parada) was prized above all metals. The Indians seem to have been the first to use mercury as a medicine. It was even mentioned as such by Caraka and Susruta both for external and internal use. Quicksilver was supposed to have superhuman powers . . . Quicksilver was used against all possible sicknesses, particularly for diseases of the skin, fevers (pox), nerve disorders, lung diseases and later against syphilis." In the absence of any dis-cussion or documentation of dates this claim of priority cannot have unquali-fied acceptance. Singer's suggestion (Singer and Underwood 1962) that mer-cury preparations may have been of Indian origin is purely conjectural.

A recent work by Filliozat (1964) illustrates very well some of the reasons for confusion in dating early Indian medical writings. This author states that the "principal and oldest texts of the classical Indian medicine are the

Samhitas, the corpus said to be of Bhela, Caraka and Susruta" but that the origins of these are lost in an ancient maze of mysticism, mythology and magic.

The difficulties in interpreting events in early Indian medical history are similar to those of China, namely, uncertainty of dates and confusion of fact with mythology. It is therefore impossible, at least on the basis of present knowledge, to establish a date or period when mercury in any form was first used in Indian medicine.

Greece

The only important source of information on the possible use of mercury in the medical practice of classical Greece is the Hippocratic "writings." (This designation is being used in order to avoid the controversy over what is properly embraced in the "genuine" works of Hippocrates and what more suitably should be spoken of as the "Hippocratic Corpus.")

Most medical historians who have discussed the Hippocratic materia medica either fail to mention mercury or state that it was not used by Hippocrates or by members of his school (Sigerist 1967; Schmidt 1924; Haeser 1875-82; Neuberger and Pagel 1902-5; Dieterich 1837). This conclusion is true if what is meant is quicksilver (argyros chytos), but may not be true if it is meant to exclude all forms of mercury.

One reason for presenting a detailed discussion of the Greek word *miltos* in Chapter 5 and in the Appendix is that this term appears at least twice in Hippocratic prescriptions. In the Littré edition these prescriptions appear in Volume 6, pages 427 and 459 respectively. Both are ointments, the first for treating burns and the second for ulcers or fistulas. The burn ointment is made up of old pork fat, wax, oil, incense, lotos scrapings, *miltos*, and arum leaves cooked in wine and oil; the prescription for fistulas is simply *miltos* and honey.

Neuberger and Pagel include "Rothel" in their list of mineral drugs used by Hippocrates. Presumably they have translated *miltos* in this way, which indicates that they considered it to be ruddle or iron oxide.

In his book *Materia Medica Hippocrates*, Raudnitz mentions both *Kinnabaris* and *Miltos*. The former he describes as *Sanguis draconis*, a vegetable resin, but he avoids entering into the controversy over the meaning of *miltos*, simply using a Latinized form. (The work is in Latin.)

Even more interesting than the Hippocratic use of *miltos* is that of *Kinnabaris*. Raudnitz, who based his work on the Foesius edition of Hippocrates (Geneva: 1657), says about Kinnabaris: *"Sanguis draconis, resina est ex fructibus Calami Draconis Willd., fam. Palmae, patr. Sumatra: quam Hipp. ad balanos pueros purgandi cum oleo lentisco et adipe anserino usurpavit* (635)."

Littré used a different text from that followed by Raudnitz and this is probably the reason for differences between these two scholars. Raudnitz, for example, gives a Hippocratic prescription containing *miltos* to be used against puerperal uterine hemorrhage, an application which is not found in Littré. On

the other hand, the word *Kinnabari* in the Greek text used by Littré is translated by him as "Sang-dragon" which places him in agreement with Raudnitz
as to the vegetable origin of this material. The use of kinnabari mentioned by
Littré is similar to that described above by Raudnitz in that it is mixed with
goose grease and made into suppositories for treating respiratory troubles in
infants. This seems more reasonable than the "balanos" of Raudnitz. Littré
comments that some scholars consider this passage to be apocryphal but that
he does not.

It is difficult to understand why both Raudnitz and Littré translated the
Greek *kinnabari* as dragon's blood and neither commented on discussions
found in Pliny and elsewhere on the meaning of this term. The preponderance
of evidence is that the Greek *kinnabari* was in fact mercury sulfide (Latin
minium) and not the same as the Latin *cinnabaris* which is a resinous material
derived from trees, *sanguis draconis* or dragon's blood.

Dierbach, in *Die Arzneimittel des Hippokrates*, devotes two pages to a discussion of the meaning of *miltos* and concludes that it is quite possible that
this name was given to a variety of entirely different substances. There can be
no quarreling with his opinion on *miltos*. His discussion of "Kinnabaris," however, is quite a different story.

"Kinnabaris," according to Dierbach, is "Drachenblut" and consequently is
listed under plants and is classified as a strong astringent. Hippocrates, he says,
used it in suppositories (Stuhlzapfgenmasse) and in this respect Dierbach is in
agreement with Littré and Raudnitz. This leads to speculation as to the extent
to which the three scholars influenced each other, particularly since they all
made the same mistake. The difficulty seems to arise as a result of putting an
"s" on the end of *Kinnabari*, resulting in a Latin rather than a Greek translation.

As his authority for concluding that "kinnabaris" is dragon's blood,
Dierbach cites the Latin translation of Dioscorides by Sarracenus, and gives
the spelling as *cinnabaris*. In the original Greek, the word is *kinnabari*, which
is not the same as the Latin *cinnabaris*. The latter is almost certainly dragon's
blood and the former mercury sulfide. Dierbach apparently chose to follow
Sarracenus into error and to ignore Theophrastus and Pliny who were quite
clear on the nature of these compounds. A careful reading of Dioscorides in
the original Greek would have alerted him to the danger.

A case for the Hippocratic use of mercury sulfide (modern cinnabar) can
be built on these points:

1. Kinnabari almost certainly, and miltos possibly, meant mercury sulfide
   (cinnabar) in the time of Hippocrates.
2. Cinnabar was known in Asia Minor, in India and on the Island of Syra
   at least a thousand years before the time of Hippocrates.
3. There was active trade between India and Greece as well as within the
   Grecian world before and at the time of Hippocrates.
4. The materia medica of Hippocrates certainly contained several drugs
   which were imported from India, such as cardamon, sesame, and cinna-

mon, thus demonstrating a trade in pharmaceuticals between the two
countries.

5. Mercury sulfide (cinnabar) was definitely known in the Greek world in
the days of Aristotle and Theophrastus, a mere hundred years or so
after the time of Hippocrates and during the days when his school still
flourished.

6. Hippocratic prescriptions made use of both kinnabari and miltos.

Indirect evidence such as that outlined above proves nothing but it does
raise a real possibility that mercury sulfide was used by the Father of Medicine
or by his disciples. The problem is more literary than chemical since much depends on whether or not the passage describing cinnabar suppositories is in
fact a part of the Hippocratic writings.

Rome

Two Roman writers of the 1st century A. D., Pliny and Celsus, have left
evidence which is well nigh incontrovertible that mercury sulfide (cinnabar,
minium) was used in the medicine of that period. Pliny was definitely not a
physician and the medical qualifications of Celsus have been questioned. The
latter, nevertheless, has been described by Castiglioni as "surely the most
powerful and intelligent mind in the medical history of classical Italy" and as
"the greatest of Latin medical writers." One of Celsus' best known contributions to medical science was his description of the four cardinal signs of inflammation so familiar to all medical students: Calor, Dolor, Rubor and
Tumor. His *De Medicina* was the first book on general medicine to be printed
with movable type, its editio princeps having been published in Florence in
1478.

Gibbon (n. d., Vol. 1, p. 316) referred to the 37 books of the *Historia
Naturalis* as "that immense register, where Pliny has deposited the discoveries,
the arts, and the errors of mankind . . ." In the light of present-day knowledge
it is not difficult to separate most of the discoveries and the arts from the errors. Pliny's accounts of quicksilver, minium and cinnabar are for the most
part credible, particularly their medicinal applications, and it must be remem-
bered that Pliny's *minium* is our cinnabar or mercury sulfide and that his
cinnabaris is the resin otherwise known as dragon's blood.

In Book XIII, in the section on unguents, perfumes and cosmetics Pliny
states that *minium* is often employed to impart color to these preparations
and that its mixture with oil of roses and oil of saffron is a practice of recent
origin.

Misuse of *minium* afforded Pliny an opportunity to lash out at incompe-
tent physicians of his day. He is highly critical of those who do not understand the art of compounding theriacs and mithradatics and consequently
prescribe *minium* in place of *cinnabari Indicae.*

Among the many prescriptions appearing in the *De Medicina* of Celsus
some call for *minium* and others for *minium Sinopicum.* There is no way of

knowing with certainty whether these terms refer to the same or to different substances. Book V contains lists of drugs classified according to their therapeutic effects and Section five of this book mentions about 50 substances which have a purging action, including *minium* and *lapis haematites*. This might be interpreted as signifying that Celsus distinguished between derivatives of quicksilver and those of iron. The next section contains a long list of erodents (caustics), including *minium Sinopicum,* but not *haematites*. This is followed by a list of exedents (destroyers of tissue) which includes *lapis haematites* but not *minium* or *minium Sinopicum,* and still another list of caustics which contains none of these. *Minium* and *minium Sinopicum* never appear in the same prescription, which suggests that they are the same thing.

Several of Celsus' uses of *minium* or *minium Sinopicum* are of special interest from the historical point of view. One (Book IV. 22) calls for a mixture of *minium* and salt water to be administered as an enema in the treatment of intestinal cancer. This represents an early internal, but not oral, use of the drug. Oral administration is suggested by the inclusion of *minium* among the medicines which have a purging effect (Book V. 5). The local application of a salve containing *Sinopic minium* is recommended in the treatment of ulceration of the genitals (Book VI. 18). This could be an early use of a mercurial against venereal infections. *Minium* in the treatment of trachoma (Book VI. 6) may reflect the influence of Egyptian medicine, and its use in ointments the influence of Hippocrates, for, as Singer has pointed out (Singer and Underwood 1962), in the writing of *De Re Medica* (sic): "Many of its phrases are closely reminiscent of the Hippocratic Collection."

The question of proper interpretation of *minium* might be raised if one wished to cast doubt on the use of mercury by Celsus. Since he apparently distinguished between haematite and *minium* and since his well-known contemporary Pliny clearly used *minium* to signify mercury sulfide, it is not unreasonable to conclude that Celsus also meant mercury sulfide when he spoke of *minium*.

A third important contributor to the pharmaceutical literature of the first century A. D. was Pedanios Dioscorides of Anazarba. Having been born in Asia Minor, Dioscorides is generally considered to have been Greek, and his *De Materia Medica,* originally written in Greek, was not translated into Latin until the fifth century. He is strongly identified with the Roman world by reason of his military service in the armies of Nero.

His *Materia Medica,* also known as the *Greek Herbal,* is a sort of pharmacopoeia in five books containing all available information on drugs from the vegetable, animal and mineral kingdoms. Several chapters in his Book V are devoted to "metallic stones," including *Kinnabari* and *Udrarguros.* Dioscorides points out clearly that something which he calls *Ammium* is the same as *Kinnabari* and that it comes from Spain, thereby almost certainly establishing its identity as mercury sulfide. He comments on the poisonous properties of the vapours of *ammium* and of *udrarguros,* noting that milk is an effective antidote against poisoning by the latter.

Dioscorides often uses language which bears a strong resemblance to that employed by Pliny. This may have been due to these two contemporaries being familiar with each other's work, but possibly is a reflection of their both drawing on a common source of information, the writings of Theophrastus. The present-day practice of plagiarism in the preparation of textbooks may thus have been in vogue 2000 years ago.

Medicinal uses of mercurials do not occupy a prominent place in the *Materia Medica* of Dioscorides. In the chapter on *Kinnabari* he says that ". . . it hath the same virtue that Haematitis hath, being good for eye-medicines and that more effectually, for it is more binding and blood-stanching, and being taken with Cerat it heals burnings and ye breaking out of pustules" (1934 translation).

Of quicksilver (*hydrargyrum-udrarguros*) Dioscorides writes that when ingested it eats "through ye inward parts, by its weight. But this is helped by much milk being dranck, or wine with wormwood, or the decoction of smallach, or with the seed of Horminium, or with Origanum, or with Hyssop with wine" (1655 translation).

Alexandrian School

For a period of about five hundred years, starting in the fourth century B. C., the principal seat of medical learning was the so-called Alexandrian School. Sudhoff made an extensive study of a number of papyri dating from that period and published a monograph on the subject in 1909, with a chapter devoted to perfumes and drugs. Several mineral drugs are mentioned, but nothing is said about quicksilver, cinnabar, minium, miltos, ochre or any substance that is in any way related to mercury. This negative evidence does not rule out the possibility that mercurials were used in that time and place, but it strongly suggests that they could not have occupied a prominent position.

Medieval Medicine

There is no evidence that Galen used any form of mercury therapeutically. On the contrary, he considered mercurials to be poisonous, thus establishing himself as one of the first anti-mercurialists. The authoritative position which Galen held in the medical world for more than a thousand years undoubtedly did much to keep mercury in disrepute during that time. Although mercury did come into prominent use in European medicine at the end of the 15th century, it was not until Paracelsus publicly burned the books of Galen and Avicenna in Basle in 1527 that the new iatrochemistry began to offer a serious challenge to Galen's doctrines.

Galen died about 200 A. D., but his all-embracing influence on medical thought and practice persisted for some 1500 years after his death, thus spanning the period traditionally called medieval and extending for some time thereafter. In spite of the slavish devotion to Galen on the part of most of those who followed him, quicksilver did not go into total eclipse, as can be seen from the writings of prominent medieval physicians. Nothing new was added, however, the medical men, like the encyclopedists, being little more than compilers and copyists.

Among the medical writers of the fourth century, Oribasius (325-403) is the best known. Since he probably derived most of his information from Hippocrates, Dioscorides and Galen (Campbell 1926) it is not surprising to find something, but not much, about quicksilver in his writings. Oribasius mentions both *hydrargyros* (Vol. 4, p. 628) and *kinnabari* (Vol. 2, p. 715). The latter he describes as acrid, with astringent properties.

A few lines are given to *argentum vivum* by Aetius of Amida (502-575) (Book 13, Chap. 79, p. 392) but he does not mention *kinnabari* or any other form of mercury. His contemporary, Alexander of Tralles (525-605) on the other hand, recommends *kinnabari* for the treatment of gout (Vol. 2, p. 557) but say nothing about *hydrargyrum*. A century later Paul of Aegina (625-690) devotes a short paragraph to *argentum vivum* and another to *kinnabari* in words which suggest the influence of Oribasius.

These few brief notes reflect the unimportant position occupied by mercurials in early medieval medicine. It was known, but not widely used.

Possibly as a result of the growing influence of Arab medicine after the 10th century, mercury gradually became more acceptable in the Latin West. Of great importance in the transition was the famous School of Salerno which had its beginnings in the ninth century (Singer and Underwood 1962). Among the physicians and scholars who were attracted to this cosmopolitan institution was Constantine the African (c. 1010-1087) who had studied medicine in Tunisia. He has been credited with being the first to introduce the Arabic *materia medica* into Europe (Multhauf 1966) thus initiating a move toward iatrochemistry which was to reach a climax with Paracelsus 500 years later. Both Constantine and his contemporary at Salerno, Nicolaus Praepositus (fl. c. 1050) included *argentum vivum* in prescriptions used for a variety of skin diseases (Dieterich 1837).

Mercury ointments were supposedly prescribed by Roger of Palermo (fl. c. 1170-1200) for treating chronic diseases of the skin (Singer and Underwood 1962), and in the 13th century, Theodoric of Cervia (1205-1298) and Arnold of Villanova (1240-1311) joined the ranks of the early mercurialists. The latter is better known as an alchemist than as a physician, another example of the close relationships between the two professions during the Middle Ages.

A summary of drugs and prescriptions known to physicians in the 13th century was compiled by John of St. Amand (fl. 13th cent.) and published in his *Areola* or compendium (Joannes 1893). This encyclopedic work, by a man whom Neuberger (1910-1925) called "One of the most learned physi-

cians of the day" quotes extensively from Galen and Avicenna and to a lesser
degree from Aristotle, Hippocrates, Alexander of Tralles and from the Arab
physicians Rhazes, Serapion, Averroes, Isaac Judaeus, Haly Abbas, Mesue, Sr.,
and Alkindus. No mention is made of Pliny, Dioscorides, Scribonius Largus or
Celsus, which means that as far as quicksilver is concerned there are important
omissions.

Numerous references to quicksilver (*argentum vivum*), its uses and effects,
both beneficial and harmful, are found in the *Areola.* For the most part,
prescriptions containing mercury are for external use in the treatment of
scabies, lice and the itch. He states that *argentum vivum* can be used in small
doses or mixed with other medicines and it will not be harmful. Among the
harmful effects, John mentions that the vapors of *argentum vivum* will destroy the eyesight and likewise will destroy hearing. Bad breath, paralysis, and
tremors are other toxic effects listed. Quicksilver is not included in the lists of
medicines which are bad for the stomach, which harm the liver, act as vermi-
fuges or stimulate the flow of urine.

Saint Thomas Aquinas was not the only pupil of Albertus Magnus to
distinguish himself in his own right. Less well known, but more important in
the history of quicksilver, is Pope John XXI, 187th occupant of the Throne of
Saint Peter (Kirsch 1913). Born in Lisbon about 1210 and enthroned in
September 1276, John was killed by the collapse of a faulty roof about eight
months after his elevation. After studying with Albertus, Petrus Hispanus, as
John XXI was then called, continued his interest in medicine and was appointed
physician-in-ordinary to Gregory X (St. Gregory) in 1272. Before that, however, he had been professor of medicine at the University of Siena and while
in that position compiled a collection of prescriptions which he published in
his *Thesaurus Pauperum,* in which mercury is mentioned frequently in ointments to treat body lice and other skin diseases. John XXI represents the
only instance in history of a professor of medicine progressing to the papacy.

Neuberger (1910-1925 translation) states that Guy de Chauliac
(c. 1300-1367) describes the treatment of chronic ulcers by the application
of a plate of lead coated with quicksilver.

Islamic Medicine

Galen's death (c. 200 A. D.) can be used as a convenient milestone or
turning point in the history of medicine. Coinciding as it did with the beginnings of the disintegration of the Roman Empire it marked the end of what
Singer has called "the observational period of antiquity" and signaled the
beginning of an era when "the medical writers became mere compilers from
the works of former authors." Had it not been for Arabic writers, some of the
works of Hippocrates, Galen and perhaps others would have been forever lost.
This, of course, would have resulted in even larger gaps than now exist in the

history of medicine, and consequently in the history of quicksilver. Arabian contributions, although strongly influenced by Galen (Campbell 1926), were not limited merely to the preservation of the older medical knowledge; they also included some important original observations on the part of a number of prominent physicians. Mercury did not go unnoticed.

As noted above, Hippocrates probably used mercury sulfide, but Galen did not use mercury at all. Surviving documents show that Arab physicians were familiar with the *Materia Medica* of Dioscorides, but there are no records to indicate whether or not they knew Celsus. Contact between Arab and Indian medicine is known to have existed at least as early as the latter part of the eighth century A. D. since there were Indian physicians in Baghdad during the reign of Haroun-al-Rashid, who was Caliph from 786 to 809. By that time the alchemical-medicinal use of mercurials had become well established in India. The possibility of contacts between Arab and Chinese medicine before or during the "Golden Age" of Arabian culture (750–850 A. D.) both directly and perhaps indirectly through India has been suggested. On the other hand, there is no evidence of direct transmission of the medical knowledge of ancient Assyria, Persia or Egypt to the Arab physicians (Campbell 1926).

Any information on quicksilver which the medical practitioners of Islam may have gotten from Galen or from early medieval physicians such as Oribasius, Aetius, Alexander, or Paulus would certainly not have encouraged them to use this material. If an explanation for the prominence of mercury in Arab medicine is to be sought, it is more likely to be found in the East than in the West. The intimate relationship between Islamic medicine and alchemy parallels that which had developed in India starting in the fourth or fifth century and strongly suggests that the latter influenced the former. Neuberger (1910–25 translation) refers to a ninth century Arab physician, Abu Mansur Muwaffak ben Ali Harawi, commonly known simply as Mansur, as ". . . the author of a *materia medica,* the importance of which lies in the fact that not only Indian remedies, but also Indian medical principles are given . . . ." This work is said to describe 585 drugs, of which 75 are of mineral origin.

According to Campbell (1926) there were at least 400 Arabic medical authors of repute among whom a dozen or so achieved pre-eminence. Chaucer knew of Haly, Serapion, Razis, Avicen, Averroes and Damascien (1919).

One of the earliest of the Arab physicians to prescribe quicksilver was al-Kindī (c. 800–870 A. D.). His *Medical formulary* or *Aqrābādhīn* (al-Kindī 1966 translation) contains a prescription calling for aquatic costus, male frankincense, castoreum, musk and pure quicksilver. These drugs are boiled after sieving with the mercury and introduced into the urethra as a remedy against urethral discharge and calculi. This remedy, as well as most of those prescribed by al-Kindī, is said to cure the disease "with the help of Allah." The use of a mercurial in the treatment of what seems to be gonorrhea is interesting in relation to the early confusion between this disease and lues venerea (syphilis).

A later Arab formulary, that of al-Samarqandi (d. ca. 1222) lists vermilion and quicksilver among the useful drugs (1967 translation).

The first recorded account of animal experimentation on the toxicity of quicksilver comes from Rhazes (Abu Bekr Muhammad ibn Zakariya Ar-Razi [ca. 850–ca. 923]): "I myself gave an ape quicksilver to drink and I have only observed the effects which I have just mentioned (pains in the belly and intestines). I found out about these pains by conjecture when the ape twisted about and clutched at his belly with his hands and his mouth." (Iskandar 1959).

Rhazes says that he does not believe quicksilver to be very harmful to humans, except for the symptoms mentioned above. "It is, " he states, "passed out unchanged, especially if the patient moves about" (Iskandar 1959). He disagrees with those ancient physicians who claim that the effects of ingesting quicksilver are like those from litharge (lead oxide). But: "As to calcinated mercury, especially the quality obtained by sublimation, it is very harmful and indeed fatal. Its sharpness excites very severe pains about the belly, causes colic and bloody stools."

According to Rhazes, quicksilver poured into the ear will cause severe injury.

> When mercury is poured into the ear, severe pains accompanied by delirium and convulsions occur. Moreover, there will be a feeling of a heavy weight within the side (of the head) into which the mercury is poured. It is likely, however, that no bad effects are brought about by pouring mercury into the ear, if the mercury happens to flow back (immediately). But if some mercury reaches the narrow passages of the ear, bad effects should be expected, for some physicians inform me that they have seen such patients smitten by epilepsy followed by apoplexy (Iskandar 1959).

The Jew, Moses Maimonides [= Moses ben Maimon] (1135–1208) is important in Arab medicine. His extensive writings include a *Treatise on Poisons and Their Antidotes* which is disappointing as far as quicksilver is concerned. Maimonides discusses lead, arsenic, copper (verdigris), and cobalt but does not mention mercury.

Best known among the Arab physicians is the Persian Abu Ali al-Husain ben Abdallah ibn Sina, familiarly called Avicenna (980–1037). His writings were voluminous and included the *Canon Medicinae* which ". . . served as the main textbook of medicine, both among the Arabic-speaking peoples and in the Latin West, until the 17th century" (Singer and Underwood 1962). Unfortunately only a small part of this encyclopaedic work has been translated into English (Gruner 1930), thus limiting its usefulness in historiography.

Avicenna has been credited with being the first to describe mercurial tremor (Proksch 1874) and he also lists weakness, foul breath, blindness, and deafness as toxic manifestations caused by quicksilver (Browne 1962). The *Canon* devotes major attention to diseases of the skin and male generative organs and presumably Avicenna prescribed mercury in some form for these ailments. He is said to have recommended the use of mercurials, including sublimate, only as an external remedy (La thérapeutique d'Avicenne 1917; Sprengel 1815-20 translation). According to Sprengel, Avicenna regarded corrosive sublimate as the most violent of all poisons.

REFERENCES

Aetius Amidenus. 1533. *Aetii Antiocheni medici: De cognoscendis et curandis morbis sermones sex.* Jam primum in lucem editi, interprete Janus Cornarius. De ponderibus et mensuris, ex Paulus Aegineta, eodem interprete. Basel: In Officina Frobeniana.

Alexander, Trallianus. 1878-79. *Alexander von Tralles: Original-text und übersetzung nebst einer einleitenden Abhandlung.* Trans. Dr. T. Puschmann. 2 vols. Vienna: W. Braumüller.

Browne, E. G. 1962. *Arabian medicine.* Cambridge: Cambridge University Press.

Campbell, D. 1926. *Arabian medicine and its influence on the middle ages.* 2 vols. London: Kegan Paul, Trench, Trubner and Co.

Castiglioni, A. 1941. *A history of medicine.* Trans. and ed. E. B. Krumbhaar. New York: A. A. Knopf.

Celsus, A. C. 1935-38. *De medicina.* Trans. W. G. Spencer. 3 vols. London: W. Heinemann.

Chaucer, G. 1919. *Complete works.* Ed. W. W. Skeat. Oxford: Oxford University Press.

Clendening, L. 1960. *Source book of medical history.* New York: Dover Publications.

Dawson, W. R. 1964. *The beginnings: Egypt and Assyria.* New York: Hafner Publishing Co.

Deines, H. von, and Grapow, H. 1959. *Wörterbuch der Ägyptischen Drogennamen.* Berlin: Akademie Verlag.

Dierbach, J. H. 1824. *Die Arzneimittellehre des Hippokrates.* Heidelberg.

Dieterich, G. L. 1837. *Die Merkurialkrankheit in allen ihren Formen.* Leipzig: Wigand.

Dioscorides, P. 1934. *The Greek herbal.* Englished by John Goodyer, A. D. 1655. Edited and first printed, A. D. 1933, by Robert Gunther. Oxford: At the University Press.

Ebers Papyrus. 1852. *Papyrus Ebers: Die Maasse und das Kapitel über die Augenkrankheiten.* Leipzig: Sächsische Akademie der Wissenschaften.

Ebers Papyrus. 1875. *Papyros Ebers: Das hermetische Buch über die Arzneimittel der alten Ägypter in hieratischer Schrift.* Herausgegeben mit Inhalts Angabe und Einleitung versehen, von G. Ebers. Mit hieroglyphisch-lateinischem Glossar von L. Stern. 2 vols. Leipzig: W. Engelmann.

Ebers Papyrus. 1890. *Papyros Ebers: Das älteste Buch über Heilkunde.* Aus dem aegyptischem zum erstenmal vollständig übers, von dr. med. H. Joachim. Berlin: G. Reimer.

Ebers Papyrus. 1931. *Papyrus Ebers.* Trans. from the German version by C. P. Bryan. New York: D. Appleton and Co.

Ebers Papyrus. 1937. Trans. B. Ebbell from W. Wreszinski's hieroglyphic transcript. Copenhagen: Levin and Munksgaard.

Elgood, C. 1934. *Medicine in Persia.* New York: Paul B. Hoeber.

Filliozat, J. 1964. *The classical doctrine of Indian medicine.* Delhi: Munshiram Manoharlal.

Ghalioungui, P. 1963. *Magic and medical science in ancient Egypt.* London: Hodder and Stoughton.

Gibbon, E. N. d. *The decline and fall of the Roman Empire.* New York: Modern Library.

Gruner, O. C. 1930. *A treatise on the canon of medicine of Avicenna.* London: Luzac and Co.

Haeser, H. 1875-82. *Lehrbuch der Geschichte der Medicin und der epidemischen Krankheiten.* 3d rev. 3 vols. Jena.

Hippocrates. 1839-61. *Oeuvres complètes.* Trans. and ed. E. Littré. 10 vols. Paris: J. B. Baillière.

Iskandar, A. Z. 1959. A study of Ar-Razi's medical writings with selected texts and English translations. Ph. D. Dissertation, 1959, University of Oxford.

Joannes, de Sancto Amando. 1893. *Die Areola des Johannes de Sancto Amando (13. Jahrhundert)*. Ed. J. L. Pagel. Berlin.

al-Kindi. 1966. *The medical formulary or Aqrābādhin of al-Kindi*. Trans. M. Levey, with a study of its materia medica. Madison: University of Wisconsin Press.

Kirsch, J. P. 1913. In *The Catholic encyclopedia*, vol. 8, pp. 429–31. New York: Encyclopedia Press.

Leake, C. D. 1952. *The old Egyptian medical papyri*. Lawrence, Kansas: University of Kansas Press.

Meyer, E. S. C. von. 1905. *Geschichte der Chemie von den altesten Zeiten bis zu Gegenwart*. Leipzig: Veit.

Morse, W. R. 1938. *Chinese medicine*. New York: Paul B. Hoeber.

Moses ben Maimon. 1966. *Treatise on poisons and their antidotes*. Ed. S. Muntner. Philadelphia: J. B. Lippincott Co.

Multhauf, R. P. 1966. *The origins of chemistry*. London: Oldbourne Book Co.

Neuberger, M. 1910–25. *History of medicine*. Trans. E. Playfair. Vol. 1., London: Henry Frowde, Oxford University Press; vol. 2, pt. 1, Humphrey Milford, Oxford: Oxford University Press.

Neuberger, M., and Pagel, J. 1902–05. *Handbuch der Geschichte der Medizin*. 3 vols. Jena: Gustav Fischer.

Oribasius. 1851–76. *Oeuvres*. Trans. U. C. Bussemaker and C. Daremberg. 6 vols. Paris.

Partington, J. R. 1935. *Origins and development of applied chemistry*. London: Longmans, Green and Co.

Paulus Aegineta. 1844–47. *The seven books of Paulus Aegineta*. Trans. with commentary by Francis Adams. 3 vols. London: Sydenham Society.

Plinius Secundus, C. 1938–63. *Natural history*. Trans. H. Rackham, W. H. S. Jones and D. E. Eichholz. 10 vols. London: W. Heinemann.

Proksch, J. K. 1874. *Der Antimercurialismus in der Syphilis-Therapie*. Erlangen.

Raudnitz, J. M. 1843. *Materia medica Hippocratis*. Dresden.

Ray, P. 1956. *History of chemistry in ancient and medieval India,* incorporating *The history of Hindu chemistry,* by A. P. C. Ray. Calcutta: Indian Chemical Society.

al-Samarqandi, Najib al-Din Muḥammad ibn 'Ali. 1967. *The medical formulary of al-Samarqandi*. Trans. M. Levey and N. al-Khaledy. Philadelphia: University of Pennsylvania Press.

Schmidt, A. 1924. *Drogen und Drogenhandel im Altertum*. Leipzig: Barth.

Sigerist, H. E. 1961. *A history of medicine*. Vol. 2. New York: Oxford University Press.

Sigerist, H. E. 1967. *Primitive and archaic medicine*. New York: Oxford University Press.

Singer, G., and Underwood, E. A. 1962. *A short history of medicine*. 2d ed. New York: Oxford University Press.

Sprengel, K. 1815–20. *Histoire de la médecine*. Trans. from the 2d ed. A. J. L. Jourdan. 9 vols. Paris.

Stern, L. 1875. Glossarium hieroglyphicum. In *Papyros Ebers,* ed. G. Ebers, 2 vols. Leipzig: W. Engelmann.

Sudhoff, K. 1909. Ärztliches aus griechischen Papyrus-Urkunden. *Studien Gesch. Med.* 5/6:1–296.

La thérapeutique d'Avicenna. 1917. *Rev. Gen. Clin. Therap.* 31 (suppl.) :916–17, 989–91.

Thompson, R. C. 1936. *A dictionary of Assyrian chemistry and geology*. Oxford: Clarendon Press.

Veith, I. 1967. *Huang Ti Nei Ching Su Wen: The Yellow Emperor's classic of internal medicine*. Baltimore: Williams and Wilkins Co.

ADDITIONAL READINGS

Aretaeus. 1856. *The extant works of Aretaeus, the Cappadocian*. Ed. and trans. Francis Adams. London: Sydenham Society.

Hoernle, A. F. R. 1907. *Studies in the medicine of ancient India*. Oxford: Oxford University Press.

Huard, P., and Ming Wong. 1968. *Chinese Medicine*. Trans. Bernard Fielding. London: World University Library.

Marsili, A. 1956. *Scribonio Largo Ricetta.* Pisa: Edizioni "Omnia Medica."
Partington, J. R. 1965. *A short history of chemistry.* 3rd ed. New York: Harper and Row, Harper Torchbooks.
Schonack. W. 1913. *Die Rezepte des Scribonius Largus.* Jena: Gustav Fischer.
Simon, F. A. 1860. *Geschichte und Schicksale der Inunktionskur.* Hamburg.
Stillman, J. M. 1960. *The story of alchemy and early chemistry.* New York: Dover Publications.
Suṣruta. 1897. The Suṣruta-samhitā. Trans. by Dr. A. F. R. Hoernle. Calcutta.
Wang Chi Min. 1926. China's contribution to medicine in the past. *Ann. Med. Hist.* 8:192-201.
Zimmer, H. R. 1948. *Hindu Medicine.* Baltimore: Johns Hopkins Press.

15

Mercury and Syphilis

Much of the super-abundant literature on the history of syphilis and on the treatment of this disease with mercury has been based on an assumption that early writers had been correct in their diagnoses. As is well known, lues venerea has been properly called "the great imitator," and even with modern diagnostic techniques it is not always easy to establish a diagnosis. From time to time the question has been raised about confusing syphilis with leprosy, particularly prior to the 16th century. Leprosy, too, has protean manifestations and presents its own problems of diagnosis. When sanitation and simple bodily cleanliness are deficient, secondary infections of the skin and intercurrent systemic infections are very likely to be superimposed on individuals suffering from any chronic wasting disease, thus adding further difficulties to diagnosis. Gonorrhea was not conclusively distinguished from syphilis until near the middle of the 19th century, and it required the painstaking studies of Kussmaul, published in 1865, to establish that the late manifestations of lues were not due to over-treatment with mercury. With few exceptions, clinical descriptions of disease pictures before the 17th or even the 18th centuries are not sufficiently precise to permit of anything more than speculation on the part of latter-day diagnosticians.

Another frequently-made assumption is that mercury has definite and "specific" action in curing syphilis. It is true that there are records of many patients who were presumed to have had the disease and who showed improvement or even cure following one or more courses of treatment with quicksilver. Modern standards of clinical investigation would demand something more than this type of *post hoc* assumption, particularly in view of the fact that many persons with proven syphilis recover spontaneously and never develop detectable late manifestations.

In the ensuing discussion, when syphilis or one of its synonyms is mentioned, the word is understood to be qualified by "alleged" or "presumed."

Early History

As early as 1615 the literature on the treatment of syphilis with mercury had reached sufficient proportions to attract the scornful attention of Miguel Cervantes. (At that time the disease was variously called morbus gallicus; French, Spanish and Neapolitan disease; the pox or great pox; lues venerea, etc.). In *Don Quixote*, Part II, Book 3, Chapter XXII, the following passage appears: "Virgilius [Polydorus] forgot to tell us who was the first man in the world to have a cold in the head, or the first to take inunctions for the French disease, all of which I explain most accurately, citing the authority of more than twenty-four authors." Sancho replies: "Tell me, sir, . . . tell me if you can, since you know everything, who was the first man that ever scratched his head?"

These words of Cervantes apparently had no effect on physician-authors of the subsequent two and one-half centuries, for Proksch, in 1874, was impelled to remark: "To know who was the first to recommend and use this treatment against syphilis is much less important than one-tenth the cost of the paper that has been filled with scribbling on the subject."

Inextricably interwoven into the question of mercury therapy for the pox is the even more troublesome problem of the early history of the disease itself. A typical 19th century comment is that of Haeser who said: "There is hardly any question in historical pathology so frequently investigated and answered in so many different ways as that dealing with the origin of syphilis in antiquity."

That the situation had shown no change in 1969 is illustrated by the remark of Crosby that ". . . the matter of the origin of syphilis is doubtlessly the most controversial subject in all medical historiography." The crux of the controversy has to do with whether or not syphilis was introduced into Europe by members of Columbus's crew on their return from the New World in 1493.

Students of ancient Chinese medicine have claimed that chancres were recognized in China as early as the seventh century A. D. and that they were treated with mercury fumigations, but this view is not widely accepted.

During the period of the Crusades (1096-1270 A. D.) many of the crusaders returned to Europe with what was called leprosy and some of them who were treated with the mercurial Unguentum Saracenum were allegedly cured. The composition of Unguentum Saracenum given by Guy de Chauliac (d. 1368) (Neuberger 1910-25 translation) was:

Euphorb. et Lythargyri	ana lib. ss
Staphid. agriae	quartam ss
Argenti vivi	quartam I
Axunguiae porci veteris	lib. I

Of course, mercury ointment is effective in the treatment of a variety of skin diseases, particularly those caused by parasites. A successful therapeutic test with mercury does not prove that the disease treated was lues venerea.

Fifteenth and Sixteenth Centuries

Printed books dealing with syphilis began to appear during the closing
years of the 15th century (Sudhoff 1912b and 1925). At first a mere trickle,
the volume continued to increase over the years, becoming a veritable flood
in the 19th century (Proksch 1891). Of particular interest is a work by Konrad
Schellig or Schelling (1448–1508), variously called *Consilium in morbum gal-
licum, Consilium in pustulas malas* and *De morbo gallico.* The author was a
professor at Heidelberg and personal physician to Count-Palatinate Philip
(1448–1508).

Schellig's book is the first known printed work in which mercury is mentioned as a possible form of treatment for syphilis. The date of this book is
uncertain, but it must have been published no later than 1496, and possibly
as early as 1488. An introductory letter to the book speaks of syphilis as a
sickness which ". . . is not, as the vulgar think, a new disease, but in prior
years has been often observed . . . ." Mercury therapy is alluded to briefly on
the last page of the book with a warning that if it is used as an ointment or
for inunctions, great caution should be observed. Nothing is said about internal administration.

Sudhoff believes that Schellig's book was published in 1495 or 1496
(Sudhoff 1912a and 1925; Richter 1910). His opinion is based largely on similarity of wording of Jakob Wimpheling's introductory letter to Schellig's
book and a pronouncement emanating from the Council of Worms held in
1495 (Holstein 1889). (Relevant to the date of Schelling's book are the following references: Trustees of the British Museum 1963, p. xxx; Polain, 1932,
vol. 3, p. 685; Reichling 1905–14, Vol. 7, p. 40; Goff 1964, p. 551; and
Rogers [Bodleian Library] 1969, personal communication; and G. D. Painter
[British Museum] 1969, personal communication.)

Shortly after Schellig, Joseph Grunpeck (1470–1531) in 1496 recommended the use of mercurial ointment for treating syphilis. Five different
forms of ointment using metallic mercury are mentioned in the writings of
Caspare Torella (fl. c. 1500) dating from 1497. An ointment containing argentum vivum extinctum was prescribed in 1497 by Johannes Widmann
(1440–1524), and in the same year Carridino Gilino (fl. 1497) recommended
adding sublimate to the metallic mercury that was used for inunctions. The
pre-fifteen-hundred literature was enriched by contributions from Bartolo-
maeus Steber (d. 1506), Natale Montesauro (c. 1500), and Antonio Scanaroli
(c. 1500) all writing in 1498.

Pedro Pintor (1423–1503) who practiced medicine in Rome has recorded
the case histories of several distinguished patients whom he treated for lues
venerea and on some of whom he used mercury. His clientele included the
Borgia Pope, Alexander VI, and other members of the Borgia family, but his
results were not always favorable. One of the victims of his treatment was the
Cardinal of Segovia, but this did not seem to discourage a number of high-
ranking churchmen from seeking his services.

Other physicians of the late fifteenth and early sixteenth centuries whose names are connected with mercurial anti-luetic therapy are: Antonio Benivieni (c. 1440–1502), Marinus Brocardus (fl. 1530), Juan Almenar (fl. c. 1500), Giacomo Cataneo (fl. c. 1500) (an early anti-mercurialist), and Giovanni de Vigo (c. 1460–c. 1520) (who put live frogs along with quicksilver in the mix from which his ointment was made).

The term "syphilis" originated in a poem published in 1530 by Hieronymus Fracastorius (1478–1553). A first version of his poem, in two books, was completed in 1525 (Fracastoro 1930 translation), and was dedicated to the author's friend and patron, the scholarly Cardinal Pietro Bembo (1470–1547). This version of *De Morbo Gallico* is not extant (as far as is known), but the Cardinal is known to have been critical of part of the poem. At Bembo's behest, Fracastorius produced a revised form of his poem in 1530 (the form that is known today), the revision including a new Book III which glorifies the hyacus tree and the guaiac treatment for lues (Fracastoro 1934 and 1935 translations). For reasons which are not clear, guaiac was more acceptable to the Church than mercury, being known as lignum sanctum or Holy Wood. Although the name "syphilis" was first presented to the world in 1530, or possibly 1525, it does not appear commonly in medical literature until near the end of the 18th century.

Fracastorius, creator of "Syphilis," was not only a poet of note, he was also a physician of great distinction. His views on therapeutic uses of mercury and its compounds are indicative of theory and practice in the 1500's.

In the *De Morbo Gallico* mercurial drugs are mentioned: "First of all mix styrax, cinnabar and minium . . . etc." There is no way of knowing whether his use of *cinnabrium* and *minium* in juxtaposition meant that he was thinking of two different materials or if the choice of words was merely for the sake of the meter. He goes on to say that treating the entire body with this preparation will get rid of lues but is a bit severe and may cause choking and difficult breathing; therefore it should be used only locally on those parts of the body where ulcers are present.

A little later Fracastorius says: "But most people do better by dissolving [or "thoroughly resolving"] everything in quicksilver [i.e., to clear up all pustules] , since the power inherent in this is wonderful" (Fracastoro 1935 translation).

The story of Ilceus, to which Cardinal Bembo supposedly objected, follows the introduction of mercury. The patient is cured by bathing three times in a stream of quicksilver, the healing fluid being poured over his body by the nymph Lipare who was in charge of the stream. Soon word of the cure spread far and wide, and an ointment made of mercury and grease, even though it was recognized as being a bit messy, became the favorite treatment for the dread disease. There can be little doubt that in this allegory Fracastorius was expressing his own views.

Syphilus the shepherd is introduced in the third book of the poem, the part which was a later addition to the original version. Ilceus had been stricken

as punishment for killing a deer sacred to Diana, and Syphilus because of ir-
reverence toward the Sun. The former was cured with quicksilver, the latter
with guaiac. Had Cardinal Bembo not insisted on having the third book added
to the poem there may have been no Syphilus for whom,to name the disease.
Perhaps it would have been given a name derived from Ilceus.

Although Fracastorius, ostensibly to please Bembo, had extolled the virtues of guaiac in the revised version of *De Morbo Gallico*, he never abandoned
his belief in the "miranda vis" of quicksilver. When he wrote as he says, "not
as a poet but as a physician," he again discussed the treatment of syphilis, this
time in his *De contagione et contagiosis morbia et eorum curatione libri III.*
F. H. Garrison says this work "contains the first scientific statement of the
true nature of contagion, of infection, of disease germs and the modes of
transmission of infectious diseases" (Fracastoro 1930 translation).

The *De Contagione*, published in 1546, contains a detailed description of
guaiac and its uses, since of all remedies this was in most general use. It is
quite clear, however, that Fracastorius prefers the use of mercury in difficult
or advanced cases of lues, although he accepts the efficacy of guaiac in milder
forms of the disease. Fracastorius the physician gives detailed instructions for
the preparation and application of mercury inunctions as well as procedures
for fumigations, but he condemns the oral administration of mercury in any
form.

In the last chapter of *De Contagione*, which deals with elephantiasis,
Fracastorius says, "But for this disease, as for the French sickness, the most
effective remedy is quicksilver." (This sounds something like Galileo saying
under his breath "but it does move.")

Never loath to express his views on practically anything, Benvenuto Cellini
(1500-1571) in his *Autobiography* (1924 translation) admits, or perhaps
boasts, of having been a victim of "the sickness," and although he does not
relate that he had been treated with mercury, he has a few trenchant comments to make about a physician of his time, Giacomo Berengario of Carpi
(1470-1530), familiarly known as Giacomo Carpi, who gained fame and fortune through his mercury therapy. Giacomo was a client of Cellini rather than
the other way around, and this is what the silversmith says of the physician:

> There arrived in Rome a surgeon of the highest renown, who was called
> Maestro Giacomo da Carpi. This able man, in the course of his practice,
> undertook the most desperate cases of the so-called French disease. In
> Rome this kind of illness is very partial to the priests, and especially to the
> richest of them. When, therefore, Maestro Giacomo had made his talents
> known, he professed to work miracles in the treatment of such cases by
> means of certain fumigations; but he only undertook a cure after stipulat-
> ing his fees, which he reckoned not by tens, but by hundreds of crowns. He
> was a great connoisseur in the arts of design . . . .

The town of Carpi produced in addition to Giacomo, a second illustrious
physician, Bernardino Ramazzini (1633-1714), who became known as the
Father of Industrial Medicine. Ramazzini credited his fellow townsman with

being the first to treat syphilis with mercury inunctions, but this was probably a matter of local pride getting the better of historical accuracy.

Fallopius, in his treatise on the French disease (1564), speaks of Giacomo Carpi as one who was better known for his wealth than for his skill. He relates that Giacomo became a very rich man through his treatment of the Morbus Gallicus by mercury inunctions and that although he killed a great many of his patients, he cured the majority.

During the sixteenth century the treatment of lues venerea with mercury was largely in the hands of the barber-surgeons. They were soon joined by some of the alchemists and an array of charlatans. Berengario was not the only one who found wealth in this pursuit and thus it has been said that the alchemists realized their dream of turning mercury to gold. There is a story of a sixteenth century barber-surgeon in Paris who was seen praying at the statue of Charles VIII. When it was pointed out to him that the statue was not that of a saint he replied that he knew what he was doing . . . . "This man has given me an income of 7000 livres." (The allusion is to the belief that the Morbus Gallicus had been spread throughout Europe by the men of Charles's army and their camp-followers during and after the invasion of Italy and the campaign at Naples in 1495.)

Philippus Theophrastus Aureolus Bombastus von Hohenheim (1493–1541), is a formidable name. It belonged to a formidable personage. Philippus was his given Christian name, Theophrastus was given to the boy in honor of Aristotle's famous pupil of that name, Aureolus is the equivalent of Goldilocks—because of his blond hair, Bombastus von Hohenheim was his father's name (preceded by Wilhelm). The origin of the appellation "Paracelsus," by which he is commonly known, is obscure. One theory is that it was assumed by the man himself to signify that he excelled or went beyond the great Celsus, and another is that it was bestowed as an honor by one of his few admirers, Trimethius, Abbot of Spannheim, who was his mentor in alchemy.

Paracelsus has been called everything from charlatan to genius. These terms are not necessarily mutually exclusive. No one denies that he played a significant role in the history of medicine, particularly in the field of iatrochemistry. He was a strong advocate of the use of mercury in syphilis. A contemporary of Martin Luther, Paracelsus had a strong religious background and was imbued with the spirit of reform which pervaded the period. He had the greatest scorn for the slavish acceptance of the teachings of Aristotle, Galen, and Avicenna which had dominated medical practice for centuries. Paracelsus opened the door to wide acceptance of the use of mercury for treating lues, as well as to other applications of chemistry to medicine. The seventeenth and eighteenth centuries were to see the full effect of this reform.

Some historians have credited Paracelsus with having been the first to prescribe mercury internally in the treatment of lues venerea, but this is open to question. According to Fallopius, Berengario of Carpi was the first to use this mode of administration, a practice which Fallopius condemned. Alessandro

Benedetti (1460-1525) may also have administered mercury internally as the red oxide in a syrup at a time which antedated Paracelsus (Hauser 1875-82).

As pointed out by Walter Pagel in his book on Paracelsus, syphilis was the main medical problem of the first half of the 16th century. The controversy over the relative merits of mercury and guaiac was in full tilt when Paracelsus entered the lists as a champion of mercury. His writings on the subject were extensive and characteristically forceful, attacking von Hutten (see below) and the "impostors" and quacks who used the Holy Wood. His first book on guaiac was followed by three on the "französischen Krankheit" [French disease] and then eight more on the same subject. Not only did Paracelsus attack those physicians who prescribed decoctions of guaiac but he also was critical of those who used mercury in excess.

Quite naturally, Paracelsus's enemies reacted to the challenge, using, among other things, the weapon of censorship. It has been suggested (Debus 1966) that the Fugger banking house had a hand in the suppression of Paracelsus' books because of the revenue they were deriving from a monopoly on the importation of guaiac wood from the New World. However, at that time, the Augsburg financiers had just become deeply involved in the production of quicksilver at Almadén (Chap. 4) so it should not have made too much difference to them which form of treatment, guaiac or mercury, gained ascendancy.

A contemporary of Fracastorius and Paracelsus, the French physician Jean Fernel (1497-1558), is less familiar to present-day readers than he deserves to be (Sherrington 1946). He popularized the term lues venerea (veneral plague) although he was not the first to use this appellation. It is more meaningful than the fanciful "syphilis" of Fracastorius and surely in France, lues venerea was a welcome alternative to the opprobrious morbus Gallicus.

Fernel was a highly successful practitioner of medicine. At least one of Fernel's royal patients was a victim, and a fatal victim at that, of lues venerea. Francis I succumbed to the ravages of *Treponema pallidum* in 1547. This event established Fernel as an outstanding anti-mercurialist since he refused to treat the king with mercury. There is a story that another one of the court physicians, Antonio Gallus (b.?-1550) remarked of the king's illness, "He got it in the same way as his subjects, and, as with them, mercury will take it away."

A study of Fernel's writings reveals that he did not reject the use of mercury without what seemed to him to be valid reasons. He was familiar with the experience of Berengario of Carpi and was apparently more impressed with the failures than with the alleged successes. He also knew the tragic story of Ulrich von Hutten (1488-1523) with its condemnation of mercury and plea for the use of guaiac instead. Fernel was more concerned with the frequent toxic side effects of mercury than with its benefits. A majority of his contemporaries did not agree with him, but he was by no means alone in his anti-mercurial stand.

Doctor François Rabelais (c. 1494-1553) was a contemporary of his compatriot Fernel and may have shared the latter's anti-mercurial views. The "Abstractor of the Quintessence," in his *Account of the Inestimable Life of the Great Gargantua, and of the Heroic Deeds, Sayings and Marvellous Voyages of his Son the Good Pantagruel* (Book I, Chap. LVIII), gives an excellent clinical description of the effects of mercury therapy: "But what shall I say of those poor men that are plagued with the pox and the gout? O how often have we seen them, even immediately after they are anointed and thoroughly greased, till their faces glister like the key-hole of a powdering tub, their teeth dance like the jacks of a little pair of organs or virginals, when they are played upon, and that they foamed from their very throats like a boar, which the mongrel mastiff-hounds have driven in . . . ."

This passage is one of the earliest descriptions of iatrogenic mercury intoxications and is one of two places in which Rabelais speaks of the powdering tub—a large wooden vat in which meat was salted for purposes of preservation; it was large enough to contain the body of a man and so was used for sweating and fumigating syphilitic patients (Figures XV-1 and 2). This method of treatment, in modified forms, was employed until the end of the nineteenth century. Shakespeare mentions the tub several times, and some of the Restoration dramatists make it clear that it was a familiar part of everyday life in the latter part of the 17th century.

Several additional 16th century physicians are worthy of mention. Johannes Benedictus (1483-1564) is believed by some historians to have been

Figure XV-1 Treatment of Syphilis at the beginning of the 16th Century. This 17th century engraving shows the patient sitting in a barrel containing cinnabar (mercuric sulfide) which was then heated. (From *Folke Henschen, Sjukdomarnas Historia*. Stockholm: Albert Bonniers Forlag. 1962.)

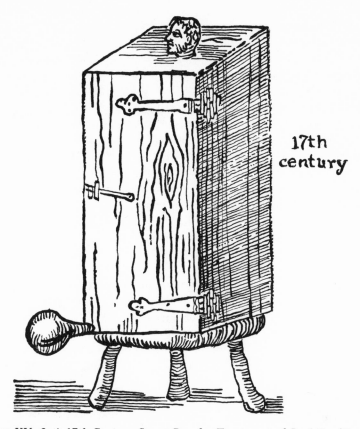

Figure XV-2 A 17th Century Sweat Box for Treatment of Syphilis. (From G. Bugger [ed.], *Das Buch der grossen Chemiker*, Vol. I. Berlin: Verlag Chemie. 1929–30.)

the first to use mercury internally in the treatment of lues. Jacques de Bethencourt was the first French physician to publish a scientific treatise on syphilis. It was he who first used the term "morbus venereus" which was later modified and popularized by Fernel. A number of other names appear in the 16th century literature on mercury and syphilis. Their contributions are discussed in standard works such as those of Proksch, Haeser, Simon, and Neuberger and Pagel.

Seventeenth Century and Later

The dominant position of mercury in anti-luetic therapy in the 17th century is reflected in the number of articles on the subject published during that period. The exhaustive bibliography compiled by Proksch (1891) contains a total of 65 entries dealing with syphilis therapy in the 17th century and of these, 43 are on mercury in one or another form. This relatively small number of medical articles is surprising in view of the prominence of the pox in non-medical literature, particularly in the theater of the Restoration period in England.

It is interesting to contemplate the large volume of syphilis literature contributed by non-medical writers and the small amount coming from doctors in the golden days of the Restoration. Perhaps many of the doctors were somewhat illiterate, and knew it, and left the writing to professional writers.

Mercury retained its pre-eminence in the treatment of syphilis through the 18th century. Proksch's compilation lists a total of 517 articles on anti-luetic therapy from 1700 to 1799, and of these there were 382 dealing with the use of mercury. Published accounts of adverse toxic effects rose from two in the 17th century to twenty-one in the 18th century. In the 19th century, something like 3,000 articles on the treatment of syphilis are listed by Proksch for the period 1800–1889. About one-third of these deal with mercury, and there are close to 400 reports of intoxication resulting from its use.

The first paper dealing with the subcutaneous injection of mercury was published in 1826. The variety of compounds used is impressive: mercury metal, calomel, sublimate, iodide, perchlorate, cyanate, albuminate, peptonate, formamide, ammoniated peptonate, bloodserum mercury, oxide, carbolate, salicylate, grey ointment, benzoate, thymolate, iodo-tannate, succinamide, thymol-acetate and cinnabar. The introduction of injection therapy was followed by a prompt and dramatic decrease in the number of anti-mercurial articles in the literature.

Oral preparations used during the nineteenth century included most of those that were given by injections as well as mercury acetate, ammoniate, tannate, urate and chromate.

Mercury retained its leading role in the treatment of syphilis well into the twentieth century and was still being used at least as late as 1948 (Merklen, Goury-Lafont, and Nezelhof 1948) in spite of the introduction of Salvarsan and related arsenical drugs after 1910 and of penicillin in the early 1940's. No less an authority than Jonathan Hutchinson, just before his death in 1913 wrote: "The small dose plan begun early and continued without intermission over a very long period is so efficient and so free from all drawbacks that its advocates are almost excusable if they show no zeal in the search for other remedies (cited in Power and Murphy 1914).

Hutchinson was not the last of the mercurialists. The second edition of J. E. Moore's standard text, *The Modern Treatment of Syphilis*, published in 1941, contains a detailed account of the use of the inunction treatment, accompanied by a statement that "The administration of mercury by inunction is the best method of use of the drug."

Anti-mercurialism and Syphilis

Opposition to the use of mercury in the treatment of lues venerea arose very early, in fact, almost as soon as the drug was put to that use. Most medical writers before the closing years of the sixteenth century held that quicksilver in any form was too toxic to be used medicinally, especially when given internally.

Konrad Schellig (Schelling) was certainly among the first to leave a record of treating lues with mercury. The first non-medical writer to take a stand against the administration of mercury in lues was Sebastian Brant (1458-1521) who is best known as author of *The Ship of Fools* (Narrenschiff). His anti-mercurial views are expressed in his *De Scorra Pestilentiali, sive Mala de Frantzos*, published in 1496.

The best known and by far the most influential of the early antimercurialists was Ulrich Ritter von Hutten (1488-1523). He was not a physician but he had undergone the inunction treatment. The fact that he obtained no relief might raise the question as to whether or not he actually had lues venerea (if the efficacy of quicksilver is accepted). His prompt cure when he switched from mercury to guaiac increases the grounds for doubt about the true nature of von Hutten's ailment.

The anti-mercurial, pro-guaiac influence of von Hutten stemmed from his prominence in public affairs. He played an important role in the early days of the Protestant Reformation as an ardent supporter of Martin Luther and was concerned not only with religious but also with political reform, being one of the leaders in the abortive revolt of the imperial knights in 1522-1523. Von Hutten was a forceful writer and for a time was poet laureate at the court of the Emperor Maximilian I of the Holy Roman Empire. With such credentials there is small wonder that his word carried great weight in what can be called not just a reformation but a revolution in syphilis therapy. The antimercurial vogue is reflected in the publication of 45 articles on the guaiac treatment of lues between 1517 and 1600 as compared with eight on mercury. For nearly a century the Holy Wood held sway over quicksilver, thanks largely to von Hutten. The mercurialists would have been justified in saying: A pox on him!

Guaiac was not the only agent to challenge mercury, but it was the most important during a period of more than 400 years. At least ten other metals (Table XV-1) and 22 materials of plant origin (Table XV-2) were used against syphilis.

Mercury clearly deserves an honored place in the annals of medical history, particularly in syphilology. It was the first drug to be used for the purpose of killing pathogenic microorganisms in the human body, antedating the sulfonamides and antibiotics by more than 400 years. During most of that period it was held to be one of three or four "specifics" available to the medical profession. No other drug has such a long and distinguished, if at times stormy, record. Whether or not this reputation was deserved is uncertain.

TABLE XV-1

**Metals Other Than Mercury That Have
Been Used In Treating Lues**

Antimony	Iron
Arsenic	Lead
Bismuth	Platinum
Copper	Silver
Gold	Zinc

SOURCE: Proksch 1891.

Did Mercury Cure Syphilis?

The use of mercury in the treatment of syphilis may have been the most colossal hoax ever perpetrated in the history of a profession which has never been free of hoaxes. Starting at least as early as the last decade of the 15th century and continuing well into the 20th, the cult of mercurialism held sway among venereologists. True, it was challenged from time to time, the very vigor of the attacks and violence of the counter-attacks attesting to the emotional rather than scientific aura generated by the mercurial vapors. The case can be summed up very briefly:

1. Persons presumed to have syphilis were treated with mercury.
2. Some of these persons recovered or appeared to recover.
3. Ergo—mercury cures syphilis.

Such "reasoning" is so patently fallacious that it is almost unnecessary to present any further evidence to discredit this form of treatment. Ample evidence to support a sceptical point of view is readily available.

Primary and secondary lesions of syphilis are known to clear up spontaneously, often in a matter of a few weeks after their appearance. Small wonder that a treatment which is continued for months or years is credited with ef-

TABLE XV-2

**Plants Other than Guaiac That Have Been
Used in Treating Lues**

Aconitum	Opium
Agave	Pulsatilla
Asclepias	Saponaria
Astragalus	Sarsaparilla
Begonia	Sassafras
Chelidonium	Smilax
China wood	Stramonium
Cicuta	Tayuya
Clematis	Theer
Guaco	Walnut
Lobelia	etc.
Mezereum	

SOURCE: Proksch 1891.

fecting a cure: the disease disappeared at some time after therapy was begun. Continuation of the unpleasant treatment long after the subsidence of the outward manifestations may well have been motivated by a Calvinistic sadism to discourage the miscreant from further immoral activities.

Prior to the identification of *Treponema pallidum* and the development of the Wassermann reaction, both during the first decade of the 20th century, the diagnosis of syphilis was often difficult and at times virtually impossible. The early stages of lues are relatively benign and are definitely self-limited. Late syphilis is characterized by periods of latency lasting for many years, often exhibiting no overt manifestations up to the time of death of the infected person. A treatment lasting for a year or more would certainly allow ample time for the initial outward signs to disappear and for latency to supervene. These features of the disease introduce formidable difficulties when it comes to evaluating the efficacy of therapeutic measures. The long-held conviction that mercury could effect a cure might lead to ethical objections if an investigator deliberately withheld treatment in order to conduct a controlled experiment; however, at least one such study has been undertaken (see below).

The multitude of therapeutic measures employed by later venereologists is evidence that no single form of treatment, including mercury, was unquestioned. Proksch (1891) lists 128 articles published between 1800 and 1887 describing non-mercurial methods of anti-syphilitic therapy.

During British military operations in Portugal in 1812, William Fergusson, Inspector General of Hospitals to the Portuguese Army, had an opportunity to make some important observations on the treatment of syphilis. He noted that the Portuguese soldiers who contracted the disease generally received no treatment, mercurial or otherwise, while the infected British soldiers were treated vigorously with mercury. As a result of experience over a period of two years, Fergusson concluded that the Portuguese recovered from syphilis more quickly and completely than the British and that among the former, the disease ". . . exhausts itself, and ceases spontaneously." This study has the characteristics of controlled clinical investigation, with a treated and an untreated population. The findings and conclusions were to be confirmed by Norwegian investigators a century later (Bruusgaard 1929).

Not only among avowed anti-mercurialists do questions about the efficacy of mercury find expression. Thus, Tongue in 1801 writes: "I should ask why, after having destroyed the venereal virus at its source; after having cured the original venereal ulcers, we should still see syphilitic affections in other parts of the body?" These are the words of an observer with an inquiring mind, albeit ensnared by the types of *post hoc* reasoning so common in his time and still all-too-common in more recent medical writings. Under mercurial treatment the local lesions disappeared, this being viewed as cause and effect. Spontaneous resolution of the manifestations of early syphilis was not widely recognized so that disappearance was attributed to the treatment. Yet, the disease was not eradicated since late manifestations appeared elsewhere in the body. Had anyone followed Tongue's query to its logical conclusion a century of unfortunate syphilitics might have been spared the discomforts of mercury therapy.

So strong was the hold of mercury on the medical profession that when it failed to effect the expected cure of syphilis all sorts of excuses were offered. When mercury did not cure syphilis it was because it had been improperly administered, either too little or too much, or some other fault of the treater, never of the treatment (Colles 1881). Frequently, the patient was blamed for not continuing under therapy.

Doubts as to the efficacy of available therapy for syphilis prompted C. P. M. Boeck (1845–1917) of Oslo, Norway, to initiate his classic studies. During the 20-year period 1891–1910, Boeck deliberately withheld treatment from 1,978 patients with primary and secondary syphilitic lesions. All of these patients were hospitalized as long as infective processes were present, thus protecting the community from further spread of the disease. Some of the cases were free of outward manifestations of syphilis within a month, others took longer, up to a year, the average being 3.6 months (Gjestland 1955).

Two follow-up studies were made of Boeck's patients, the first by E. Bruusgaard in 1929 and the second by Danbolt and Gjestland between 1948 and 1951 (Gjestland 1955). The results of these studies have been summarized by Clark and Danbolt (1964) who report that "It was estimated that between 60 and 70 out of every 100 of these patients went through life with a minimum of inconvenience despite no treatment for early syphilis." (This is certainly more than can be said for patients who were subjected to the mercury treatments.) Clark and Danbolt note that significant follow-up information was obtained on about 80 per cent of the original study group.

Comments on the Boeck-Bruusgaard material by J. E. Moore (1941), a leading 20th century authority on syphilis, are of interest: "Within the last fifty years of the 19th century there were not lacking such eminent clinicians as Lesser and Boeck to dispute the 'curative' effect of mercury, and to contend that the apparent 'cures' from its use represented only the natural evolution of the disease and its tendency toward spontaneous 'cure.'" Moore goes on to say that: "Numerous studies have appeared to prove decisively that the results of treatment, expressed in any terms one desires, but especially in terms of ultimate clinical outcome, are better if arsphenamine and mercury (or bismuth) are both employed than if arsphenamine is used alone." This does not sound like a very high recommendation for mercury, and elsewhere Moore says that: "The clinical observations of centuries, indicating that the drug will 'cure' syphilis or at least markedly reduce the incidence of grave late complications, have not been explained, *nor even largely corroborated....*" (Emphasis added.)

Certainly the number of articles published from the 16th through the 19th centuries in which mercury is recommended for the treatment of syphilis is many times greater than that of works in which other forms of treatment, or no treatment at all, is espoused. This in and of itself should not convince any scientifically-minded person that mercury was necessarily of any value to the patient. That it could cause serious and even fatal effects is amply documented. While it may be unfair to apply late 20th century criteria in judging the physicians and therapies of earlier times there is no reason why modern concepts cannot be applied in retrospect as a basis for present judgments. This can be expressed in another way in the form of a simple question: Knowing what we now know about mercury and about the natural history of syphilis, would we use the drug for that disease? The answer based on proven danger and unproven benefit, must be in the negative, and even if physicians wanted to employ this form of therapy there is a strong probability that the Food and Drug Administration (or its equivalent in countries other than the United States) would not permit it.

REFERENCES

Bruusgaard, E. 1929. Über das Schicksal der nicht spezifisch behandelten Luetiker. *Arch. Dermatol. Syphil.* 157:309–32.

Cellini, B. 1924. *The life of Benvenuto Cellini.* Trans. John Addington Symonds. New York: Charles Scribner's Sons.

Clark, E. G., and Danbolt, N. 1964. The Oslo Study of the natural course of untreated syphilis. *Med. Clin. N. A.* 48:613–23.

Colles, A. 1881. *Selections from the works of Abraham Colles.* Ed. R. McDonnell. London: New Sydenham Society.

Crosby, A. W., Jr. 1969. The early history of syphilis: A reappraisal. *Am. Anthropol.* 71:218–27.

Debus, A. G. 1966. *The English Paracelsians.* New York: Watts.

Fallopio, G. 1564. *De morbo gallico.* 2 pts. Patavii.

Fergusson, W. 1812. Observations on the venereal disease in Portugal. *Med. Chir. Trans. London* 1:1–5.

Fracastoro, G. 1930. *Hieronymi Fracastorii De contagione et contagiosis morbis et eorum curatione libri III.* Trans. and ed. Wilmer Cave Wright. New York: Putnam.

Fracastoro, G. 1934. *The sinister shepherd.* A translation of Girolamo Fracastoro's *Syphilidis sive de morbo gallico libri tres*, by William Van Wyck. Los Angeles: Primavera Press.

Fracastoro, G. 1935. *Syphilis or The French disease.* Trans., notes and appendix by Heneage Wynne-Finch, and introd. by James Johnston Abraham. London: William Heinemann Medical Books.

Goff, F. R. 1964. *Incunabula in American libraries.* New York: Bibliographic Society of America.

Gjestland, T. 1955. *The Oslo Study of untreated syphilis: An epidemiologic investigation of the natural course of untreated syphilis based on a restudy of the Boeck-Bruusgaard material.* Oslo: Akademisk Forlag.

Haeser, H. 1875–82. *Lehrbuch der Geschichte der Medicin und der epidemischen Krankheiten.* 3d rev. 3 vols. Jena.

Holstein, H. 1889. Ein Wimpfeling-Codex. *Z. Litt.-Gesch. Ren.-Litt.* 2:213–15.

Merklen, F. P.; Goury-Lafont, M.; and Nezelhof, C. 1948. Penicillin and ascorbic acid in prevention and therapy of mercurial and bismuth stomatitis in syphilis. *Paris Med.* 38: 166–68.

Moore, J. E. 1941. *The modern treatment of syphilis.* 2d ed. Springfield, Ill.: Charles C. Thomas.

Neuberger, M. 1910–1925. *History of medicine.* Trans. E. Playfair. 2 vols. London: Oxford University Press.

Neuberger, M., and Pagel, J. 1902–05. *Handbuch der Geschichte der Medizin.* 3 vols. Jena: Gustav Fischer.

Pagel, W. 1958. *Paracelsus.* Basel: S. Karger.

Polain, M. L. 1932. In *Catalogue des livres imprimés au quinzième siècle des Bibliothèques de Belgique,* vol. 3. Brussels.

Power, D., and Murphy, J. K., eds. 1914. *A system of syphilis.* Oxford Medical Publications, 2d ed., vol. 1. London: Oxford University Press.

Proksch, J. K. 1874. *Der Antimercurialismus in der Syphilis-Therapie.* Erlangen.

Proksch, J. K. 1891. *Die Litteratur ueber die venerischen Krankheiten.* Bonn.

Reichling, D. 1905–14. Appendices ad *Hainii-Copingeri Reportorium bibliographicum.* 7 vols. Munich: I. Rosenthal.

Richter, P. 1910. Uber Conrad Schellig und sein "consilium in pustulas malas." *Arch. Gesch. Med.* 3:135–40.

Sherrington, C. 1946. *The endeavour of Jean Fernel.* Cambridge: At the University Press.

Simon, F. A. 1860. *Geschichte und Schicksale der Inunktionskur.* Hamburg.

Sudhoff, K. 1912a. *Graphische und Typographische Erstlinge der Syphilisliteratur aus den Jahren 1495 und 1496.* Munich: C. Kuhn.

Sudhoff, K. 1912b. *Aus der frühgeschichte der syphilis.* Studien zur Geschichte de Medizin, vol. 9. Leipzig: J. A. Barth.

Sudhoff, K. 1925. *The earliest printed literature on syphilis.* Adapted by Charles Singer. Florence: R. Lier and Co.

Tongue, J. 1801. An experimental inquiry into the modus operandi of mercury, in curing the lues venerea. In *An inaugural dissertation,* M. D. dissertation, University of Pennsylvania, 1801. Philadelphia: Printed for the author.

Trustees of the British Museum. 1963. *Catalogue of books printed in the 15th century now in the British Museum.* Reprint of 1913 ed. London.

ADDITIONAL READINGS

Cellini, B. 1845. *Memoirs of Benvenuto Cellini: A Florentine artist: Written by himself.* With notes and observations of G. P. Carpani from the edition printed in Milan in 1806. Trans. Thomas Roscoe, Esq. 2 vols. New York: Wiley and Putnam.

Donne, J. 1941. *Ignatius: His conclave or His enthronization in a late election to hell.* Reproduced in facsimile from the edition of 1611. New York: Columbia University Press, The Facsimile Text Society.

Kussmaul, A. 1861. *Untersuchungen über den konstitutionellen Merkurialismus und sein Verhältnis zur konstitutionellen Syphilis.* Wurzburg.

Sydenham, T. 1848–50. *The Works.* Trans. from the Latin ed. of Greenhill by R. G. Latham. 2 vols. London: Sydenham Society.

Voress, H. E., and Smelcer, N. K. 1957. *Mercury toxicity: A bibliography of published literature.* Oak Ridge, Tennessee: U. S. Atomic Energy Commission, Technical Information Service Extension.

16

Mercury in 16th and 17th Century Medicine

Sixteenth Century

For the most part, the history of quicksilver in 16th century medical practice is inextricably interwoven with the history of syphilis. The battle between the mercurialists and the anti-mercurialists which had been joined before 1500 continued throughout the century and beyond.

The traditional view among 16th century physicians was that mercury in any form taken internally was poisonous; even external uses did not have universal acceptance. This distrust is reflected in a decree of the Augsburg Senate in 1582 which warned apothecaries not to prepare or sell substances known to be dangerous, including all preparations of mercury.

Important information on mercurial drugs and prescriptions can be found in the 16th century pharmacopoeias, of which that of Augsburg, the *Pharmacopoeia Augustana,* is the best known. The date of the first appearance of this work is in doubt, but it is generally thought to have been published in 1564 (Crosland 1962; Augsburg Pharmacopoeia 1927). It strongly influenced later pharmacopoeias in Germany and elsewhere.

All early pharmacopoeias were based largely, but not exclusively, on the traditional medical teachings of Hippocrates and Galen and consequently contain no listings or prescriptions of mercurials for internal use. This is true of the Augustana, but this work does list a number of mercury-containing compounds for external application:

Aqua cum Mercurio. This is made from mercury sublimate and aqua fortis (now nitric acid but of uncertain identity in the 16th century).
Unguentum Rubeum, sive de Minio. The meaning of minium is uncertain but it probably is mercury sulfide since litharge and plumbum ustum are used elsewhere to designate lead oxides.
Unguentum ad Pediculos. This ointment contains argentum vivum.

Unguentum Mercuriatum cum Theriaca. An ointment containing argentum vivum extinctum, i. e. intimately mixed with fat.
Emplastrum Nervinum Vigonis. This prescription calls for 25 ingredients including earthworms and minium.

So intimately has the name of Fracastorius been associated with syphilis that his more important medical treatise *De Contagione, Contagiosis Morbis et eorum Curatione* tends to be relegated to a position of secondary importance. This work, first published in 1546, covers a number of contagious diseases. For at least one of these, elephantiasis, Fracastorius recommends the use of mercury therapy (Chap. 15).

Just what disease entity was meant by elephantiasis or elephantia is not clear. Fracastorius points out that it is sometimes erroneously confused with syphilis. Its identification with leprosy was accepted by Fracastorius and is plausible today, but the name was probably applied to other diseases including schistosomiasis. In any event, "The medicaments that are effective against this contagion are: nitre, sulphur, hellebore, cedar, incense, iris, quicksilver (argentum vivum), turpentine and liquid styrax; mix them with fat, and I advise you to use the fat of the bull, the fox, the bear or the lion . . . . [and] for this disease, as for the French Sickness, the most effective remedy is quicksilver" (Fracastoro 1930 translation).

Information on attitudes toward, and uses of, mercury in medicine can be found elsewhere than in the medical literature. As a matter of fact, if a well rounded picture is to be obtained other sources *must* be consulted. One such source is the apothecaries.

There could be no more eloquent spokesman for the apothecaries of the 16th century than Thibault Lespleigney (1496-1567), Apothicaire à Tours, whose *Promptuaire des Médicines Simples en Rithme Joieuse* tells the story about all drugs, Galenical and otherwise, including quicksilver.

In order to be sure not to offend either the clergy or his fellow apothecaries, Lespleigney precedes the Promptuaire by a "Rondeau à La Vierge Mère, Marie Rayne de Virginité" and a "Prologue par lequel ledict auteur dedie ledict Promptuaire aulx appoticaires de ladict ville de Tours." Then for double assurance he adds a "Ballade à La Mère de Jésus" at the end. These gestures toward the Church reflect the usual 16th century caution in scientific writing, particularly if any new ideas are to be introduced. The Promptuaire contains 165 chapters, all in rhyming couplets. Chapter 158 (Lespleigney 1899), and a rough translation, goes as follows:

Vermillon. Cap. 158

Vermillon en grec est sercog,	The Greek vermilion won renown,
Rouge comme creste de cog,	For brilliant red like rooster's crown,
Cinabrion dict aultrement,	Its other name is cinnabar,
Aulx boutons donne allegement,	Clears up pimples, leaves no scar,
Car d'iseulx est desiccatif,	It dries things up and does it well,
De flux de sang restractif,	The bloody flux yields to its spell,

Utille aulx serotz et colires.   Useful in several kinds of dressings.
Use en, si sante tu desires.   Use it, and enjoy health's blessings.

In these eight short lines Lespleigney provides a great deal of information about the contemporary status of mercury compounds. His chapter on "Vermillon" coming immediately before that on "Vif Argent" indicates that in his mind the two were associated with each other. Yet he suffered from the same confusion in terminology which has plagued many writers before and since. In describing the action of cinnabar as astringent he gives it the principal characteristic of dragon's blood or cinnabaris. Internal use of the drug for dysentery confirms this, in that he must have believed it to be non-toxic. Finally, the author shows that he held cinnabar in high esteem for external as well as internal use.

Lespleigney's chapter on quicksilver (1899, Chap. 159) is considerably longer than the one on cinnabar. In it he says of quicksilver

VIF ARGENT

HYDRARGYROS la grecque gent
Dict ce que disons vif argent,
Froict et humide au degré quart,   Hydrargyros, the word in Greek
Traict de mine par subtil art;   Of quicksilver is used to speak,
Non obstant, si par tout veulx lire,   Cold and wet to the fourth degree
Trouveras que aulcuns veulent dire   'Tis drawn from mines with subtlety;
Qu'il sort naturel d'une terre.   For anyone who wants to know,
Qui tient telle opinion erre.   A little reading soon will show
Quant à parler de sa vertu,   It's quite at home in firma terra;
Je n'en donne pas ung festu;   If you don't think so you're in error.
Car, conbien qu'elle soit vigoreuse.   Its virtues trouble you, mayhap,
Sa vigueur est trop rigoreuse.   For them I wouldn't give a rap;
Aussy qu'il est rare es usaiges   For just as it is vigourous
Des medecins expers et saiges!   Its vigour is too rigourous.
Il penetre, dissoult, consomme   A doctor who knows how it's used
En mondifiant; c'est la somme.   Is one not easily confused.
Gallien n'en faict pas grant cas,   Consumes, dissolves and penetrates
Car, luy vivant, ne regnoit pas   It modifies all kinds of states.
La maladie impatience!   Galen gives it scanty mention
Aussy tel art n'est pas science   [Plants and dung got his attention].
Liberalle, mays cirurgicque.   Surgeons understand me well,
Les expers én telle pratticque   For art from science they can tell.
Enfendent ássez ma parolle.   The great pox they all know by heart,
Ce n'est pas la grosse verolle,   They have the facts before I start.
C'est la hyddeuse maladye: It comes from choosing beds unknown
Entendez sans que je le dye.   And plugging holes best left alone;
Elle prent d'avoir trop mal cousché   From turning in without a light
Et d'avoir le trou mal bousché,   And trusting sound instead of sight.
Par faulte d'y porter chandelle   Repentance comes a bit too late;
Et s'estre endormy au chant d'elle;   Now you've got the story straight.
Puys c'est tard, si on s'en repent.   Why, like a freshly butchered calf
Voila dont tout le mal despent,   Must we be tied and bent in half,
Parquoy il fault, comme une beste,   Plunged in a fire worse than Hell,
Depuys les piedz jusque à la teste   Roasted till we're done quite well.

Lié, garotté comme ung veau,   No matter how you kick or bite,
Estre plongé en ung fourneau   Or sweat, you cannot set things right.
Plus cruel que n'est purgattoire,   Prayers will bring no benefit,
Tant le faict est criminatorie;   Song and dance won't help a bit;
Puys, deussiez vous mordre ou ruer,   Only more of being rubbed,
Sy fault il là dedens suer   Sends you off all fresh and scrubbed.
Et faire dure penitence,   Thus the drug dissolves your pain
Chanter fault et mener la dance,   With cures so swift you can't complain.
Davantaige estre bien frotté;   It's always comforting to know
On s'en va frays et descrotté: You've got some property to show;
Voyla la vertu de la droggue.   You really have not lost a thing,
Le feu puisse brusler la boggue,   You're in the class of duke and king!
Le chasteignier et la chateigne!   Does this not give you consolation?
On ne voyt homme qui s'en pleigne,   Yes; but still my tribulation
Car il y a quelque confort.   Fills my heart with pangs of conscience.
On en a tousjours quelque apport.   Woe to him who has no patience.
Communement on n'y per rien,
Car c'est le mal des gens de bien
En tous degrez et tous estatz,
De nobles, princes et prelatz.
N'esse pas consolation?
Oy; mays tribulation
Donne remors de conscience.
Mal vit qui ne prent patience.

This "rithme joieuse" tells the whole story of quicksilver and its more important uses as well as its Greek name, its properties of cold and wet, its source, and its principal pharmacological actions, its great potency, and Galen's low opinion of it. Advice on how to avoid catching the great pox is fortified with a vivid description of the horrors of treatment when this advice is not followed. The fumigation treatment (Estre plongé en ung fourneau) is worse than inunctions, but the latter is pretty bad, too. By way of consolation it is pointed out that a cure will be accomplished and that those who take the cure are in the distinguished company of nobles, princes, and prelates.

Seventeenth Century

During the 17th century, pharmacopoeias were published in many European cities. The treatment of syphilis continued to be the principal application of mercurial drugs and the mercurialist-anti-mercurialist controversy continued unabated. Relevant material can be found in both medical and popular writings, the pox doctor having become a special object of satire.

The first Pharmacopoeia Londinensis, prepared by the Royal College of Physicians, was published in 1618. It carries several entries relating to quicksilver, not all of which are easily comprehensible. One of the listings is:

Argentum Vivum, sive Hydrargyrum Crudum
     Praecipitatum, Pulvis Angelicus
     Sublimatum, Mercurius sublim.

From the accompanying text it is not possible to be certain which of the
compounds are related to quicksilver and which, if any, to silver. There is no
clarification as to the nature of Pulvis Angelicus but it probably is Argentum
Vivum Praecipitatum. The "angelic" name suggests an alchemical influence.

Elsewhere in the 1618 London Pharmacopoeia there is a group of
compounds which are clearly derived from quicksilver:

Mercurius Dulcis
Mercurius Dulcis Praecipitatus
Turbith minerale (crude mercury and oil of vitriol)
Mercurius sublimatus

The influence of Pliny and other early writers can be seen in the entries
dealing with compounds of quicksilver and sulfur.

Cinnabaris, Minium nativum, factitum ex Argento vivo & sulphure
Minium, eiusque vena, Anthrax e quo fit Argentum vivum in Hydra

There are no prescriptions in this pharmacopoeia calling for internal use of
mercury or its compounds.

The Augsburg Pharmacopoeia of 1622 lists argentum vivum under poisons.
There are no prescriptions for internal administration, but there are prescriptions for two forms of mercurial ointments: *unguentum pediculorum* and
unguentum mercuriale. The latter has 15 ingredients, including "mercurii vivi
mundat," the nature of which is not explained. The revised edition of 1657
contains directions for preparing the internationally famous *Emplastrum de
Ranis cum Mercurio,* which was not present in the 1622 edition.

Among the 49 minerals and stones listed in the Amsterdam Pharmacopoeia
of 1636 can be found argentum, cinnabaris, minium and sublimatum. A section entitled "De Pulver. Chymic." includes two prescriptions containing mercury which apparently are intended for internal use. This pharmacopoeia is
certainly among the first to give official approval for the internal use of
mercurials.

The Brussels Pharmacopoeia of 1671 has a section entitled "Materia
Argenti vivi, vena minii, sive cinabri vocatur" which contains several descrip-
tive paragraphs on mercury and some of its compounds. This pharmacopoeia
lists more varieties of mercurials than the other 17th century works, including
ing: mercurius regeneratus, mercurius dulcis, mercurius ex cinabro praecipitatus ruber, mercurius praecipitatus ex cinabrio albus, magisterium mercurii
rubicundissimum, panacea ex mercurio, and Turphetum minerale. It is not
clear which, if any, of these are intended for internal use but the listing of
ointments in a separate section suggests that the compounds just mentioned
were to be taken internally.

Quite similar to the Brussels Pharmacopoeia is that of Lille published in
1694. The latter lists one compound which is not found in the former,
namely, Aetiops mineralis, which is made from "mercurii vivi" and "florum
sulphuris," in other words, artificial cinnabar.

Valencia produced a pharmacopoeia in 1698 which, among other things,
contains two formulas for ointments to be used in treating the morbus gallicus. One of these is simply quicksilver and pork fat, precursor to "blue
mass."

Among the prominent physicians of the 17th century a few may be chosen
to illustrate contemporary views on the use of mercury (Feissinger 1897).
Daniel Sennert (1572-1637) who served for a time as Professor of Medicine
at the University of Wittemberg, was a follower of Paracelsus in the use of
mercury for treating syphilis, but in other respects he adhered to the tenets of
Galen. Jean Baptiste van Helmont (1577-1644) was a strong believer in the
use of mercuric chloride and Gregory Horst (1578-1638), sometime Professor of Medicine at Giessen, favored the use of the aurum vitae (sublimate
and gold) of Paracelsus. On the other hand, Nicholas Tulp (1593-1674),
immortalized by Rembrandt as the demonstrator in The Anatomy Lesson, believed that mercurial salivation did more harm than good.

In England, Thomas Sydenham (1624-1689) with his strong naturopathic
leanings, held mercury in low regard. His views are disputed by James
Primrose (1580-1659) who says that ". . . the moderne Physicians have found
by experience, that it may safely be administered, if it doe not exceed due
measure. For no medicine is taken in excessive quantitie without hurt to the
body" (1651). To emphasize the dangers of taking too much mercury
Primrose cites the story ". . . of a certaine Druggist, who being in a burning
Fever, instead of a glasse of water to quench his thirst, unhappily chanced to
light on a glasse full of Quicksilver, and drinking up a great quantity thereof,
he dyed congealed, insomuch as when his dead body was opened, the Physicians found the blood about his throat congealed and frozen." In spite of this
dire picture, Primrose concludes his essay by saying that "The most excellent
remedy of all is quicksilver, either taken crude by it self, from a scruple to a
few drammes, or else first quenched with the juyce of Lemmons, but then the
does must be lesse, because being quenched, it stayes longer in the belly."

Textbooks of chemistry such as that of Nicholas Lemery are a rich source
of information about mercurial drugs used in the 17th century. The title page
of Harris's translation of the fifth edition of Lemery's text (1686) advises the
reader that the book contains "An easie Method of Preparing those Chymical
Medicins which are used in PHYSICK. with Curious Remarks and Useful Dis-
courses upon each preparation, for the benefit of such who desire to be instructed in the Knowledge of this Art." "Quicksilver," it is said, "is one of
the greatest remedies we have in Physick, when it is used as it should be, but
full as dangerous, when it happens into the hands of Quacks, who use it upon
all occasions for all sorts of Diseases, and give it indifferently to all sorts of

persons." This seems to have been the sentiment of the more resonable physicians of that, as well as of other times. (See also Chap. 12.)

In a treatise by Lodovico Bellinzani (fl. c. 1650) published in Rome in 1648, in addition to the well known uses of mercury for worms, skin diseases, and syphilis, Bellinzani mentions its administration, presumably by mouth, to assist parturition in cases of difficult labor. Perhaps this idea is the basis for the mistaken belief that abortion can be produced by swallowing mercury.

Views on the position of the medicinal uses of quicksilver during the 17th century can be found in well-known non-medical books. Robert Burton (1577-1640) for example, in *The Anatomy of Melancholy,* originally published in 1621, relates that "Matthiolus approves of Potable Gold [and] Mercury. . .," but Molière (J. B. Poquelin, 1622-1673) was less enthusiastic about the virtues of Potable Gold (see Poquelin, 1666).

John Donne (1572-1631) a few years earlier (see 1941 edition) had condemned Paracelsus and the entire concept of iatrochemistry, in which quicksilver occupied a prominent place.

Restoration plays, particularly those of Thomas Shadwell, Philip Massinger, Charles Sedley, and Mrs. Aphra Behn, are full of the French disease and of broad lampoons of French doctors who treat it (Silvette 1967). Mercury therapy, either by rubbing or tubbing, seems to have been a fact of everyday life, especially in high society. The employment of foreign doctors to treat the pox suggests that English physicians were not interested in this aspect of practice. Mercury itself was not attacked or ridiculed.

The use of rubbing and tubbing as the standard forms of treatment for syphilis has been recorded in the familiar nursery rhyme:

> Rub-a-dub
> Three men in a tub . . . .

Mercury's capacity for stimulating the composition of verse has been mentioned in the work of Thibault Lespleigney. A counterpart in 17th century England is found in the person of John Woodall (1556?-1643) who composed a poem entitled: "In Laudem Mercurii: or IN PRAISE of Quicksilver or Mercurie" (1639, p.228 ff.). A general idea of the nature of the work can be gotten from the first four stanzas:

> Whereto shall I thy worth compare,
> whose actions so admired are?
> No medicine knowne is like to thee,
> in strength, in vertue and degree.

> Thou to each Artist wise art found,
> a secret rare, ye sage and sound,
> And valiantly though pla'st thy part,
> to cheere up many a doleful heart.

Yet makest thy patient seeme like death,
with ugly face, with stinking breath:
But thou to health him soone restores,
although he have a thousand sores.

The perfectst cure proceeds from thee,
for Pox, for Gout, for Leprosie,
For scab, for itch, of any sort,
These cures with thee are but a sport.

Guaiac remained on the scene as a rival of quicksilver at least until the end
of the 17th century. Bernardino Ramazzini (1633-1714) used both of these
drugs as did his contemporary and countryman Carlo Musitano (1635-1714).
The latter, who was both a gynecologist and a priest (Feissinger 1897), prescribed instillations of calomel mixed with plaintain for urethritis.

REFERENCES

Augsburg Pharmacopoeia. 1927. A facsimile of the first edition of the *Pharmacopoeia
Augustana* with introductory essays by T. Husemann. Madison: State Historical
Society of Wisconsin.
Bellinzani, L. 1648. *Il mercurio estinto resuscitato.* Rome.
Burton, R. 1948. *The anatomy of melancholy.* Ed. Floyd Dell and Paul Jordan-Smith.
New York: Tudor Publishing Co.
Crosland, M. P. 1962. *Historical studies in the language of chemistry.* London:
Heinemann Educational Books.
Donne, J. 1941. *Ignatius: His conclave.* Reproduced in facsimile from the edition of
1611. New York: Columbia University Press, The Facsimile Text Society.
Feissinger, C. 1897. *La thérapeutique des vieux maîtres.* Paris.
Fracastoro, G. 1930. *Hieronymi Fracastorii De contagione et contagiosis morbis et
eorum curatione libri III.* Trans. and ed. Wilmer Cave Wright. New York: Putnam.
Lemery, N. 1686. *A course in chymistry.* 2d ed., enl. and trans. Walter Harris from the
Fifth Edition in the French. London.
Lespleigney, T. 1899. *Promptuaire des medecines simples; en rithme joieuse.* Nouvelle ed.
Dr. Paul Dorveaux. Paris: H. Welter.
Poquelin, J. B. [Molière]. 1666. *Le médicin malgré lui.* Paris.
Primrose, J. 1651. *Popular errours or The errours of the people in physick.* London.
Silvette, H. 1967. *The doctor on the stage.* Knoxville: University of Tennessee Press.
Woodall, J. 1639. *The surgeons mate or Military and domestic surgery.* London.

ADDITIONAL READINGS

Agricola, G. 1588. *De natura fossilium.* Bk. 8. Trans. from the 1st Latin ed. of 1556 by
Mark Chance Bandy and Jean A. Bandy. New York: Geological Society of America.
Bauderon, B. 1588. *Paraphrase sur la Pharmacopée.* Lyons.
Beguin, J. 1620. *Les élémens de chymie.* Paris.
Freind, J. 1726. *The history of physick from the time of Galen to the beginning of the
16th century.* 3rd. ed., pt. 1. London.
Thorndike, L. 1923-58. *A history of magic and experimental science.* 8 vols. New York:
Columbia University Press.

17

Mercury in 18th and 19th Century Medicine

Mercury in one form or another continued to be the treatment of choice for syphilis throughout the 18th century but it also found many other uses. The growing list of mercurials available to the physician is illustrated in the second edition of Quincy's pharmacopoeia, published in 1719.

Mercurius — Quicksilver
Arcanum Joviale — Amalgam of tin and mercury treated with spirit of nitre and wine
Mercurius Sublimatus — White sublimate [mercuric chloride]
Mercurius Sublimatus Dulcis — [Mercurous chloride, calomel]
Mercurius Resuscitatus — Mercury revived [recovered] from cinnabar
Mercurius Praecipitatus ruber — Red precipitate [red oxide]
The Prince's Powder — Red precipitate treated with spirit of wine
Panacaea Mercurii rubra — made from red precipitate
Panacaea Mercurii alba — made from calomel
Mercurius Praecipitatus albus — white precipitate
Turpethum Minerale (Turbith Mineral) — yellow precipitate [oxide]
Mercurius Praecipitatus viridis — Green precipitate
Arcanum Corallinum — made from red precipitate
Aethiops Mineral — mercury and sulfur
Cinnabar Nativum — natural cinnabar [Mercuric sulfide]
Cinnabar Factitium — Artificial cinnabar

Obviously the choice of product was great enough to suit the taste of doctor and patient, even when it came to color. One of the most popular patent medicines of the period was Townsend's Mixture, made of red mercuric oxide, potassium iodide, syrup of orange, tincture of cardamon and water, developed by a clergyman, Joseph Townsend (1739–1816). The combination of mercury and iodide survived well into the 20th century in the form of the so-called mixed treatment for syphilis.

239

The article on mercury in James's three volume medical dictionary of 1745 is one of the longest in the entire work, occupying 13 double columns of folio. In addition to medicinal preparations and uses, it covers many other aspects of quicksilver including sources, properties and ancient knowledge of the metal. James held mercury in high esteem, for he says that:

> Quicksilver judiciously administered is . . . undoubtedly a most excellent Medicine; it opens the Pores, small Vessels, and Ducts of the Glands, resolves obstructed Humours, attenuates those which are too thick and viscid, especially the Lympha; and dissipates Concretions, even in the remotest Parts of the Body. On all these Accounts it is found to be of singular Service in Tumours, swelled Glands, scirrous Spleen, Mesentery, or Liver, Ganglions, Strumae, and other such Diseases. It likewise blunts the acrimony of the Fluids, and hence performs Wonders in Venereal Tumours, Buboes, and Ulcers, in cutaneous Pustules, Scabs, and other Affections of the Skin . . . Quicksilver is a great Enemy to all Sorts of Vermin, as well as to Worms; and it suddenly kills or banishes them, being applied in an Ointment to any Parts of the Body where they are found.

But, says James:

> By an injudicious Use of it, whether outwardly apply'd, or inwardly taken, the Nerves are, likewise, affected, weakened, corrupted, and contracted; whence Tremblings, Spasms, Palsies, and too great an Attenuation of the Fluids, which often brings on a fatal Salivation, Ulcers in the Mouth and Throat, and incurable Looseness.

Appearing as they did in a standard reference work, the views of James probably reflected the accepted position of mercurials in the materia medica of 18th century Britain. In France, the great syphilographer Jean Astruc (1684-1766) added a touch of elegance, his favorite preparation for internal use being mercury triturated with crawfish eyes. His countryman François Chopart (1743-1795) showed his originality by treating cancer of the breast with mercury sublimate, a centigram daily taken internally (cited in Feissinger 1897).

One of the most spirited controversies over the use of mercury in medicine centered around the person of Dr. Thomas Dover (1660-1743). Known as the Quicksilver Doctor, Dover had a successful career not only as a physician but also as a privateer. On one of his expeditions he participated in the capture of the city of Guayaquil and later happened to put in at the island of Juan Fernandez where he found the exiled Alexander Selkirk and returned him to England. (Selkirk's adventures have been immortalized by Daniel Defoe in his story of Robinson Crusoe.) Dover is best known to physicians of the 20th century as originator of Dover's Powder, a mixture of opium and ipecac. In his own day, however, Dover was notorious for his uninhibited use of mercury (Dewhurst 1957).

Dover's experiences in the use of mercury are described in his book *The Ancient Physician's Legacy to his Country* first published in 1732. The second edition appeared in 1733 and included a supplementary treatise on the wonders of mercury by "the Learned Belloste" which probably was added to counter the attacks which the first edition had engendered. A reading of a few of Dover's case reports provides some evidence of why he should have been attacked and even ridiculed. The epithet Quicksilver Doctor had been applied in scorn. Says Dover:

> I desire to know, Why I am called the Quicksilver Doctor, by way of Derision? Pray do not you, Gentlemen Physicians, Surgeons, and Apothecaries, prescribe it almost every Day of your lives? I aver, you do. Only you disguise it; and I give it in such an open honest Manner, that my patients cannot be deceived in taking it. Let me ask you, What is your Aethiops Mineral? Is it not Quicksilver ground to a black Powder, with Brimstone? and in as great Esteem with you as any of your Medicines?

The list of diseases which Dr. Dover cured with mercury does not quite include everything from A to Z but it does embrace practically every known malady from Apoplexy to Worms. For the most part he prescribed the metal itself, as in treating consumption where he found that "crude Mercury is the most beneficial Thing for the Lungs, taking one Ounce every Morning." To promote conception, Dover recommended "An Ounce may be taken once a Day for a Month or two." Usually his doses were smaller, although he pre- scribed a pound or a pound and a half for the cure of intestinal obstruction.

Dover seems to have enjoyed a fashionable practice and was familiar with the medical literature. He cites a number of authorities, including Sydenham, Boerhaave, Willis, Freind, Cheyne, and Boyle. His admiration for the "learned Belloste," however, suggests that he chose to quote from those physicicans who supported his views on mercury. Belloste, for example, held that "Mercury, whose Virtues I here publish, is a Miracle of Nature, and the greatest Gift of Providence in the whole Materia Medica." Compared with him, Dover was an ultra-conservative. Belloste's abandon in prescribing mer- cury saw its high point in his acceptance of the advice of Marianus Sanctus that no less than four Pounds of mercury be taken for intestinal obstruction (Dover 1733). This makes Dover's pound and a half look almost homeopathic in size.

Dover's arrogance and his aggressive personality invited controversy; his opponents were not slow in accepting his challenge. There followed a war of words which lasted for nearly 50 years and gave rise to some of the most colorful polemics in the history of medicine, worthy of Dr. Johnson's "age of quarrelers and detractors."

Among the first to attack Dover was Daniel Turner (1667–1740) (1733) who made a cause célèbre of the death of the prominent actor Barton Booth, allegedly as a result of Dover's treatment with quicksilver. (Turner was awarded his M. D. degree by Yale College in 1723, thereby receiving the first medical diploma granted by an American school.)

Another of Dover's more formidable antagonists was H. Bradley, whose *Treatise on Mercury* was published anonymously in 1733. Countering the extravagant claims of Dover and Belloste with equally strong language, Bradley says in his preface that "If Men were to search the whole Dispensary, they could not pitch upon a Drug more improper for a Panacea, or Universal Medicine, than Mercury; it being very precarious in its Operation." At the same time, Bradley, who shows a wide familiarity with all aspects of mercury, does not entirely condemn its proper use. To support his cautious stand, he cites all sorts of cases of poisoning, both therapeutic and occupational. One of his interesting comments is "Some say a Jew from Lisbon discovered its Use in the Venereal Disease by chance."

Bradley may have been the first to call attention to a type of complication that was to attract attention 200 years later, following the introduction of the Miller-Abbott tube (Chap. 11). He pointed out that ". . . if some of the Mercury should be lodged in the Appendix Vermicularis, or blind Gut . . . it may there in time acquire a corrosive Nature, and destroy the Substance of that Bowel . . . ."

Dover and Belloste were by no means alone in their espousal of the marvels of mercury. They were vigorously supported by a number of prominent practitioners of their time, including George Cheyne (1671-1743) whose words appear in the Preface. Hyperbole and extravagance of expression were characteristic of the period, while the status of medical practice had reached the low point so poignantly painted by Samuel Wesley:

> When Radcliffe fell, afflicted Physic cried
> How vain my power! and languished at his side.
> When Freind expired, deep-struck, her hair she tore
> And speechless fainted, and revived no more
> Her flowing grief no further could extend;
> She mourns for Radcliffe, but she dies with Freind.

An antimercurialist might be tempted to find causal relationship between the rise of quicksilver and the decline of medicine. Analogous medical faddism was known before and has been known since the 18th century.

The excesses and absurdities of the mercurialists of the 18th century did not escape the attention of the great English satirist, Thomas Rowlandson (1756-1827). Chosen as the "good guy" in one of Rowlandson's engravings was Mr. I. Swainson, an arch-enemy of quicksilver, who in 1797 published in London a tract "Mercury stark naked. A series of letters addressed to Dr. Beddoes, stripping that poisonous mineral of its medical pretensions, and showing that it perpetuates, increases and multiplies all the diseases for which it is administered; and while it may sustain worthless branches of medical practice, the use of it is an opprobrium to the scientific and moral character of the profession." Ironically, the building at 21 Frith Street in Soho, Swainson's stronghold, depicted by Rowlandson, (Fig. XVII-1), housed a brothel in 1969.

Figure XVII-1 Eighteenth Century advocates of mercury therapy face Mr. Swainson, a foe of mercury therapy, in front of his shop in London.

Enthusiasm for quicksilver spread to the New World. The *Columbian Centinel* of Boston, on September 8, 1798, published a lengthy communication signed by Doctors Isaac Rand and John Warren, dealing with the ravages of yellow fever.

> It is with the highest degree of pleasure [say the writers] that we communicate to the public, our hopes that after proper evacuations, the use of calomel, may be found to answer . . . important purposes. This medicine has accordingly been used with much success in 15 patients within 18 days, all of whom excepting one, have recovered; or have past the dangerous period. It has not been given in the usual doses, for the purpose of an evacuation by the intestines, but in small doses of one, two or three grains, every hour or two, so as to produce a salivation as soon as possible; . . . and in every instance as the salivation came on, the disease has abated.
>
> Coinciding in sentiment, respecting the use of mercury, so as to produce a salivation, we with pleasure mention the learned Dr. Rush of Philadelphia, but the Method is more explicitly and highly recommended by James Clark, M. D. F. R. S. E. in a treatise on the yellow fever, as it appeared in the Island of Dominica, in the years 1793, 94, 95 and 96.

(The same page of the *Centinel* carries a story about Buonaparte's capture of Malta and about the Kings's troops killing 17,000 Irish rebels in a single engagement.)

Nineteenth Century

Therapeutic uses of mercurial drugs during the 19th century resulted in the production of a literature of prodigious proportions. Proksch's compilation (1891) lists 1,121 titles on mercury in syphilis published between 1800 and 1889, and the number of accounts of other uses is of similar magnitude. And this was long before the days of publish or perish logorrhea. Nothing short of a work of encyclopedic size could begin to present a review of this voluminous mass. Selected samples must suffice to portray the role of mercury in 19th century medicine.

One of the leading mercurialists of the early years of the 19th century was James Curry (?-d. 1819), known to his associates as "Calomel Curry." As a teacher at Guy's Hospital, Curry numbered among his students Charles Turner Thackrah and his classmate, the poet John Keats. When the latter contracted a venereal disease, probably gonorrhea, he wrote to a friend in October 1817 that "The little Mercury I have taken has corrected the poison and improved my Health. . . ." (Cited in Hale-White 1938). His use of mercury can be explained by the influence of Curry who recommended this medicine for practically all ailments (1809). Had Keats's trouble been syphilis he would almost certainly have used more than a "little Mercury."

The second volume of medical theses published by Caldwell in Philadelphia in 1806 contains several dissertations which give insight into the status of mercury in early 19th century American medicine. One of the theses (Stuart 1806) says:

"Mercury has not only eliminated the venereal virus, humbled the obstinacy of dropsy, broke the enchantment of epilepsy, and subdued an innumerable list of diseases, equally inimical to life, but now compels malignant fever to own its sway."

The author's definition of a malignant fever includes yellow fever, and Dr. Rush is quoted as saying of this disease that "effects of mercury, in every case where salivation was induced were salutary." In order to produce a salivation, mercury may be administered by several means:

1. By the mouth.
2. By the gums.
3. By frictions.
4. By shoes or socks impregnated with the ointment.
5 By ointment in the form of clysters.
6. By fumigations.

Elsewhere in this collection of dissertations, Tongue (1806) mentions the administration of mercury to infants by rubbing it into animals from which milk is obtained for infant feeding. This procedure is taken from Swediaur who describes its application to milk supplies from humans as well as from other animals.

An early 19th century evaluation of mercury is given in the following letter (Townsend 1811):

A Case of Phthisis Pulmonalis cured by Mercury.
by Dr. L. Townsend.
Philadelphia, August 1st, 1810.

Dear Sir,

Permit me to address this note to you for publication, in your very useful periodical pamphlet. If there be a specific in medicine (which I entirely deny) mercury is exclusively entitled to this character. However, certain it is, that by this hero of the Materia Medica, there are more diseases cured, than by all the formidable lists of remedies contained in medical volumes beside: mercury has not only eliminated the venereal virus, humbled the obstinacy of dropsy, but compels diseases of the utmost malignancy to own its sovereign sway — it is the regula regulans (and if I may be allowed the expression) the panacea of medicine—it disappoints and blasts the hopes of the grisly tyrant death, and arrests from his insatiate grasp the unhappy victims of disease. There is, perhaps, no malady, to which the suffering system of man is subjected, more distressing and painful, than Phthisis Pulmonalis; it has long been a desideratum of the first magnitude, to discover a remedy adapted to the cure of this dreadful, this horrid disease — and thanks to the indefatigable researches of a Rush, for the discovery of this herculean medicine.

Some of these words bear a striking resemblance to those used a few years earlier in the thesis quoted above.

Dr. Townsend goes on to describe how he had cured George Shively of Kensington who had been ". . . labouring under phthisis pulmonalis of 12 months duration." The treatment included venesection and calomel. That a reputable medical periodical accepted Dr. Townsend's letter for publication suggests that the views expressed did not do violence to the beliefs of that time and place. And Philadelphia was the center of medical learning in America during the post-revolutionary period!

Claims as to the efficacy of mercury in treating a variety of diseases other than syphilis were not the exclusive province of rabid 18th century mercurialists such as Dover and Belloste or of enthusiasts like Dr. Townsend. Abraham Colles (1773-1843), of Colles fracture fame, was noted for his objectivity and yet he was an ardent, if not rabid, admirer of mercury. In 1837 he wrote:

To the observations on the use of mercury, in Syphilis I have added some observations on the use of this medicine in the treatment of diseases not venereal. In this second part of the work I have studiously avoided any mention of its use in the diseases in which its efficacy is generally known and long established; and I have designedly and strictly confined myself to those in which it has been only very rarely employed, but in which I have had the good fortune to have found it a most active medicine, and a most speedy remedy.

Colles found mercury ". . . an invaluable medicine in derangements of the brain and nervous system accompanied with paralysis of the voluntary muscles" such as hemiplegia, paraplegia and total paralysis. In addition, it was useful or actually curative in treating epilepsy, gastrodynia [bellyache], ulcers of the tongue and extremities, and chronic tumors of the abdomen. In none of these conditions was syphilis implicated as a cause.

Mercury's versatility, in fact its omnipotence, is nowhere better illustrated than in the published version of Latham's *Lectures on Subjects Connected with Clinical Medicine: Comprising Diseases of the Heart* (1847). Here it is found that the metal is useful not only because of its general antiphlogistic properties but also because of its specific action in several diseases of the heart, including endocarditis and pericarditis. A direct relationship was shown between the ease with which salivation is produced and the rapidity of recovery. That this is not the idle boasting of a quack is attested by the credentials of Dr. Latham: Fellow of the Royal College of Physicians, late Physician to St. Bartholomew's Hospital and Physician Extraordinary to the Queen. One can only say God save the Queen!

Nineteenth century pharmacopoeias, formularies and dispensatories provide information on the preparations of mercury (and other drugs) available to the physician along with usual doses and uses. Medical textbooks reflect the utilization of the compounds in various diseases. Toxicology texts describe improper or excessive use as well as therapeutic accidents and incidentally give additional insight into how the drugs were being employed. The book *On Poisons,* by A. S. Taylor, published in 1875 is a good example of its genre because it reflects the situation in both Great Britain and the United States during the middle years of the 19th century.

Taylor states that metallic mercury is not commonly regarded as a poison, that it has been prescribed and taken in large doses by patients suffering from intestinal obstruction and that in some cases as much as two pounds has been administered. He emphasizes the need for caution, however by citing the case of a girl who swallowed four and a half ounces of mercury for the purpose of producing an abortion and who developed a mercurial tremor. Blue pill, in which mercury is mixed with a fatty base, and gray powder, which is mercury with chalk, are among the mercurial drugs noted as being in common use and which can cause untoward reactions.

Corrosive sublimate (mercuric chloride, bichloride of mercury) is discussed at great length by Taylor, principally in relation to its high toxicity. He describes its external use in the treatment of skin disorders such as ringworm and mentions that it is present in some quack medicines.

Calomel (mercurous chloride) ". . . although commonly regarded as a mild medicine, is capable of destroying life, even in comparatively small doses." By this he means six to 12 grains.

Other mercury compounds mentioned by Taylor are ammoniated mercury (white precipitate) used in ointments, red oxide (red precipitate), cinnabar, cyanide of mercury, mercury sulfate (turpeth mineral), nitrates of mercury and mercuric methide (methyl mercury). Except for ammoniated mercury, none of these materials had medicinal uses.

A tradition established by Molière in France and by the Restoration dramatists in England found expression in America during the first decade of the 19th century in the form of *The Mercuriad,* a tragi-comedy in five acts, by an anonymous author (A Friend to Mankind 1807). It had the sub-title, *Spanish Practice of Physic,* and was described by its creator as ". . . a burlesque on the excessive use, and an exposition of the malignant effects of MERCURY, Introduced into the modern practice of Physic," and was signed "A Friend to Mankind."

If quicksilver is taken as an example it is clear that the practice of medicine in the 19th century was more closely related to past than to future times. Alchemical ideas had not completely disappeared, and ancient humoral doctrines were still very much alive. Bleeding and purging retained their sadistic sway; extreme measures were taken in the sick room and then attacked and defended with extreme vigor. Pharmacology, with its beginnings early in the century, had achieved the status of a recognized discipline before the century's end, but no one showed any great interest in examining the accepted dogma about the virtues of mercury. No more was known about it in 1900 than in 1800, or perhaps in 1700.

REFERENCES

Bradley, H. 1733. *Physical and philosophical remarks on Dr. Dover's late pamphlet, entitled, "The ancient physician's legacy to his country."* London: C. Rivington.

Caldwell, C., ed. 1806. *Medical theses, selected from the inaugural dissertations published and defended by the graduates in medicine of the University of Pennsylvania, and other medical schools, etc.* Philadelphia.

Colles, A. 1881. *Selections from the works of Abraham Colles.* Ed. R. McDonnell. London.

Curry, J. 1809. *Examination of the prejudices commonly entertained against mercury as beneficially applicable to most hepatic complaints.* London.

Dewhurst, K. 1957. *The quicksilver doctor: Life and times of Thomas Dover.* Bristol: John Wright and Sons.

Dover, T. 1733. *The ancient physician's legacy to his country.* 2d ed. London.

Feissinger, C. 1897. *La thérapeutique des vieux maîtres.* Paris.

"A Friend to Mankind." 1807. *The Mercuriad.* Lansingburgh, N. Y.

Hale-White, W. 1938. *Keats as doctor and patient.* London: Oxford University Press.

James, R. 1745. *A medicinal dictionary.* 3 vols. London.

Latham, P. M. 1847. *Diseases of the heart.* Philadelphia.

Proksch, J. K. 1891. Die Litteratur ueber die venerischen Krankheiten. Bonn.

Quincy, J. 1719. *Pharmacopoeia officinalis et extemporanea:* or *A compleat English dispensatory, in four parts.* 2d ed. London.

Stuart, J. 1806. A dissertation on the salutary effects of mercury in malignant fevers. In *Medical theses,* ed. C. Caldwell. Philadelphia.

Swediaur, F. 1815. *A complete treatise on the symptoms, effects, nature and treatment of syphilis.* Translated from the fourth French edition by Thomas T. Hewson. Philadelphia.

Taylor, A. S. 1875. *On poisons.* 3d American ed. from the 3d English. Philadelphia: Henry C. Lea.

Tongue, J. 1806. An experimental inquiry into the modus operandi of mercury in curing the lues venerea. In *Medical theses,* ed. C. Caldwell. Philadelphia.

Townsend, L. 1811. A case of phthisis pulmonalis cured by mercury. *Philadelphia Med. Museum* 1:166–67.

Turner, D. 1733. *"The ancient physician's legacy" impartially surveyed.* London.

ADDITIONAL READINGS

A Country Physician. 1733. *Remarks upon the late scurrilous pamphlet entitled "A review of the quicksilver controversy."* London.

Dangerfield, H. P. 1806. An experimental essay on cutaneous absorption. In *Medical theses,* ed. C. Caldwell. Philadelphia.

A Mercurialist. 1733. *A short review of the quicksilver controversy.* London.

18

Mercury in 20th Century Medicine

As a remedy for syphilis, mercury finally disappeared from the scene just before the midpoint of the 20th century. In dermatology, another and much older use has managed to survive, possibly because the minor skin ailments for which mercury ointments are used have attracted less attention than lues venerea. For a time, mercurial diuretics occupied a prominent place in the materia medica but their status has been reduced by the chlorothiazides. Several mercurial antiseptics are listed in the most recent pharmacopoeias, and a related use is found in contraceptive jellies. Radioactive mercury has been introduced as a means for localizing brain tumors and for studying kidney function. All in all, mercury's role in medicine has become far from impressive, and there are valid reasons for reducing it still further.

Diuretics

Early History of Mercury as a Diuretic

One of the disease complexes discussed by Paracelsus in his *Elf Tractat* (Bombastus 1922-31) is dropsy (Wassersucht). Mercury is not mentioned in the section dealing with the treatment of this condition, but in a succeeding paragraph it is stated that mercurial drugs, the nature of which is not specified, are the best form of treatment. As is true in much of the writing of Paracelsus, the entire account of dropsy is a mixture of mumbo jumbo and sense. For reasons which are not entirely clear, mercury is said to be beneficial because Wassersucht is caused by the influence of the stars. Thus, with the right answer and the wrong reasons, Paracelsus becomes an early proponent of the use of mercury as a diuretic. This probably is the basis for statements such as "Calomel (mercurous chloride) was used as a diuretic by

249

Paracelsus in the 16th century . . . . " (Goodman and Gilman 1959) and "The medicinal use of mercurials began in the 16th century with the use of calomel as a diuretic" (Friedman 1957). The second of these quotations sounds very much like a paraphrase of the first but no reference is given with either. That calomel was the mercurial used by Paracelsus is a matter of inference. Certainly it is not true that "the medicinal use of mercurials began in the 16th century" and it is unlikely that any controlled observations were made at that time to prove diuretic efficacy. Simple clinical results, however, may have been sufficient to demonstrate the diuretic action of calomel. Nothing in the formularies, dispensatories and pharmacopoeias of the 16th or 17th centuries indicates that mercury compounds were widely used as diuretics during that epoch, although Woodall, in a poem "In Laudem Mercurii" published in 1639, says:

> Sweat to provoke, thou goest before,
> and urine thou canst move good store . . . .

He also speaks of "mercurius Diaureticus." On the other hand, Record, in 1651 lists 47 "Medicines which doe provoke Urine" and does not include mercury on his list.

In the 18th century Boerhaave, who was distinguished both as a clinician and as a chemist, gave his blessing to the use of mercury as a diuretic when he wrote: "It is serviceable in the dropsy, as well as in the venereal disease, and also in the most obstinate diseases of the glands" (1753 translation, Vol. 2, p. 311).

A comprehensive treatise on the ill effects of mercury published in 1837 (Dieterich) includes a discussion of "Urorrhoea mercurialis. Merkurieller Harnfluss." The author states that this mercurial diuresis is one of the rare complications resulting from the use of mercury, in fact he could find in the literature only two cases of this sort. Both of the cases had been reported by J. D. Schlichting, one in 1744 and the other in 1748, and one of the patients had experienced marked salivation and diarrhoea in addition to the diuresis. The latter, as a side effect of the mercury treatment of syphilis, was to initiate a new era of diuretic therapy nearly 200 years later (see below).

The ancient tradition of herbals, bestiaries and lapidaries found a 19th century exponent in the person of John Stephenson, M. D., F. R. S., whose *Medical Zoology and Mineralogy* was published in 1832. Four naturally occurring species of mercury are listed, followed by some of the mercury compounds that are included in the London and Edinburgh pharmacopoeias. Stephenson notes that mercury is useful in the different forms of dropsy and that it "operates powerfully as a diuretic."

Popular mercurial formulations mentioned by Stephenson are De Velno's vegetable syrup (corrosive sublimate in a decoction of woods), Spilsbury's drops (sublimate and gentian) and Gowland's lotion (sublimate and bitter almonds). The first of these bears the same name as the remedy sold by the antimercurialist, Mr. Swainson (Fig. XVII-1).

Among those who have been credited with discovering or re-discovering the diuretic value of mercurials, the name of A. P. W. Philip has appeared (Friedman 1957). This British physician was a prolific writer with strong convictions on many aspects of medicine, physiology, anatomy, pathology and other subjects. His views must have commanded respect since he was a Fellow of the Royal Society and often read papers before that body. In 1834 he published a monograph of 60 pages *"On the Influence of Minute Doses of Mercury. . . ."* By minute doses he means one-half grain of Blue Pill three or four times a day. His thesis is that small doses of any drug act as a stimulant while large doses act as a sedative or depressant. According to Philip "The most remarkable of the effects peculiar to mercury, is its influence on the liver." But since, in minute doses, it has ". . . a power of exciting the various secreting surfaces . . . [it] must prove a means of relief in many states of disease, especially those attended with a general failure of power of these surfaces . . . [as] in the various forms for the dropsy . . . ." The matter-of-fact way in which Philip writes of the value of mercury in dropsy and his failure to claim credit for discovering this action strongly suggests that the diuretic effect of mercury was well known in his time.

The work of Jendrassik (1886) is sometimes cited as a significant event in the history of mercurial diuretics (Friedman 1957; Shoemaker 1957). In a report published in 1886, Jendrassik gives detailed clinical histories of eight patients, most of whom had congestive heart failure, who were treated with various drugs, including calomel. He points out that prior to his time all accounts of the successful use of calomel for reducing edema had involved the use of other drugs in combination with the mercurial. He, too, used combinations, but compared the efficacy of calomel alone with that of combined therapy, concluding that this drug alone was just about as good as the combinations when given in doses of 0.20 grams two to four times a day. His article includes no references to earlier work.

Mercury Diuretics in the 20th Century

Like the famous story of Rhazes and the ape, that of the "discovery" of the organo-mercurial diuretics has been told often and not always accurately. In both cases original sources are readily accessible; in the case of the diuretics one of the principals, Dr. A. Vogl, published an account of the event about 30 years after its occurrence (1950).

As a third year medical student working in the Wenckebach Clinic in Vienna in 1919, Vogl had been instructed by an attending physician, Dr. P. Saxl, to give one cc of mercury salicylate parenterally to a syphilitic patient. The student did not completely understand the order and requested the hospital pharmacist to make up 10 cc of mercury salicylate in water. After several days the pharmacist advised him that the mercurial was not soluble in water but could be prepared in an oil base. This resulted in further delay. In the meantime a recently discharged Austrian army doctor who was at the hospital as an observer told Vogl that he had just received some samples of a new parenteral mercurial anti-luetic drug called Novasurol, and suggested that he try it while waiting for the mercury salicylate. Vogl accepted the suggestion and was struck by the marked diuresis experienced by the patient. He promptly repeated the treatment on a second luetic patient and got the same result of diuresis. His preceptor, Dr. Saxl, extended the observations to about 60 patients, with and without syphilis, with and without edema. He published the results in 1920. Saxl also tested the diuretic effect of mercury salicylate and found it to be effective in edematous patients, but very much less so than Novasurol. The story was retold by Dr. Vogl in 1969 in a slightly different and more colorful version (Vogl 1970).

The announcement of this discovery was followed by widespread and almost indiscriminate use of Novasurol for producing diuresis. Although Saxl had found no contra-indications and no toxic effects, Vogl states that there were many cases of severe mercury poisoning in Vienna during the months which followed the introduction of this new parenteral mercurial diuretic. Reports of fatalities attributed to its use began to appear in the medical literature in the early 1930's and continued through the early 1950's (Voress and Smelcer 1957). Reviews published in 1948 (Kaufman) and 1949 (Stanley) describe 32 and 27 fatal reactions, respectively, but there was some duplication of cases in the two reports. All of the deaths following the use of these diuretics may not have been due to the drug since, as pointed out by Kaufman and emphasized by E. A. Brown (1957), many of the patients were in a moribund condition at the time of administration. The use of Novasurol as a homicidal agent also has been described (Leschke 1934 translation).

Intensive investigation on the part of the pharmaceutical industry resulted in progressive improvement in the mercurial diuretics, with elimination of the toxic effects without loss of therapeutic efficacy. Even after the advent of the chlorothiazide diuretics around 1960, the organomercurials retained a place in the materia medica, with eight listed in the 1967 edition of the United States Dispensatory (*Dispensatory of the United States*) and a like number in the 1968 edition of the *Merck Index*.

Successful treatment of dropsy, particularly with mercurials, may be accomplished by means other than diuresis. Substantial quantities of fluid can be removed from the body as a result of salivation, diarrhoea and sweating. The last of these is probably not very important, although Dieterich (1837) describes a case in which "Hidrosis mercurialis" resulted in significant fluid loss. Salivation, with the production of a quart or more a day, was not at all uncommon during the aggressive treatment of syphilis with a mercurial drug. The saliva was not ordinarily swallowed but discharged into a receptacle so that it could be measured. This, in fact, was one way of judging the adequacy of treatment. Diarrhoea was and is a common result of the ingestion of some forms of mercury and this may contribute to the dehydrating effect of calomel and corrosive sublimate. Diuresis and dehydration are not necessarily synonymous, the former being only one of several ways in which the latter can be produced.

Theories of Diuretic Action

An understanding of the diuretic action of mercury offered no difficulties to the physicians of the early part of the 18th century, as attested by the words of John Quincy, M. D. (1730). The properties of mercury, says Quincy, can be explained by geometric laws ". . . that upon the Division of solid Spheres, their Gravities decrease in a triplicate Proportion of their Diameters; but the Superficies [= surface — *ed*.] only in a duplicate." The repeated division of mercury globules renders them lighter and lighter and it is this which ". . . occasions such prodigious Alterations, in rendering the animal Fluids thinner, and breaking open the secretory Passages."

Two hundred and fifty years after Dr. John Quincy had given his explanation of how mercury produces increased urine flow, the question still was unresolved. The problem is succinctly stated by Baer and Beyer in a comprehensive review published in 1966: "Since the original disclosure of the diuretic propensity of the antisyphilitic organmercurial Novasurol by Saxl & Heilig (in 1920), the literature on the site and mode of action of these agents has been as conflicting as the recognition of their utility had been uniform."

As early as 1928 it had been demonstrated that the site of action was on the kidney (Govaerts) and this was later confirmed (Bartram 1932; Pitts 1958). Attempts at more specific localization led to the conclusion that the proximal portion of the renal tubule is the principal but not exclusive site, and that the distal convoluted tubules as well as the upper collecting tubules may also be involved (Giebisch 1958; Giebisch and Windhager 1959; Giebisch 1960; Windhager and Giebisch 1961). In regard to the mode of action of the organomercurial diuretics, available evidence points to depression of active transport of sodium rather than to decreased permeability of the tubules to chloride (Weston 1957; Jamison 1961). Inhibition of adenosine triphosphatase may also play a role (Jones, Norris, and Landon 1963; Mason and Saba 1969). Studies on the sub-cellular location of the site action point to the microsomes and mitochondria (Bergstrand et al. 1961), but this has yet to be confirmed (Baer and Beyer 1966).

Difficulties in understanding the action of organomercurial diuretics have not been limited to their physiology and pharmacology; the chemistry has also presented problems. One of the complicating features is, as stated by Hughes (1957) that "The biological effect of mercury has been found to vary according to the nature of the chemical compound administered." This is certainly true, but it tells only a part of the story. Variation in responses can also depend on the route of absorption or administration and on the size of the dose as well as on a number of "host factors," some of which can be determined and some of which cannot.

Ointments

Of the hundreds of prescriptions in the present-day materia medica there is probably none which has had as long a history as mercury salves and ointments. They may possibly have been used in ancient Egypt and Assyria, and it is almost certain that Hippocrates prescribed them. There is no doubt that mercurial salves were employed by Celsus. Galen, with all his antipathy toward quicksilver, and in spite of his great influence, brought about only a temporary and partial cessation in thier use. Many of the leading Arab and other medieval physicians prescribed mercury ointments, thus initiating a period of continuous use lasting for at least a thousand years. The earlist pharmacopoeias, formularies and dispensatories, as well as the most recent, all contain listings of mercury ointments. This is a truly amazing record!

Modern medicine prides itself on its scientific approach to questions of diagnosis and therapy, condemning what is scornfully called empiricism (a term which is closely associated with quackery in the minds of many physicians). Governmental agencies, particularly in the United States, have properly taken steps to protect the public by insisting that drugs, or combinations of drugs, be of proven safety and efficacy before they may be placed on the market. This principle was recognized as early as the 14th century by Bernard de Gourdon (died c. 1320) who taught at Montpellier that ". . . if we want to test some drugs on the human body, we should first experiment on birds, then on dumb animals, then in hospitals, then on Franciscans, and so on progressively, lest it should prove to be poisonous and so fatal" (McVaugh 1969). Mercury ointments have eluded both the scientific eye of the physician and the watchful eye of the government official.

No drug can be used completely without fear of adverse reactions. Decisions for or against use are based on considerations of risk versus benefit. The use of mercury ointments is attended by serious and frequent risks. The literature, starting with the writings of Albucasis and Avicenna, contains many reports of systemic and local untoward effects, the latter being both irritative and allergic. Skin lesions for which mercury ointments are customarily prescribed are not a threat to life and consequently do not call for therapy which may indeed be life-threatening.

What about efficacy? Present practice demands that efficacy of any drug be demonstrated through controlled experiments, first on lower animals and ultimately on humans. Matched groups of subjects, treated and untreated, are compared, great pains being taken to see that the only variable is the presence or absence of the drug under test, and that the observer and subject do not know who is receiving the drug and who is a "control." A thorough search of the literature has failed to turn up any evidence that anything resembling a controlled clinical evaluation of the efficacy of the mercurial ointments has ever been undertaken. On the other hand, a bibliography (Voress and Smelcer 1957) covering the literature on mercury toxicity for the period 1903-1955 lists no less than 13 reports of serious reactions to mercury ointments, with at least one fatality. Three instances in which nephrosis developed following the use of ammoniated mercury ointment were reported by Becker in 1962 in an article in which he cites several cases in addition to his own. Another case of nephrosis was reported from the Mayo Clinic in 1967 (Silverberg, McCall, and Hunt). No doubt the adverse reactions which are recognized and reported represent only a fraction of those which actually occur.

Some of the published articles on the testing of mercury ointments contain nothing more than formulas and directions for preparing a superior product (Franklin 1928; Jordan et al. 1932) while others describe methods for assaying the amount of mercury present (Allport 1928). One study, designed to ascertain why some formulations were better than others, led to the conclusion that the presence in the ointments of gelatin and a higher proportion of water were responsible for the greater bacteriostatic properties when the preparations were tested on agar cultures (Schiller 1938).

Research workers at the U. S. Food and Drug Administration proposed an assay method depending on the amount of mercury found in the kidneys of test animals (Laug, Vos, and Kunze 1947). This would be an index of absorption and availability of mercury as a systemic rather than as a local remedy, rather surprising in 1947 when inunction therapy should long since have passed into history. These investigators do state, however, that skin permeability ". . . is not necessarily the sole criterion of therapeutic efficiency," but they do not mention any other criteria.

In an article in 1933, the reason for the continued popularity of one mercury ointment was raised:

> How did yellow oxide of mercury salve—fine old panacea for all the ills of the eye and adjacent anatomic structures, known to every physician, surgeon, specialist, nurse, optician, housewife and grandmother—obtain its undisputed place in the pharmacopoeia and in the hearts of our ailing countrymen? . . . . Long ago abandoned by most experts, why does it tower above the one other preparation embalmed in the memories of generations of medical students and share its place only with a bastard silver agglomeration [argyrol] as "something good for the eyes"? (Hosford and McKenney 1933)

These words could be repeated in 1972 with undiminished cogency. There is still no answer to the question other than a persistence of medieval theory. At one time the formula was a family secret and its efficacy in treating assorted diseases of the eye was related to the constitution of the patient. In California a century ago mercuric oxide ointment was spoken of as "Barkan's Golden California Eye Salve" and Dr. Barkan resisted the urging of his friends to patent the remedy. When subjected to bacteriological testing, the yellow oxide of mercury ointment was found to be little more effective than its petrolatum base. In actual practice, petrolatum alone has been shown to be as effective as the mercury ointment in treating infections of the eyelids, and the same is true in the treatment of pediculosis (Hosford and McKenney 1933).

The survival of mercury ointments in the materia medica invites reflection. How have they survived in the face of known danger and unknown efficacy?

How many other drugs are in a similar position? Is empiricism in medicine as reprehensible as some scientifically-minded physicians maintain? Who, if anyone, would be or should be interested in performing a controlled study of the efficacy of mercury ointment?

If the use of salves which contain quicksilver is justified on the basis that the Egyptians, Assyrians, Arabs and medieval physicians held them in high esteem the users will be practicing empirical medicine and should recognize what they are doing. If mercury is one of many accepted drugs of known danger and unknown efficacy, the pharmacopoeias are in need of some drastic revisions. Governmental regulatory agencies, in their zeal to keep new, unproven drugs from being sold, should not overlook the continuing danger from older drugs of known toxicity and unproven value.

Quicksilver has always led a charmed life, surrounded by mystery, magic and mysticism. Perhaps that is why it continues to command the faith of practicing physicians, themselves descended from an ancient priesthood.

Antiseptics

Along with ointments and diuretics, organic mercurial "antiseptics" represent the third type of mercury compound in current (1972) medicinal use. Introduced about 1920, these materials gained almost immediate popularity in the home medicine chest and in the hospital operating room, principally as a substitute for tincture of iodine. Perhaps the most striking feature of some of the organomercurial "antiseptics," in addition to their brilliant colors, is the difference in the manner in which they are described by their manufacturers and by the authors of standard textbooks of pharmacology. The former speak of them as highly effective agents with strong antiseptic properties, while the latter use such terms as "feebly active" (Goodman and Gilman 1965) and refer to them as bacteriostatic rather than bacteriocidal.

A study of three of these popular organomercurial compounds, undertaken in 1948, points out that bacteriostasis but not killing may occur, and this casts serious doubt on the efficacy of organomercurial "antiseptics" (Morton, North, and Engley 1948). The summary of the report states that "The organomercurial compounds 'mercurochrome,' 'merthiolate,' and 'metaphen' as supplied in aqueous solutions on the market, possess many shortcomings as disinfectants."

At a symposium on antiseptics held in 1949 (Davis 1950), a spokesman of the pharmaceutical industry ". . . presented evidence to show that some of the mercurials are just as capable of preventing infection with clostridium tetani spores as are common germicides . . ." (Brewer 1950), while a representative of the Chemical Corps of the U. S. Army, maintained that ". . . three representative mercurial compounds . . . did not fulfill the requirements of an antiseptic in that they failed to render bacteria incapable of causing infection" (Engley 1950). At least one reason why mercurial compounds "have not enjoyed a peaceful career," as Koch said (Engley 1950), is brought out by this little contretemps.

Pointing out that ". . . the effectiveness of antiseptics as applied to the human body has never been established clearly, and that the proper specific laboratory tests to reflect degree of utility have not yet been devised," Powell and Culbertson (1950) undertook to remedy the situation. They found that two proprietary organomercurial "antiseptics" ". . . remained *in situ* for as long as eight hours in titers stronger than those necessary for suppression of staphylococcal growth." They thus combined *in vivo* application with *in vitro* testing of potency but failed to demonstrate what could be expected under conditions of actual clinical use. The cloud of uncertainty, in fact, still hangs over the mercurial "antiseptics."

The organomercurial "antiseptics" illustrate the way in which the pharmaceutical industry undertakes extensive research in an effort to develop products that combine efficacy with low toxicity. Mercuric chloride was believed to have strong germicidal activity but its toxicity imposed limitations on its usefulness. A desideratum would be to find a compound that would take advantage of the presumed potency of mercury in killing microbes and at the same time not kill the person on whom the agent is used. Some of the organomercurial products resulting from pharmaceutical research, although of little clinical value as germicides, have achieved and retained great popularity, which emphasizes the familiar American folkway that promotional advertising can be more effective than scientific evaluation when it comes to the use of drugs.

The extent to which the magic of mercury has been exploited is illustrated by its use in throat tablets which are "Suggested as an aid in the relief of hoarseness, and minor throat irritations and for counteracting the foul breath often attending same." These tablets, each containing 0.3 milligrams of phenylmercuric nitrate, were being sold in the 1960's. Another interesting use is found in the product of a well-known manufacturer of toothbrushes advertised as "Germ Fighter Brand," which according to its label "Stays antibacterial in use up to four months." The bristles are stated to have as an active ingredient a mercuric complex containing 0.15 percent mercury — which is actually too little to do either good or harm. Action against the manufacturer was instituted in 1970 by the U. S. Federal Trade Commission, with charges of misleading advertising.

Contraceptives

Unwanted pregnancy has been a sociomedical problem at least as far back as the time of Hippocrates, a fact which is attested by the injunction in the

Hippocratic Oath against physicians providing women with means for producing abortion. Mercury has played a prominent, and at times tragic, role
in the prevention of pregnancy and in the related area of inducing abortion.
Mercuric chloride has been the form most commonly used, usually applied
locally in the vagina or uterus (Voress and Smelcer 1957). Oral self-
administration of sublimate with the objective of producing abortion has been
reported (Leschke 1934 translation), although the account does not make it
entirely clear that abortion rather than suicide was the objective. Metallic
mercury, too, has been taken by mouth as an abortifacient (Taylor 1875, p.
351).

Mercury compounds designed to kill spermatazoa without killing or
injuring the user came into use in the 1940's, phenylmercuric acetate being
the most popular mercurial for this purpose. Vaginal contraceptive jellies,
including those containing mercury, lost some of their popularity with the
advent of "the pill," but they were still on the market in 1970 (Physicians'
Desk Reference 1969), and are reasonably effective. Very often users have
not been aware of the presence of mercury and this has resulted in the occurrence of puzzling and distressing allergic reactions to one or the other marital
partner. It has also resulted in the finding of elevated values for mercury in
the urine or blood of supposedly "normal" individuals who were unaware
that they had been using a mercurial preparation.

REFERENCES

Allport, N. L. 1928. A new method for the assay of ointments of mercuric oxide and
    ammoniated mercury. *Quart. J. Pharm.* 1:23-27.
Baer, J. E., and Beyer, K. H. 1966. Renal pharmacology. *Ann. Rev. Pharmacol.* 6:
    261-92.
Bartram, E. A. 1932. Experimental observations on the effect of various diuretics when
    injected directly into one renal artery of the dog. *J. Clin. Invest.* 11:1197-1219.
Becker, C. G.; Becker, E. L.; Maher, J. F.; and Schreiner, G. E. 1962. Nephrotic
    syndrome after contact with mercury. *Arch. Internal Med.* 110:178-86.
Bergstrand, A.; Friberg, L.; Mendel, L.; and Odeblad, E. 1961. Studies on the excretion
    of mercury in the kidneys. *Acta Pathol. Microbiol. Scand. Suppl.* 144(51):115-17.
Boerhaave, H. 1753. *A new method of chemistry.* Trans. Peter Shaw. 3d ed., vol. 2.
    London.
Bombastus von Hohenheim, P. A. T. [Paracelsus]. 1922-31. *Sämtliche Werke.* Ed. K.
    Sudhoff. Munich: R. Oldenburg.
Brewer, J. H. 1950. Mercurials as antiseptics. *Ann. N. Y. Acad. Sci.* 50:211-19.
Brown, E. A. 1957. Reactions to the organomercurial compounds. *Ann. N. Y. Acad. Sci.*
    65:545-52.
Davis, H. L., ed. 1950. *Conference on mechanism and evaluation of antiseptics,* held by
    the Section on Biology of the New York Academy of Sciences, 28-29 October
    1949, New York City. *Ann. N. Y. Acad. Sci.* 53:1-219.
Dieterich, G. L. 1837. *Die Merkurialkrankheit in allen ihren Formen.* Leipzig.
Dispensatory of the United States. 1967. 26th ed. Philadelphia: J. B. Lippincott Co.
Engley, F. B., Jr. 1950. Evaluation of mercurial compounds as antiseptics. *Ann. N. Y.
    Acad. Sci.* 53:197-206.
Franklin, J. H. 1928. Mercury ointment. *Quart. J. Pharm.* 1:347-50.
Friedman, H. L. 1957. Relationship between chemical structure and biological activity
    in mercurial compounds. *Ann. N. Y. Acad. Sci.* 65:461-70.

Giebisch, G. 1958. Electrical potential measurements on single nephrons of necturus. *J. Cellular Comp. Physiol.* 51:221–39.

_____. 1960. Measurements of electrical potentials and ion fluxes on single renal tubules. *Circulation* 21:879–91.

Giebisch, G., and Windhager, E. E. 1959. Chloride fluxes across single proximal tubules of Necturus kidney. *Federation Proc.* 18:52.

Goodman, L., and Gilman, A. 1955. *The pharmacological basis of therapeutics.* 2d ed. New York: Macmillan Co.

_____. 1965. *The pharmacological basis of therapeutics.* 3d ed. New York: Macmillan Co.

Govaerts, P. 1928. Origine rénale ou tissulaire de la diurèse par un composé mercuriel organique. *Compt. Rend. Soc. Biol.* 99:647–49.

Hosford, G. N., and McKenney, J. P. 1933. Ointment of yellow mercuric oxide (Pagenstecher's Ointment): Its use and abuse. *J. Am. Med. Assoc.* 100:17–19.

Hughes, W. L. 1957. A physiochemical rationale for the biological activity of mercury and its compounds. *Ann. N. Y. Acad. Sci.* 65:454–60.

Jamison, R. L. 1961. The action of a mercurial diuretic on active sodium transport, electrical potential and permeability to chloride of the isolated toad bladder. *J. Pharmacol. Exp. Therap.* 133:1–6.

Jendrassik, E. 1886. Das Calomel als Diureticum. *Deut. Arch. Klin. Med.* 38:499–524.

Jones, V. D.; Norris, J. L.; and Landon, E. J. 1963. Interaction of a rat-kidney endoplasmic reticulum fraction with glycolytic enzymes. *Biochem. Biophys. Acta* 71: 277–84.

Jordan, C. B. et al. 1932. Mercury ointment. *J. Am. Pharm. Assoc.* 21:1018–22.

Kaufman, R. E. 1948. Immediate fatalities after intravenous mercurial diuretics. *Ann. Internal Med.* 28:1040–47.

Laug, E. P.; Vos, E. A.; and Kunze, F. M. 1947. Biologic assay of mercury ointments. *J. Am. Pharm. Assoc.* 36:14–15.

Leschke, E. 1934. *Clinical toxicology.* Trans. C. P. Stewart and O. Dorrer. Baltimore: William Wood and Co.

Mason, R. G., and Saba, S. R. 1969. Platelet ATPase activities: Part II. *Am. J. Pathol.* 55:215–24.

McVaugh, M. R. 1969. Quantified medical theory and practice at 14th century Montpellier. *Bull. Hist. Med.* 5:397–413.

Merck Index. 1968. Ed. Paul G. Stecher. 8th ed. Rahway, N. J.: Merck and Co.

Morton, H. E.; North, L. I.; and Engley, F. B. 1948. The bacteriostatic and bacteriocidal actions of some mercurial compounds on hemolytic streptococci. *J. Am. Med. Assoc.* 136:36–41.

Philip, A. P. W. 1834. *On the influence of minute doses of mercury.* London.

Physicians' desk reference. 1969. 23 ed. Oradell, N. J.: Medical Economics, Inc.

Pitts, R. F. 1958. Some reflections on mechanisms of action of diuretics. *Am. J. Med.* 24:745–63.

Powell, H. M., and Culbertson, C. G. 1950. Assay of antiseptics at different times after application to human skin. *Ann. N. Y. Acad. Sci.* 53:207–10.

Quincy, J. 1730. *Lexicon physico-medicum* or *A new medicinal dictionary.* 4 ed. London.

Record, R. 1651. *The urinal of physick.* London.

Saxl, P., and Heilig, R. 1920. Über die diuretische Wirkung von Novasurol und anderen Quecksilberpräparaten. *Wien. Klin. Wochschr.* 33:943.

Schiller, F. W. 1938. The bactericidal effectiveness of the improved calomel ointment. *Am. J. Pharm.* 110:289–96.

Shoemaker, H. A. 1957. The pharmacology of mercury and its compounds. *Ann. N. Y. Acad. Sci.* 65:504–10.

Silverberg, D. S.; McCall, J. T.; and Hunt, J. C. 1967. Nephrotic syndrome with the use of ammoniated mercury. *Arch. Internal Med.* 120:581–86.

Stanley, T. E. 1949. An analysis of 27 reported fatalities immediately following the injection of a mercurial diuretic. *Virginia Med. Monthly* 76:416–19.

Stephenson, J. 1832. *Medical zoology and mineralogy.* London.

Taylor, A. S. 1875. *On poisons in relation to medical jurisprudence and medicine.* 3d American ed. Philadelphia: Henry C. Lea.

Vogl, A. 1950. The discovery of the organic mercurial diuretics. *Am. Heart J.* 39: 881–83.

————. 1970. On clinical medicine. The Pirquet Lecture, delivered 14 May 1969. *Bull. N. Y. Acad. Med.* 46:39–60.

Voress, H. E., and Smelcer, N. K. 1957. *Mercury toxicity: A bibliography of published literature.* Oak Ridge, Tenn.: U. S. Atomic Energy Commission, Technical Information Service Extension.

Weston, R. E. 1957. The mode and mechanism of mercurial diuresis in normal subjects and edematous cardiac patients. *Ann. N. Y. Acad. Sci.* 65:576–600.

Windhager, E. E., and Giebisch, G. 1961. Micropuncture study of renal tubular transfer of sodium chloride in the rat. *Am. J. Physiol.* 200:581–90.

Woodall, J. 1639. *The surgeons mate or Military and domestic surgery.* London.

19

Occupational Poisoning

Had a little book by Ulrich Ellenbog (d. 1499) of Augsburg been published even within 25 years of the time it was written it would have been the only known incunabulum in industrial toxicology. For unknown reasons, Ellenbog's *Von den gifftigen besen Tempffen,* written in 1473, was not printed until 1543.

The precaution, common in the 15th, 16th, and 17th centuries, of delaying publication of a work until after the death of the author could hardly have been necessary in the case of Ellenbog's book since it contains nothing of a controversial nature that might have caused trouble with the church authorities. Ellenbog states that the work was written by his own hand and it was no doubt circulated in manuscript form. Being addressed to the goldsmiths of Augsburg it perhaps led prospective publishers to believe that it was of only local interest and hence would not justify the cost of printing. Recognition of its wider applicability came later, but in any case the number of copies produced was probably small, since not more than one or two are known to have survived. A facsimile edition was published in 1927 and an English translation in 1932.

In Koelsch's opinion, Ellenbog's book is the first work on industrial hygiene in the literature of the world, and the claim has been made that it contains the first description of occupational mercury poisoning (Ellenbog 1927). The relevant passage is: "This vapour of quicksilver, silver and lead is a cold poison, for it maketh heaviness and tightness of the chest, burdeneth the limbs and oftimes lameth them as often one seeth in foundries where men do work with large masses and the vital inward members become burdened therefrom" (Ellenbog 1932 translation). While mercury is recognized by Ellenbog as being hazardous, he is describing the effects of metal fumes in general and not specifically those of quicksilver.

Ellenbog, who was physician to the Dean and Chapter of Augsburg Cathedral, addressed his book to "The skillful, subtle and noble craft of Goldsmithery of the Imperial City of Augsburg." Particularly perceptive is the warning that vapors ". . . through their sharpness, strength and subtelty . . . always penetrate and work more strongly than the body wherefrom they are drawn," consequently, ". . . ye shall not bend too much over this vapour but turn away therefrom and bind up the mouth" (1932 translation). The author is referring not only to the vapors of quicksilver, as mentioned above, but also to those of lead and other metals. He advises that as much as possible, work should be done in the open air and not in a closed room. Antidotes and therapeutic measures, based on contemporary concepts of hot, cold, moist, and dry properties of the poison and of the remedy are included in the treatise, but there is only brief mention of the signs and symptoms of metal poisoning. Any claim that this work, in spite of its early date, contains the earliest description of occupational mercury poisoning is open to serious challenge (Mayer 1929; Goldwater 1964).

The place of Jean Fernel in the early history of occupational mercurialism is established by what he says in his *De lue venerea* (1579, Chap. nine). Chapter seven of this book, entitled Hydrargyri vires (the strength or powers of mercury), summarizes a good deal of the knowledge about quicksilver from the time of Dioscorides up to the 16th century. Fernel mentions that mercury poisoning can cause tremors, but he describes this manifestation as occurring in the treatment of lues venerea. He relates the case history of a "Faber aurarius" (worker in gold) seen in 1556, who was made ill by mercury vapor, the main symptoms being confusion and depression (Fernel 1579, p. 60). He does not describe the work process in any detail so it is not possible to tell what other noxious vapors, if any, were present in the workplace; but Ramazzini, who cites this case (Ramazzini 1964 translation), notes that fire is used to drive off the mercury as part of the gilding process. This suggests the possibility that carbon monoxide or other toxic gases may have been present in addition to the mercury vapor and may offer a better explanation for the confusion and mental depression in Fernel's patient.

Significant contributions to the literature on occupational mercurialism during the 16th century are not numerous; the two outstanding contributors were Agricola and Paracelsus.

The three books by Paracelsus, *"Von der Bergsucht und anderen Bergkrankheiten,"* cover a wide range of topics dealing with the health hazards to which miners are exposed. Quicksilver receives occasional passing mention in Books I and II, while the third book on miners' diseases treats only of mercurialism. The title accounts for its claim to priority as the first work of monograph length dealing with a specific problem in occupational health. While it does contain a description of some of the manifestations of mercurialism, such as tremor, loss of teeth, cachexia and psychic disturbance, much of it is given over to typical Paracelsian mysticism and alchemical jargon (Bombastus von Hohenheim 1923-31, pp. 517-44).

Agricola's qualifications as both physician and metallurgist appear to good advantage in his *De Re Metallica* when he gives directions for recovering quicksilver from its ore. He includes precautions against poisoning, recommending the use of the best grade of potters' clay for making the pots in which the ore is roasted and in which the vapors are collected ". . . for if there are defects the quicksilver flies out in the fumes. If the fumes give out a very sweet odor it indicates that the quicksilver is being lost, and since it loosens the teeth, the smelters and others standing by, warned of the evil, turn their backs to the wind, which drives the fumes in the opposite direction; for this reason, the building should be open around the front and the sides, and exposed to the wind" (Agricola 1950 translation, p. 428). In this one short paragraph, Agricola displays his awareness that mercury vapors are harmful, that one of the major toxic effects is loosening of the teeth, that the escape of vapors can be detected by smell, and that toxic effects can be prevented by a combination of enclosure of the process combined with natural ventilation. All of these points have their counterparts in modern industrial hygiene.

A more detailed description of mercury poisoning is given by Agricola in his *De Natura Fossilium* which was published in 1546, ten years earlier than the better known *De Re Metallica*. After citing the statement of Dioscorides about quicksilver eating through the vital organs because of its weight, Agricola points out some inconsistencies in Galen who ". . . writes in one place that the heat of the body activates it to such a point that it kills by corrosion and in another place considers it among the substances essential to mankind. These are contradictory views since a very small quantity taken into the body attacks it violently. In still another place, he writes that no one has actually tested its strength to ascertain if a potion would kill or if it could destroy the body when placed outside." These are the words of Agricola the physician (1955 translation, p. 175). He relates the story of ". . . a depraved wife [who] gave her husband quicksilver and swallowed some herself but this was ejected from the stomach without any harm." This experiment confirmed in humans what Rhazes had found in an ape. "When mixed with other substances," says Agricola, "so that it does not corrode, when taken internally or rubbed on the skin so that the body heat is able to exert its full force, it attacks the head and causes excessive discharges, part of which flows out through the mouth, part settles in the gums and cheeks and causes them to swell." Again the physician speaks, perhaps reflecting his familiarity with the salivation treatment of what he says the Italians call the French disease and the French call the Spanish disease. This account of mercurialism in a book on mineralogy presumably was directed to physicians concerned with the health of mine and smelter workers, but it is of interest also in relation to non-occupational poisoning.

Descriptions of occupational mercury poisoning may be found in the writings of persons who are not physicians, particularly when the manifestations are so severe as to be obvious to anyone. Such was the case in the quicksilver mines at Huancavelica during the first two decades of the 17th century. Juan de Solorzano Pereira, who was in charge of the mines from 1616 to 1619, wrote in 1648 that "... sooner or later even the strongest mitayos succumbed to mercury poisoning, which entered into the very marrow of their bones and made them tremble in every limb ... ," and he goes on to say that he had never known one who had survived this condition for more than four years (cited in Whitaker 1941, p. 19). Thus one of the earliest accounts of occupational poisoning due to mercury originated in America.

Nicholas Lemery (1645-1715), writing during the latter part of the 17th century, shows his familiarity with occupational poisoning due to mercury when he says that "Those who draw it out of Mines, or work much with it, do often fall into the Palsie, by reason of Sulphurs that continually steam from it; for these Sulphurs consisting of gross parts do enter through the Pores of the Body, and fixing themselves rather in the Nerves by reason of their coldness, than in other Vessels, do stop the passage of the Spirits, and hinder their course" (1686 translation, p. 160).

The *Cours de Chymie,* first published by Lemery in 1675, was one of the most widely used chemical textbooks of its period. It went through a number of editions in French and English and was translated into Latin and into most modern languages (Partington 1965). This book, with its wide circulation, must have played a prominent part in familiarizing physicians and chemists with the occupational health hazards due to mercury. It seems strange indeed that Ramazzini, who cites many writers more obscure than Lemery, makes no mention of this work.

Ramazzini's *De Morbis Artificum*

One of the great classics of occupational medicine and, in fact of all medicine, is the *De Morbis Artificum* of Bernardino Ramazzini (1633-1714) published in 1700 (1964 translation), which earned for its author the unchallenged epithet of Father of Industrial Medicine. So comprehensive was Ramazzini's book and so great its influence that little of significance was added to the literature on occupational diseases for more than 100 years. During the entire 18th century most works on the subject were based primarily on Ramazzini. His authoritative position justifies a detailed examination of what he had to say about quicksilver as an occupational health hazard.

Occupational mercurialism is discussed or described in each of the first eight of the 40 chapters of the original edition of the *De Morbis Artificum.* An expanded version of the work, published in 1713, contains 12 additional chapters, none of which has anything to do with exposure to quicksilver.

Chapter I — Miners of Metals. A reading of this chapter strongly suggests that Ramazzini himself had little or no personal experience with mines or miners. All of his comments on the hazards of mines in general and of quick-silver mines in particular are quotations from earlier writers, including Agricola (1490-1555), Athanasius Kircher (1601-1680), Bernardo Cesi (1581-1630), Gabriel Fallopius (1523-1562) and others. Ramazzini adds nothing to what had been written previously.

Chapter II — Gilders. This chapter consists of a curious mixture of first class, first hand clinical observation along with all sorts of weird and fantastic stories. The cardinal manifestations of mercurialism are all mentioned: fetid ulcers of the mouth, salivation, loss of teeth, tremors, paralysis and psychic disturbances. Emphasis is correctly placed on the dire results of inhaling the vapors when goldsmiths heat gilded articles to drive off the mercury, and this is contrasted with its safe use orally in treating worms. Still, Ramazzini has no hesitation in repeating, and apparently accepting, a report by Fernel about a man whose brain dissolved and oozed from his eyes and who continued to live for many years, presumably because he had been protected by earlier mercury treatment for syphilis.

Chapter III — Those who give mercurial inunctions. Quite naturally this chapter combines a discussion of syphilis with the hazards to those who treat it. Ramazzini tells about a surgeon with whom he was familiar who developed diarrhea, colic, and salivation as a result of administering mercury inunctions and who avoided recurrences of the affliction by preparing the ointment for his patients and standing by while they carried out the inunctions themselves. He managed to convince his patients that this was better for them, and it certainly was better for the surgeon. Ramazzini highly recommends this approach, which, of course, is good preventive medicine, at least for the therapist.

Chapter IV — Chemists. To emphasize the hazards to which chemists are exposed, Ramazzini states that two great chemists, van Helmont and Paracelsus, became seriously ill as a result of their work. There is no suggestion that any particular chemicals were involved, but the statement about Paracelsus invites speculation as to his erratic behavior being a manifestation of mercurial erethism.

In this chapter he tells the story of a resident of the town of Finale, near Modena, who brought suit against the owner of a chemical factory where corrosive sublimate was manufactured. The plaintiff claimed that the entire neighborhood was contaminated with emanations from the factory; as a result, the death rate was greatly increased. Expert witnesses testified on both sides, and the case was decided in favor of the defendant. Ramazzini discreetly avoided taking sides in the dispute, leaving the decision, as he says, to those who are expert in natural sciences, but his account of this mercury pollution must be one of the earliest episodes in which air pollution was claimed to have had an adverse effect on health.

Chapters V and VI — Potters and Tinsmiths. These two short chapters show
that Ramazzini retained some alchemical concepts about quicksilver. He
quotes Thruston as saying that "Saturn does to Mercury what the poets say
Vulcan did to Mars" (chains and fetters it), and speaks of the "mercurial
fumes of metals."

Chapter VII — Glass Workers and Mirror Makers. This, too, is a short
chapter in which Ramazzini points out the similarity between gilding and
mirror making in regard to the harmful exposure to quicksilver.

Chapter VIII — Painters. Of interest in this chapter is Ramazzini's statement
to the effect that it is common knowledge that cinnabar is made from quicksilver. The case history of one of Fernel's patients who was a painter is cited
to emphasize the dangers of getting cinnabar on the hands and of pointing the
paint brush with the lips since ". . . it is probable that the cinnabar was carried . . . to the brain by direct communication and so to the whole nervous
system . . . ."

Without in any way detracting from the importance of Ramazzini's book,
it is of some interest to assess it in terms of its impact at the time it was written and during the 18th century, and to consider its overall historical significance. Because of its scope, the *De Morbis Artificum* well deserves being
accepted as a classic in occupational medicine. In the history of quicksilver,
however, its importance is limited. Ramazzini adds nothing to what was already well known about the occupational hazards to which mercury miners
are exposed. He does no more than confirm the previously described clinical
picture of mercurialism. On the other hand, he perpetuates a number of
myths and adds a few of his own creation. Because of his prestige and authority these fanciful notions must have received wide acceptance. In effect, because his book was so admired that nothing new seemed to be needed,
Ramazzini's influence stifled progress in his field during a period when great
advances were being made in other branches of medicine. His contributions
are analogous to those of Galen — both positive and negative.

On Saturday, May 25, 1833, *The Leeds Mercury* carried an obituary notice
recording the death, two days earlier, of Charles Turner Thackrah, age 38, "a
surgeon of this town," who had been a classmate of Keats at Guy's Hospital
in 1815. To students of occupational medicine Thackrah is known as the
author of an important book dealing with the effects on health of different
types of employment (1832).

Before the appearance of Thackrah's book, Ramazzini had been the
dominant figure in the field of occupational diseases, although a number of
significant works in medicine and chemistry had appeared in the century following Ramazzini's death. During that period the industrial revolution had
begun in England and, in fact, had resulted in major changes in the nature of

the hazards to health to which workers were exposed. Because of these changes Thackrah's approach is quite different from that of Ramazzini, being sociological rather than medical. Some of the differences can be seen in the manner in which mercury poisoning is handled in the two books. Another dissimilarity is Thackrah's concern almost exclusively with conditions in Leeds while Ramazzini encompassed all of Italy and other countries as well.

The expanded version of Ramazzini's *De Morbis Artificum* contains mention of just under 100 different occupations in its 52 chapters; specific occupations in which mercury constitutes a hazard number eight (see above). Thackrah lists about 300 occupations, of which there are only two with mercury exposures: makers of looking-glasses and water-gilders. Of the latter, the author notes that they ". . . are exposed to the same poisons as the silverers of mirrors." Health hazards to medical men do not include mercury, suggesting that mercurial inunctions, if used in Leeds in Thackrah's time, were not administered by physicians. Chemists, too, apparently were not exposed to dangerous amounts of quicksilver. Mercury exposure is not included in the list of hazards to hatters, suggesting that the French process of carroting or "sécrétage" (treatment of fur with mercuric nitrate in making felt) was not known in England in the early years of the 19th century. If the Hatter in *Alice in Wonderland* was mad as a result of mercury poisoning, the process of treating fur with mercuric nitrate in making felt must have come back to England some time between about 1820 and 1865 (See "The Mad Hatter" below). As far as mercury is concerned, Thackrah contributed nothing.

Often hailed as a classic in the early American literature on occupational medicine is an essay by B. W. McCready, M. D. published in 1837, the pretentious title of which (see McCready 1943) is not justified by the substance of the work. Much of the material is borrowed from Thackrah and has a strong sociological flavor. Very few specific occupational health hazards are discussed, and mercury exposures or poisoning are not mentioned at all.

Outstanding among the studies of occupational mercurialism during the 19th century, and in fact of all time, were those conducted by Kussmaul and published in 1861. The industrial process which served as the basis for Kussmaul's observations was the plating of glass mirrors during which sheets of tin foil were spread on a table and covered with mercury, resulting in the formation of a film of mercury-tin amalgam. The glass plate to be coated was then placed on the amalgam film and pressed down with heavy weights; and the excess mercury extruded from between the impinging surfaces. This method of making mirrors, believed to have been discovered about 1500, remained in use until near the end of the 19th century. Mirror plating by means of silver nitrate solutions has largely replaced the amalgam method except in making special mirrors such as those for astronomical and other fine instruments.

During the early part of the 19th century one of the most important centers for the manufacture of mirrors was in Fürth, Germany, and it was there that Kussmaul carried out his classic work. His finding of many cases of severe and even fatal poisoning did not result in immediate correction of the hazardous working conditions; Kober (1916) states that in 1885 more than three-quarters of sick days among mirror makers of Fürth were due to mercury poisoning.

A perusal of Kussmaul's classic on mercury poisoning leads to the conclusion that it is one of the books which are more often quoted than read. References to the work in the literature of industrial toxicology give the impression that it is concerned with little other than mercurialism among the mirror makers of Fürth. This subject is covered, and in sufficient detail to establish on a firm basis the classical picture of poisoning from the vapors of metallic mercury. But this represents only one of the many valuable features of Kussmaul's work. His review of the early history of mercury poisoning, done with admirable brevity, covers most of the important facts, usually with adequate references to the literature. His collection of case reports from the literature of the first half of the 19th century is a real tour de force. These, together with the descriptions of his own cases, embrace every conceivable form of mercurialism and innumerable types of accompanying conditions and complications.

Of particular concern to Kussmaul was the possible relationship between constitutional mercurialism and constitutional syphilis, as the title of his book indicates. This was due, as he says, to questions which had been raised about the existence of chronic mercurialism and constitutional syphilis as separate entities. The massive evidence he assembled leaves no doubt as to the reality of chronic, systemic mercury poisoning. The protean nature of its manifestations is the feature which leads to the possibility of its being confused with syphilis, for Kussmaul describes cases in which both mercury and the causative agent of lues produce similar effects on practically every organ-system of the body. The weight of evidence adduced by Kussmaul supports the conclusion that syphilis and mercurialism are separate entities, but they may occur concomitantly in the same individual, and that they also may mimic each other.

In the United States, where early interest in occupational mercurialism centered primarily in the felt-hat industry, a vivid description of the clinical manifestations of mercury poisoning is given by W. Gilman Thompson in his book, *The Occupational Diseases,* the first general textbook on its subject to be published in this country (1914).

> Workmen who acquire chronic mercurialism complain of increasing headache, lassitude, weakness, insomnia and muscular pains. They become very anemic, lose weight, and are prone to resort to alcoholic stimulants to give temporary relief to their mental and physical depression and restlessness. They have strong, gross muscular tremors of the limbs and facial muscles which they call the 'shakes.' The tremors also affect the tongue and movements of the head, and the eyes roll from side to side (Teleky). There is albuminuria. The complexion is sallow and gray and expression listless.

Chronic pharyngitis, a metallic taste in the mouth, chronic gingivitis, fetid breath, vomiting and diarrhea are also observed. There are pains in the muscles and joints. A condition of advanced marasmus and asthenia finally supervenes. The mouth becomes dry and sore, the gums ulcerate, the teeth loosen, alveolar abscesses form, salivation is abundant. There may be muscular twitching as well as tremors. I have seen the latter so intense as to resemble superficially advanced paralysis agitans and the victim is unable to convey the contents of a spoon or a glass of water to the mouth without spilling, for the tremors are increased by volitional movements, unlike paralysis agitans. From these symptoms the patient may recover after a long interval of cessation of work, but too often he returns again and again, each time becoming worse, until he becomes a hopeless invalid, with trembling lips and tongue, hesitating, whispering speech, a peculiar drooping or drawn expression, a staggering, drunken gait and condition akin to senile dementia.

Fatal occupational mercurialism from the metal was by no means unknown in this country as well as in several European nations mentioned by Thompson, particularly Britain, Germany and Italy, during the early part of the 20th century, but this is no longer the case.

A second, perhaps *the* second book on occupational diseases published in the United States is that of Kober and Hanson, which appeared in 1916. Unlike Thompson's, which represented the efforts of a single individual, this book combines the knowledge of 21 experts, including the two editors. The clinical section on mercury poisoning was written by Dr. Ludwig Teleky (1872–1957) of Vienna, based on his classic work *Gewerbliche Quecksilbervergiftung* (1912). A second section dealing with occupational exposures and prevention was written partly by Teleky and partly by Kober.

Several features of historical significance are found in this book, adding to its importance as an example of early American literature dealing with occupational health hazards. The list of industries in which mercury was used affords insight into a segment of early 20th century economics and technology.

One feature of special interest is found in the description of a number of industries and occupations in which a mercury hazard is present, particularly those which were later relegated to the past by changes in technology. Fire gilding, which in effect is gold plating, was found to be particularly dangerous as a cause of mercury poisoning. The process consists of applying a coating of mercury-gold amalgam, and then driving off the mercury by heat. It has been replaced by electroplating, which is much less hazardous. Similarly, as mentioned above, the use of mercury in the manufacture of mirrors has been supplanted by a method using silver nitrate. These newer methods are not only safer but are also probably more efficient and less expensive.

Among the important works on occupational mercurialism written during the first part of the 20th century, that of Zangger clearly deserves a place. Zangger's paper, almost extensive enough to be called a monograph, is based on 25 years of experience in the Swiss accident-insurance agency (Schweizerische Unfallversicherung). He covers a wide variety of occupational exposures and of clinical manifestations and is one of the first to emphasize the difference in response to inorganic and organic mercury compounds. Additional historical accounts of note are those published by Baader and Holstein and by the International Labour Office.

Descriptions of occupational mercury poison by Paracelsus and Agricola in the 16th century, by Ramazzini at the turn of the 18th, by Kussmaul in the middle of the 19th century and by Thompson in the 20th are strikingly similar. The triad of gingivitis, tremor, and erethism were the major manifestations, and are still recognized as such. This is not surprising in view of the fact that before the middle of the 19th century occupational exposure was almost entirely limited to the inhalation of mercury vapor, and in the fur felt hat industry to mercury nitrate, both being inorganic.

When A. W. von Hofmann in 1843 and Frankland in 1850 synthesized organic compounds of mercury they inaugurated a new era in industrial toxicology. This was dramatized within a few years by the fatal poisoning of two young laboratory assistants who were working with dimethyl mercury (Chap. 11), the first known victims of alkyl mercurials. About this time (1850), the first death due to mercury fulminate was recorded (Teleky 1955).

Further details on the toxicology of organic compounds are given in Chapters 11 and 12.

Hat Industry

On December 2, 1955 the *New York Times* ran a full-column story, with a dateline from Danbury, Connecticut and headlines: "600 Hatters Mark 1941 Nitrate Ban." The story notes that "The occasion was the 14th anniversary of the outlawing of the use of nitrate of mercury in the hat industry." This notable event had come to pass since "On December 1, 1941, the United States Public Health Service brought an end to mercury's use by hat manufacturers in 26 states through mutual agreements." Credit for this achievement was claimed in whole or in part by the Public Health Service, the hat manufacturers, and the secretary-treasurer of the local union of United Hatters, Cap and Millinery Workers. Cynics have suggested that credit for this "triumph" should be attributed to a war-time shortage of mercury. For close to a century prior to 1955 the ravages of mercurialism among hatters had been known and tolerated in the United States.

First among the major studies of mercury poisoning in the American felt-hat industry was that made by Dr. J. A. Freeman of New Jersey and reported in 1860. Freeman's findings were confirmed in a report published by the Board of Health of the State of New Jersey in 1878 (Dennis). In 1910, under the aegis of the Women's Welfare Department of the New York and New Jersey section of the National Civic Federation, Mrs. Lindon W. Bates, assisted by Miss Florence Roehm, undertook a survey of industrial mercury poisoning in the New York metropolitan area. The results were published in 1912 (Bates). Dozens of cases of severe mercurialism were found among hatters, a state of affairs which was confirmed by the New York City Department of Health in 1915 (Harris).

Additional reports on the health of hatters appeared during the 1920's, notably those of Alice Hamilton (1922a, 1922b), Wade Wright (1922) and the United States Public Health Service (Neal et al. 1937, 1941). The last are of particular significance.

Just as it is unfair for modern scientists to scoff at Pliny or at the alchemists because much of what was written many years ago now seems to be nonsense, so is it inappropriate to apply present-day standards to some of the studies made as recently as 30 or 40 years ago. This is especially true when analytical chemistry is involved. So dramatic have been the refinements in micro-chemical techniques that work which is much more than 10 years old may have to be viewed as belonging to an earlier age.

These brief remarks have been made as a preface to a discussion of the two later major studies of mercurialism in the felt-hat industry conducted by the U. S. Public Health Service during the 1930's. Because of their scope, these two studies have been widely quoted and, unfortunately, often misquoted. Some of the findings are difficult to understand and, in the light of more recent knowledge, difficult to accept (Goldwater 1964).

One anomalous finding in the Public Health Service study in the hatters' fur-cutting industry is the relative infrequency with which mercury was found in the urine of exposed workers. The levels of exposure, 0.06–0.72 mg per cubic meter of air, were such that all would have been expected to have mercury in the urine. Yet, according to the tables in the text, only 35 percent of urine specimens from 488 workers had detectable quantities of mercury. Lack of sensitivity of the analytical method employed is the most reasonable explanation for the anomalous results. (The summary of the report states that 65 percent of the specimens contained mercury.) More difficult to explain or understand is that of 33 active workers diagnosed as having mercurialism, 19 showed no mercury in the urine.

The second of the Public Health Service studies in the felt-hat industry also resulted in some puzzling findings similar to those in the first (Goldwater 1964). This is particularly unfortunate since the American Conference of Governmental Industrial Hygienists, which promulgates standards for permissible levels of exposure to potentially harmful agents in the work environment, chose to use this study as the basis for establishing limits for exposure to mercury (American Conference of Governmental Industrial Hygienists 1962). Fortunately, more reliable data are proposed for use in a revised standard promulgated in 1969 (American Conference of Governmental Industrial Hygienists 1971).

Still another large-scale investigation of the health of workers in the fur-felt industry was undertaken by the Division of Industrial Hygiene of the New

York State Department of Labor in 1936. The findings of this study are rarely quoted, probably because they were not published until 1948-1950 (Smith and Moscowitz 1948; Smith et al. 1949; Goldwater 1950; Moscowitz 1950), long after mercury had been eliminated from the making of hats in the United States.

Most extensive of the reported studies of mercurialism in the felt-hat industry is that conducted by the staff of the Clinica del Lavoro "L. Devoto" of Milan, under the direction of Vigliani (1953). This investigation covered about 1200 workers of whom 246 were diagnosed as having mercury poisoning. The monograph in which the findings are published is an outstanding contribution to the literature of occupational medicine.

In her classic work, *Industrial Poisons in the United States,* published in 1925, Alice Hamilton reviews the general subject of mercurialism in one chapter and devotes a separate chapter to the hat industry. The latter is longer than the former, reflecting the importance attached to health hazards among hatters during the first decades of the 20th century. According to Hamilton, the process of treating the fur with mercury nitrate, the so-called secretage, ". . . has been traced back to the middle of the 17th century when it was a secret in the hands of a few French workmen, evidently Huguenots; for at the revocation of the Edict of Nantes in 1685 when the Huguenots fled to England, they carried the secret with them, established the trade there, and for almost a century thereafter the French were dependent on England for their felt." This statement is difficult to reconcile with that which appears in Diderot's encyclopaedia published in 1753 to the effect that in preparing fur for making hats ". . . the pelts are rubbed with an acid solution before the fur is removed. . . ." It is also at variance with an account of secretage given by Lee (1968) in which he states that the process was introduced into England from Frankfurt around 1870. The latter is in consonance with Thackrah's failure to include mercury poisoning in his description of hazards in the British hat industry in the early part of the 19th century.

The complete story of the process of secretage has been difficult to ascertain, particularly for the period between the middle of the 18th and the middle of the 19th centuries. Some features are well documented while others are not. That the Huguenot hatters left France around 1685 is well established (Unwin 1904; Kellogg 1925; Cunningham 1897; Smiles 1868; Weiss 1854; Erman and Reclam 1782-94).

Aware of the loss to the French economy of the skills taken out of the country by Huguenot refugees, Louis XIV sent the Marquis de Bonrepaus to England to try to persuade the expatriates to return. The Marquis wrote from London in 1686 that "The other manufactures which have become established in this country, are those of hats of Caudebec, and the dressing, in the best manner of Chamois skins" (Weiss 1854). In commenting on this, Weiss says:

The manufacture of hats, indeed, was one of the finest branches of business with which the refugees endowed the English. In France, it had been almost entirely in the hands of 'the Reformed.' They, alone, possessed the secret of the composition water, which serves for the preparation of the rabbit, hare, and beaver skins, and they alone furnished to trade the fine hats of Caudebec, so much sought after in England and Holland. After the revocation, most of them retired to London, taking with them the secret of their art, which was lost to France for more than 40 years. It was not until the middle of the 18th century, that a French hatter named Mathieu, after having worked long in London, stole the secret, which had been imported by the refugees, took it to his own country, generously communicated it to the hatters of Paris, and founded a large manufactory in the Suburb of Saint Antoine. Before that fortunate theft, the French nobility and all those who prided themselves on the elegance of their dress, wore none but hats of English manufacture, and the cardinals of Rome, themselves, sent for their hats from the celebrated manufactory of Wandsworth, which had been established by the refugees.

This account by Weiss, which is entirely credible, has often been quoted and, no doubt, the quoters have in turn been quoted, as evidenced by the variations which have appeared. What is not clear is the fate of felt making in England after M. Mathieu absconded with the "secret." Thackrah's failure to note the occurrence of mercury poisoning in the early 1800's in England adds support to the assertion that mercury carroting was introduced, or rather, re-introduced after 1830.

A clue to the history of carroting in Britain is found in Taylor's book *On Poisons*, published in 1875. He describes the case of a man who was admitted to Guy's Hospital on December 10, 1863, suffering from chronic poisoning by the nitrate of mercury. "He had been for four years engaged in packing the fur of rabbits, rats, and other animals, the dried skins of which had been brushed over with a solution of nitrate of mercury." This clearly is the carroting process (secretage). If the dates are accurate it shows that carroting was being practiced in Britain in 1859.

"The Mad Hatter"

There is some question as to whether or not the Hatter in *Alice in Wonderland* (1865) was mad because of mercury poisoning. For one thing, the diagnosis may be questioned since it was made rather casually by the Cheshire Cat.

"In that direction," the Cat said, waving its right paw round, "lives a Hatter; and in that direction," waving the other paw, "lives a March Hare. Visit either you like; they're both mad." Although the ensuing episode was a Mad Tea Party, and both the March Hare and the Hatter were present, there is no further mention of the latter's madness. True, the Hatter's behavior was a bit erratic, but no more so than that of many other characters in the book.

His remark to Alice that "Your hair wants cutting" does not necessarily imply that he was thinking of using the hair to make felt (Dodgson 1946).

Gardner (Dodgson 1960) has pointed out that the phrases "mad as a hatter" and "mad as a March hare" were in common use in Carroll's time and that "mad as a hatter" may have been a play on the cockney corruption "mad as an adder." He suggests, however, that it ". . . more likely owes its origin to the fact that until recently hatters actually did go mad . . ." Writing in 1960, Gardner may be referring to the 20th century, since Thackrah failed to note the characteristic erethism of mercury poisoning in hatters in the 1830's, and the mercurial secretage may not have been reintroduced into England until the late 1850's (Taylor 1875). Between then and 1865 the symptoms of mercury poisoning in hatters could have been known but could hardly have become a by-word.

Another interesting point made by Gardner has to do with the appearance of the Hatter in Tenniel's drawings in Carroll's book. He rejects what he says was a common belief at the time that the Hatter was a burlesque of Prime Minister Gladstone. Rather, he says, "There is good reason to believe that Tenniel adopted a suggestion of Carroll's that he draw the Hatter to resemble one Theophilus Carter, a furniture dealer near Oxford . . ." who was known in the area as the Mad Hatter. He was so designated because he always wore a top hat and partly because of his eccentric ideas and behavior. He invented an alarm clock bed which awakened the sleeper by tossing him to the floor and which was exhibited at the Crystal Palace in 1851. This may have had something to do with the Hatter's preoccupation with time and with his concern over the dormouse's somnolence.

Confirmation of the theory that Lewis Carroll's "Mad Hatter" was not a victim of mercury poisoning is found in views expressed by a Director of Associated British Hat Manufacturers Limited (who wishes to remain anonymous). Among the credentials which enable this gentleman to speak with authority are the fact that his father's family have been making hats continuously since 1773 and his mother's since about 1660. His discussion of the subject of felting clearly shows that he has done considerable research into the historical aspects.

He points out, in consonance with Gardner's statement that the expression "mad as a hatter" was in common use in England during the middle of the 19th century, that the expression appears in Chapter 10 of Thackeray's *Pendennis,* which was published in 1850. Thackrah's failure to note the use of mercury nitrate in the hat industry supports the fact that it was not being used in 1830. It was not until the middle of the 1840's that the felt-hat industry as it is known today, with its use of mercury nitrate, was founded.

Our anonymous informant (personal communication 1970) has provided an interesting account of the origin of the use of a mercury compound in the making of felt.

The felting process began with camel hair in what is now a part of Turkey. The hair of camels was made into tough felt material for tents. It was discovered that the felting process was accelerated with the urine of the camel. The art was brought back to Western Europe by the followers of the Crusades. It became the habit for the workers to urinate on the fibres before felting them. The story goes that one particular workman (probably in France!) was being treated with mercury for a venereal disease. It was noticed that his fibres, after the treatment mentioned above, felted quicker and better than that of his more healthy comrades. This was the secret which gave the process the name of 'le secretage' in France. This may be an old wives tale, but it has been passed down in the Felt-hat Industry as the origin of the use of mercury.

Apocryphal or not, the story merits recording. There are other versions of the origins of felt-making, some of which are related by Hamilton (1925).

Legislation directed toward the prevention and control of occupational mercurialism has followed the pattern of labor legislation in general, although there have been special provisions made for mercury from time to time. Early laws aimed at improving the lot (and the health) of factory workers generally had to do with limiting the hours of work and the employment of women and children. Applied to occupations in which mercury was handled these limiting provisions would certainly be beneficial. Later laws prescribing minimum standards of factory hygiene and sanitation likewise reduced the possibility of occupational poisoning. Minimum wage laws, leading to improved living standards, represent another type of labor legislation that might be expected to benefit employed persons in general, including those occupationally exposed to mercury. Workmen's compensation laws have had a more direct impact.

In discussing legislation related to mercury, Biondi states that "The first statutory intervention known dates from 1665, when the hours of work in the quicksilver mines of Friuli (Idria) were reduced to six per day in consequence of the effects on health" (International Labour Office 1930–34). He also records the existence of special legislation restricting women and young children from working with mercury in the Netherlands, Japan, France, Switzerland, and Greece. Hamilton (1925) says much the same thing. In countries having workmen's compensation laws mercury poisoning is recognized as a compensable occupational disease subject to the general provisions of the laws. Great Britain enacted special Mercury Process Regulations in 1963, the culmination of a series of regulations which began when mercurial poisoning was made a notifiable disease in 1899, under the provisions of the Factories Act (Lee 1968). In France, a decree issued in 1913 and modified in 1935, was designed to protect the health of workers in the fur cutting industry in plants where the fur had been treated with mercury nitrate (Desoille 1949).

REFERENCES

Agricola, G. 1950. *De re metallica.* Trans. from the first Latin ed. of 1556 by Herbert C. Hoover and Lou Henry Hoover. New York: Dover Publications.

———. 1955. *De natura fossilium.* Trans. from the first Latin ed. of 1546 by Mark Chance Bandy and Jean A. Bandy. New York: Geological Society of America.

American Conference of Governmental Industrial Hygienists, Committee on Threshold Limit Values. 1962. *Documentation of threshold limit values.*

American Conference of Governmental Industrial Hygienists, Committee on Threshold Limit Values for Airborne Contaminants. 1971. *Documentation of the threshold limit values for substances in workroom air.* third ed. Cincinnati: American Conference of Governmental Industrial Hygienists.

Baader, E. W., and Holstein, E. 1933. *Das Quecksilber und die gewerbliche Quecksilbervergiftung.* Berlin: R. Schoetz.

Bates, L. W. 1912. *Mercury poisoning in the industries of New York City and vicinity.* National Civic Federation, New York and New Jersey Section.

Bombastus von Hohenheim, P. A. T. [Paracelsus]. 1922–31. *Sämtliche Werke.* Ed. K. Sudhoff. Munich: R. Oldenburg.

Cunningham, W. 1897. *Alien immigrants to England.* London.

Dennis, L. 1878. Hatting as affecting the health of operatives. *Rep. Board Health State New Jersey* 2:67–85.

Desoille, H. 1949. *Cours de médecine du travail.* Paris: E. Le François.

Diderot, D., and Alembert, J. le R. d'. 1751–65. *Encyclopédie ou Dictionnaire raisonné des sciences, des arts et des métiers, par une société de gens de lettres.* 17 vols. Paris.

Dodgson, C. [L. Carroll]. 1946. *Alice's adventures in Wonderland.* New York: Random House.

———. 1960. *The annotated "Alice" with an introduction and notes by Martin Gardner.* New York: Clarkson N. Potter.

Ellenbog, U. 1927. *Von den gifftigen besen Tempffen und Reuchen der Metal.* Eine gewerbe-hygienische Schrift des XV. Jahrhunderts. Wiedergabe des ersten Augsburger Druckes mit Biographie und einer medizin- und druckgeschichtlichen Würdigung von Franz Koelsch und Friedrich Zoepfl. Munich: Verlag der Münchener Drucke.

———. 1932. The treatise. Trans. C. Barnard. *Lancet* 1:270–71.

Erman, J. P., and Reclam, P. C. F. 1782–94. *Mémoires pour servir à l'histoire des réfugiés françois, dans les états du roi.* 8 vols. Berlin.

Fernel, J. 1579. *De luis venerea curatione perfectissima.* Antwerp.

Freeman, J. A. 1860. Mercurial poisoning among hatters. *Trans. Med. Soc. New Jersey* 61–64.

Goldwater, L. J. 1950. Blood studies on workers in the fur-felt hat industry. *Monthly Review (Div. Ind. Hyg. Safety St. N. Y. State Dept. Labor)* 29:1–3.

———. 1964. Occupational exposure to mercury. The Harben Lectures, 1964. *J. Roy. Inst. Public Health Hyg.* 27:279–301.

Hamilton, A. 1922a. The industrial hygiene of fur cutting and felt hat manufacture. *J. Ind. Hyg.* 4:137–53.

———. 1922b. Industrial diseases of fur cutters and hatters. *J. Ind. Hyg.* 4:219–34.

———. 1925. *Industrial poisons in the United States.* New York: Macmillan Co.

Harris, L. 1915. A clinical and sanitary study of the fur and hatters' fur trade. *New York City Dept. Health Monogr. Ser. No. 12*:1–55.

International Labour Office. 1930–34. *Occupation and health.* 2 vols. Geneva.

Kellogg, L. P. 1925. *The French regime in Wisconsin and the northwest.* Madison: State Historical Society of Wisconsin.

Kober, G. M. 1916. Mirror plating. In *Diseases of occupation and vocational hygiene,* ed. G. M. Kober and W. C. Hanson, p. 529. Philadelphia: P. Blakiston's Sons and Co.

Kober, G. M., and Hanson, W. C.. eds. 1916. *Diseases of occupation and vocational hygiene.* Philadelphia: P. Blakiston's Sons and Co.

Kussmaul, A. 1861. *Untersuchungen über den constitutionellen Mercurialismus und sein Verhältnis zur constitutionellen Syphilis.* Würzburg.

Lee, W. R. 1968. The history of the statutory control of mercury poisoning in Great Britain. *Brit. J. Ind. Med.* 25:52–62.

Lemery, N. 1686. *A course of chymistry,* 2d ed., enl. and trans. from the 5th edition in the French by Walter Harris. London.

Mayer, F. C. 1929. In *Proceedings of the 5th International Congress on Occupational Medicine,* 2–8 September 1928, Budapest.

McCready, B. W. 1943. *On the influence of trades, professions and occupations in the United States, in the production of disease.* Reprint of the New York, 1837, edition. Baltimore: Johns Hopkins Press.

Moscowitz, S. 1950. Exposure to mercury in industry: A statistical study. *Monthly Review (Div. Ind. Hyg. N. Y. State Dept. Labor)* 29:17–20.

Neal, P. A.; Flinn, R. H.; Edwards, T. I.; Reinhart, W. H.; Hough, J. W.; Dallavale, J. M.; Goldman, F. H.; Armstrong, D. W.; Gray, A. S.; Coleman, A. L.; and Postman, B. F. 1941. Mercurialism and its control in the felt-hat industry. *U. S. Public Health Bull. No. 263.*

Neal, P. A.; Jones, R. R.; Bloomfield, J. J.; Dallavalle, J. M.; and Edwards, T. I. 1937. A study of chronic mercurialism in the hatters' fur-cutting industry. *U. S. Public Health Bull. No. 234.*

Partington, J. R. 1965. *A short history of chemistry.* 3d ed. New York: Harper and Row, Harper Torchbooks.

Ramazzini, B. 1964. *Diseases of workers.* Trans. Wilmer Cave Wright. New York: Hafner Publishing Co.

Smiles, S. 1868. *The Huguenots: Their settlements, churches, and industries in England and Ireland.* New York.

Smith, A. R.; Goldwater, L. J.; Burke, W. J.; and Moscowitz, S. 1949. Mercury exposure in the fur-felt hat industry. *Monthly Review (Div. Ind. Hyg. Safety St. N. Y. State Dept. Labor)* 28:17–19, 21–24.

Smith, A. R., and Moscowitz, S. 1948. Urinary excretion of mercury. *Monthly Review (Div. Ind. Hyg. Safety St. N. Y. State Dept. Labor)* 27:45–47.

Taylor, A. S. 1875. *On poisons.* 3d American ed. from the 3d English. Philadelphia: Henry C. Lea.

Teleky, L. 1912. *Die Gewerbliche Quecksilbervergiftung.* Berlin: A. Seydel.

———. 1955. *Gewerbliche Vergiftungen.* Berlin: Springer-Verlag.

Thackrah, C. T. 1832. *The effects of arts, trades and professions on health and longevity.* 2d ed. London.

Thompson, W. G. 1914. *The occupational diseases.* New York: D. Appleton and Co.

Unwin, G. 1904. *Industrial organization in the 16th and 17th centuries.* Oxford: Clarendon Press.

Vigliani, E. C.; Baldi, G.; and Zurlo, N. 1953. *Il mercurialismo cronico nei Capellifici.* Cirie: Giovanni Capella.

Weiss, C. 1854. *History of the French Protestant refugees, from the revocation of the edict of Nantes to our days.* Trans. Henry W. Herbert. 2 vols. new York: Stringer and Townsend.

Whitaker, A. P. 1941. *The Huancavelica mercury mine.* Cambridge: Harvard University Press.

Wright, W. 1922. A clinical study of fur cutters and felt-hat workers. *J. Ind. Hyg.* 4: 296–304.

Zangger, H. 1930. Erfahrungen über Quecksilbervergiftungen. *Arch. Gewerbepathol. Gewerbehyg.* 1:539–60.

ADDITIONAL READING

Hawkins, J. H. 1917. *History of the worshipful company of the art* or *mistery of feltmakers of London.* London: Crowther and Goodman.

20

Mercury in Dentistry

Relationships between mercury and dentistry can be traced back a thousand years or even further (Bremner 1954; Guerini 1969; American Academy of Dental Science 1876; Weinberger 1948). The earliest associations are not concerned with the teeth but rather with the oral cavity as a site where manifestations of systemic disease may be found. In this respect early dentistry has much in common with modern practice, the dentist of today being constantly on the alert for oral evidence of disease elsewhere in the body (Baum 1934; McGeorge 1935; Akers 1936).

Early Uses of Mercury in Dentistry

Stomatitis as a result of mercury inunctions was noted by Albucasis in the 10th century, this being one of his many contributions to dentistry. His near contemporary, Avicenna, recorded foul breath as one of the manifestations of mercury poisoning. Corrosion of the tongue by sublimate of mercury was described by Peter of Abano, and Alessandro Benedetti (1460–1525) wrote about the harm that could be done to the teeth and gums when mercury was administered either orally or by frictions elsewhere upon the body (cited in Guerini 1969 and in Weinberger 1948). As a means of preventing these unpleasant complications Gerolamo Capivacci (d. 1589) recommended that a gold object be kept in the mouth to absorb the mercury (cited in Guerini 1969), a procedure for which a rationale has been shown to exist. Discoloration of the teeth following the administration of mercurial drugs or from the use of mercurial cosmetics by women was noted by Lazare Rivière (1589–1655) (Guerini 1969). In the 18th century one of the recognized contraindications to the extraction of painful teeth was gingivitis due to mercury (Weinberger 1948).

Relatively little is heard of mercury in the field of dentistry until the 19th century. The 1830's saw the beginning of an era during which the metal became indispensable to the dentists of the world. Its value to the victims of toothache and related ills cannot be appreciated unless one reads of the crude and painful methods previously followed in the treatment of carious teeth. Not without a struggle did mercury find a place in the cavities of human teeth and, once established, it has been obliged to face repeated threats of expulsion. Although the story begins in the 19th century it has a distinctly alchemical, almost medieval, flavor, involving a conflict between quicksilver and gold.

Prior to the advent of amalgams (alloys of mercury with other metals), various materials had been used to fill dental cavities. During the early decades of the 19th century gold was by far the most commonly used of these materials, in spite of its high cost and other drawbacks. Gold foil was pounded into cavities, a process which was time-consuming and painful. Experiments with fusible alloys of low melting point were undertaken during the 1820's with the objective of finding a combination suitable for use in cavities. Melting temperatures and expansion-contraction characteristics were the critical features. Among the combinations tried was one which was made up of bismuth, lead, tin and a small amount of mercury. This type of material was placed in the cavity in a solid state and then fused with a hot instrument, a method with obvious limitations and disadvantages. Frykholm, in his comprehensive monograph on dental amalgams (1957), mentions the use of mercury alloys as early as 1818, and quotes Lemerle as saying that an English chemist named Bell had prepared a dental amalgam from silver filings and mercury in 1819. Apparently neither of these early preparations gained permanent recognition in the dental world.

Amalgams of mercury and silver entered or re-entered the field around 1826 when a Parisian dentist by the name of Taveau began using what he called "silver paste," a combination of silver filings and mercury (Bremner 1954; Lufkin 1948). The silver apparently was obtained from coins and consequently contained some copper, a metal which was destined to come into common use in dental amalgams a century later. The particular combination used by Taveau underwent excessive expansion on setting, resulting in fractured teeth or protrusion of the fillings. Because of these unsatisfactory results the mercury-silver amalgam was soon abandoned and the older methods of filling were resumed.

Claims for priority in the use of amalgam fillings have been entered by Lindsay (1933) on behalf of Joseph Murphy of London, who Lindsay says was the first to suggest, in 1837, the use of mercury to form an amalgam, which he called a "succedaneum," (meaning "a substitute"). The discrepancy between this and the 1826 date ascribed to Taveau is explained by Singer and Underwood (1961) on the basis that Murphy was the first to make *extensive* use of the "silver paste" although he may not have been the first to use it at all. That there was no serious objection to the introduction of amalgam into Britain may have been due to the lack of any strong organization in the dental profession before 1880 when the British Dental Association was formed (Lindsay 1933).

Amalgam Disputes in the United States

Dentists in America, aware of the advantages of amalgam fillings, might well have continued an uninterrupted search for the most suitable combinations of mercury with other metals had it not been for two related events which are known in the annals of dentistry as the "Crawcour Incident" and the "Amalgam War," episodes which were of major importance, too, in the early history of professional dental organizations in the United States.

The "Crawcour Incident"

Some time during the 1830's there arrived in New York from France two brothers by the name of Crawcour who brought with them a material they called Royal Mineral Succedaneum which they advertised as a superior substance for filling teeth. By the Royal Mineral was meant gold, and the Succedaneum was a substitute in the form of a mercury amalgam. So effective were the advertising methods of the Crawcour brothers that the dental practitioners in New York and elsewhere who were committed to the use of gold felt seriously threatened. Not only were the accepted procedures being challenged but the economic welfare of many dentists was in jeopardy since it was common practice to invest heavily in the gold trade. A vigorous campaign to discredit the amalgam was mounted based on the allegation that mercury was poisonous and its use in dental fillings a threat to the public health. After a few years of harassment the Crawcour brothers fell into disrepute and left the country. The use of amalgam to fasten false teeth to the plates of artificial dentures does not seem to have been considered a threat to health (American Academy of Dental Science 1876) and was not attacked.

After the defeat of the Crawcours, some dentists considered amalgams superior to gold, and others continued to use them for their less affluent patients.

The "Amalgam War"

Anathematization and excommunication are not exclusively the weapons of religious bodies; they can be and have been used by groups whose bonds are professional and economic. They were invoked by the American Association of Dental Surgeons in 1843 against members who continued to use amalgams. Thus began the "Amalgam War." As in medieval religious controversies, the dissenters were given an opportunity to recant and could remain in good grace if they signed this pledge:

> I hereby certify it to be my opinion and firm conviction that any
> amalgam whatever . . . is unfit for the plugging of teeth or fangs and I
> pledge myself never under any circumstances to make use of it in my prac-
> tice, as a dental surgeon, and furthermore, as a member of the American
> Association of Dental Surgeons, I do subscribe and unite with them in this
> protest against the use of the same.

Members who refused to sign were expelled from the Association.

This arbitrary procedure engendered resentment and resistance; some
members were expelled but many more resigned. The shock was too great for
the newly-formed Association to withstand and although the pledge was abandoned in 1850 it was too late to save the organization. By 1856 its demise
was inevitable, but the use of mercury survived (American Academy of Dental
Science 1876). The anti-amalgam faction survived, too, led by Jonathan Taft,
who, as dean of the University of Michigan School of Dentistry, would not
permit amalgam to be used at that school up to the time of his death in 1903
(Black 1934). Either because of or in spite of his valiant stand he received
". . . practically all honors that it was possible for an appreciative profession
to confer upon him" (Ward 1934).

Development of New Amalgams and New Disputes

One of the most important events in the history of dental amalgams was
the publication of the classic work of G. V. Black (1836–1915) in 1895. Recognized as the "Grand Old Man" of late 19th and early 20th century dentistry, Black spent many years studying the properties of various amalgams,
measuring their expansion-contraction characteristics with instruments he devised and observing their behavior under conditions of use. He tried out combinations of tin, silver and copper with different amounts of mercury. He
found, among other things, that copper amalgams were not suitable for filling
teeth. This work of Black served as the basis for modern concepts and practices in the use of amalgam fillings and led to later standardization in the manufacture of dental amalgams. Specifications for mercury to be used in dental
work were drawn up and promulgated jointly by the American Dental Association and the U. S. Bureau of Standards in 1932 (Isaacs).

Exactly 100 years after the first use of mercury in dental fillings by
Taveau a new threat to the survival of amalgams appeared, this time from a
chemist. It is probably a matter of pure chance that the attack on a French
invention was led by a German, Professor A. Stock. Articles by Stock published in 1926 and 1928 are often cited as evidence that amalgam dental fillings can cause not only significant absorption of mercury but that this can
constitute a serious threat to health.

Stock's first report deals with amalgams of both silver and copper, the
latter having been in wide use in German dental clinics for patients who could

not afford the more expensive silver-mercury combination. Concern was shown for dentist as well as patient. Stock stated that nearly all dentists have increased amounts of mercury in their urine and many show evidence of intoxication, but no actual case histories or analytical data were given. He quoted from a letter he claims to have received from a German-American friend who wrote from New York that the dentists there no longer used amalgam. This was certainly not true; there are many New Yorkers (including the present writer) who had teeth filled with amalgam in the 1920's. Within a matter of weeks after the publication of his report in 1926 his conclusions were challenged by Pinkus (1926). This physician pointed out that in chemical factories and elsewhere workers had been observed for periods of 30 to 40 years under conditions of much heavier exposure than that which resulted from dental amalgams and with no evidence of mercurialism. Along with the publication of this note from Pinkus, a short rejoinder from Stock was printed in which the latter simply says that there is ample medical evidence to support his contention. This marks the beginning of a controversy which has not yet entirely died out and which might be called the Second Amalgam War.

In the second of his early reports Stock cites eight articles published in the German literature between 1926 and 1928 dealing with the health hazards from amalgam fillings. He records the finding of mercury in the urine of patients who had had teeth filled with amalgam, the values ranging from 0.1 to 4.0 micrograms per liter. Had he performed pre- as well as post-filling analyses he might have found the same levels of mercury in both since the range he cites is that of "normal" or non-exposed individuals (see Chap. 10). Hazards to dentists as well as patients are again emphasized.

The results of Stock's analyses were not immediately questioned, but his clinical interpretations were, and properly so.

P. Borinski in 1931 questioned the validity of Stock's frightening conclusions and pointed out that mercury was present in many foods and that the quantities found by Stock in the urines of dental patients were no greater than those present in non-exposed individuals. In his opinion there was no cause for alarm.

An interesting account of the amalgam controversy has been given by Leschke who, as Professor of Internal Medicine in the University of Berlin, must have had an opportunity for first-hand observation (1934 translation).

> The communications of Stock caused a great sensation when they were first published, especially in the case of a well-known Marburg professor, whose mental and physical health, after being bad for years, took a turn for the better immediately after removal of 24 amalgam tooth stoppings. A fierce discussion of the question began in the Berlin Society of Internal Medicine after a lecture by Fleischmann in 1928. Fleischmann reported seven cases in which, owing to copper amalgam tooth-stoppings, disturbances occurred in the form of headache, anaemia, diarrhoea, weakness of memory, diminution of mental working capacity, etc., which symptoms

disappeared after removal of the stoppings. Most of the patients had a lymphocytosis of 30–60 per cent (Luddicke). Eventually, these investigations were successful, through Stock's improved methods for the detection of mercury, in causing the somewhat objectionable copper amalgam tooth stoppings to be practically abandoned, and to be replaced by the safe silver amalgam stoppings. They succeeded also in drawing more attention to the possibility of mercury poisoning, especially in laboratories and dentists' rooms . . . ."

Stock's results have been carefully re-examined by P. Borinski of the Berlin Department of Health. This re-investigation has brought to light the important fact that human beings who have never come directly into contact with mercury, may excrete up to 10 μg of mercury in a day. Nearly all important foodstuffs contain traces of mercury. The mercury excretion after the fitting of copper amalgam stoppings is somewhat higher than after silver amalgam stoppings . . . .

In 1934, Stock added a highly significant but seldom noticed footnote to one of his articles: "We repeatedly emphasize that the older determinations of such small amounts of quicksilver, including even those of our own carried out according to the earlier analytical directions, merit little confidence." Thus Stock, in effect, repudiated his own earlier analyses. Stock's frankness is certainly admirable, but his arrogance is not. It is his privilege to reject his own findings and to warn others not to accept them, but he goes pretty far when he condemns the analyses of other chemists as well. It is important to know that pre-1934 Stock is not to be taken seriously. The same article contains a new method of analysis in which the mercury is distilled from the sample and condensed in a glass tube by means of liquid air. Most of Stock's more significant work was done after 1934.

American concern over the possibility of danger in handling and use of amalgams stimulated an investigation conducted by the U. S. Department of Commerce, Bureau of Standards. The report on this study was published in 1931 (Souder and Sweeney). Among the studies performed by Souder and Sweeney were measurements of the amount of mercury vapor which could evolve from the surface of amalgams and the extent to which mercury is dissolved. In both instances the quantities were small, leading to the conclusion that, "From these data it would appear that the claims for mercury poisoning, either as a vapor or as a solution from the standard amalgams passing into the body through the air or food taken into the mouth, are not justified."

Relatively few of the many articles on amalgam "poisoning" published since 1926 contain measurements of mercury concentrations in air or urine, or other useful factual data. One of the exceptions is the report by Storlazzi and Elkins, which appeared in 1941. These observers determined the concentrations of mercury in 37 urine samples from five persons, the samples having been collected at times up to 16 days after the filling of the subjects' teeth with amalgam. They found increased urinary excretion of mercury at the time of and for about a week after the insertion of the fillings but expressed doubt that the amount of mercury absorbed could cause mercurialism. They concluded that ". . . the rapid disappearance of mercury from the urine indicated that old dental fillings can, as a rule, be ignored as a source of urinary mercury." In a related study, Storlazzi and Elkins showed that the application of a popular mercurial antiseptic to the skin of axilla and toes also resulted in a transitory increase in urinary excretion of mercury (1941).

A major contribution to the amalgam question was made by Frykholm, whose monograph published in 1957 contains a complete review of earlier writings on the subject and also describes important original studies made by the author. As a result of numerous observations and experiments, Frykholm dismisses the possibility of systemic poisoning occurring from amalgam fillings but emphasizes the frequency with which allergic reactions can be caused by small amounts of mercury.

When the American Association of Dental Surgeons condemned the use of amalgam because it was detrimental to public health they gave no indication that they were worried about the possibility of mercury poisoning among dentists. This does not necessarily imply that they were more concerned about their patients than about themselves, but they seem to have been more apprehensive about the state of their pocketbooks than of their health. Not until 1909 was attention called to the potential hazard to dentists that might result from the handling of amalgams (Göthlin) and it was not for another quarter century that this threat was taken seriously. In the meantime, amalgams had gained universal acceptance.

In 1934 Leschke investigated the possible dangers of amalgams to dentists, and reported that 16 analyses of air in dentists' work rooms showed from 0.03 to 9.0 mg of mercury per cubic meter of air, and that the urine of dentists and dental assistants contained up to 0.1 mg of mercury per liter.

Grossman and Dannenberg set out in 1949 to study the mercury vapor levels in 100 dental offices and laboratories. They terminated their work after having surveyed 50 premises, since they had found uniformly low amounts, never in excess of 0.02 parts per million of mercury vapor in the air. Unfortunately they overlooked the fact that mercury might also be present in the form of aerosol or dust and that therefore the quantities present in respirable air were probably higher than those which they measured.

Reports on possible hazards from amalgams have continued to appear in the literature; a number of these are reviewed in an article published in 1968 (Joselow et al. 1968). Published works (Lintz 1935; Preussner, Klöcking, and Bast 1963; Knapp 1963) have varying degrees of usefulness; some have little substance. There seems to be no excuse, for example, for the statement in one article (McCord 1961) written by a distinguished authority on occupational diseases that "Proof is lacking, but it may be surmised that over the land at all times some thousands of dentists are the victims of mercury poisoning in varying degrees of severity. All this appears to be unknown and not even suspected." As pointed out above, the possibility of harm to dentists was very well known and has been constantly under suspicion. Not only is there no proof, but there is hardly any reason for surmising that such a dire situation exists.

The Columbia Study, 1964 - 1967

The entire question of health hazards resulting from the use of amalgam dental fillings was re-examined in some detail by a Columbia University research team during the years 1964-1967 (Joselow et al. 1968; Hoover and Goldwater 1966). One phase of the investigation involved 114 adults who were not currently under dental treatment and among whom there were 85 who had one or more amalgam fillings. The average number of fillings was 10, and the range was 1-28 fillings. None of these persons had any medical abnormalities that could be attributed to mercury poisoning and the levels of urinary mercury were similar to those found in the population at large. These findings corroborated those reported by Storlazzi and Elkins (1941).

As part of the Columbia study, the immediate effects of amalgam fillings on urinary mercury excretion were studied on 24 subjects. Analyses were made prior to treatment and then after 6, 24, 48 and 72 hours. All of the subjects did not submit urine samples at all of these intervals, but at least one post-treatment sample was obtained from all. The urinary mercury levels before and after treatment were practically the same.

In the course of this investigation it was ascertained that dental supply houses sell substantial amounts of mercurial germicides to dentists for sterilizing instruments and for preparing the mouth before operative procedures. Experimentally it was shown that painting the gums with a dilute solution of one such preparation, nitromersol, resulted in increased excretion of mercury within 24 hours. These findings pointed to the importance of considering the possibility that germicides may contribute to the presence of mercury in the urine following the filling of dental cavities with amalgams.

Dentists' occupational exposure to mercury as a result of handling or from the inhalation of vapors was also studied by the Columbia University group (Joselow et al. 1968). This phase of the research embraced 50 private practitioners, all of whom submitted urine samples for analysis and in all of whose offices atmospheric levels of mercury were measured. The time that these dentists spent working with mercury ranged from 0 to 40 hours per week, with the average being 14 hours per week. The air in the dental operating rooms studied contained from 0.002 to 0.160 mg of mercury vapor per cubic meter (average = 0.020 mg per cubic meter); and the air in the dentists' waiting rooms contained nearly as much — from 0.002 to 0.100 mg of mercury vapor per cubic meter (average = 0.018 mg per cubic meter).

In addition to measuring mercury vapor, these investigators measured particulate mercury in the air, and found that the average particulate concentration in the dental operating rooms was 0.025 mg per cubic meter—which was higher than the concentration of mercury vapor. In some cases the total mercury concentration in the air in these dental offices could be considered potentially hazardous.

The urinary excretion of mercury by the dentists was higher than that of an unexposed population. It ranged from 0 to 155 micrograms per liter, with an average of 40 micrograms per liter. None of the practitioners, however, complained of or admitted to any symptoms that could be attributed to mercurialism. It should be emphasized that this investigation revealed no more than a potential hazard. Had the danger been actual, some sign of intoxication probably would have been found in at least one dentist. Evidence presently available does not point to any significant health hazard to dentists as a result of handling amalgams.

Dentures

For a short time amalgam served as a base for fastening artificial teeth into dentures. For this purpose, however, it was soon replaced when gutta percha and vulcanized rubber came into vogue. At one time, mercury sulfide (cinnabar) also found a place in dentures, in a manner described by Taylor (1875):

> This substance is . . . sometimes employed by dentists as a coloring matter to vulcanized rubber or gutta percha for mounting artificial teeth . . . . In May, 1854, a medical man consulted me under the following circumstances. Upon the recommendation of a dentist, he had worn this red composition as a frame for false teeth, in place of gold. After some time he perceived a metallic taste in his mouth, the gums became inflamed and ulcerated, there was a great weakness and want of nervous power, with pains in the loins and an eruption on the legs. When the composition was removed, these symptoms abated.

Other similar cases are described by Taylor with citations to references in the medical literature. The toxic effects are ascribed to the gradual liberation of mercury due to contact with saliva.

Mercury may have figured indirectly in the story of the most famous false teeth in history: those of George Washington. These dentures seem to have exerted an irresistible charm over historians of medicine and dentistry. Relatively little concern has been expressed, however, about possible reasons why the Father of His Country lost his natural teeth and needed a prosthetic replacement.

It is well known that General Washington did not always enjoy the best of health and that one of the remedies he frequently used was calomel. This has led one dental historian to suggest that his loss of teeth was a result of mercurialism (Lufkin 1948). If this theory is correct it might be said that where George III and all the might of the British empire failed, mercury succeeded— in knocking the teeth out of George Washington!

REFERENCES

Akers, L. H. 1936. Ulcerative stomatitis following the use of mercury and bismuth. *J. Am. Dental Assoc.* 23:781-85.

American Academy of Dental Science. 1876. *A history of dental and oral science in America.* Philadelphia.

Baum, H. B. 1934. Occupational diseases of the mouth. *Dental Cosmos* 76:247-54.

Black, A. D. 1934. Operative dentistry: A review of the past 75 years. *Dental Cosmos* 76:43-65.

Black, G. V. 1895. An investigation of the physical characters of the human teeth in relation to their diseases, and to practical dental operations, together with the physical characters of filling materials. *Dental Cosmos* 38:353-421; 469-84; 553-69; 637-61; 737-57.

Borinski, P. 1931. Sind kleinste Quecksilber-mengen gesundheitsschädlich? *Deut. Med. Wochschr.* 57:1060-1061.

Bremner, M. D. K. 1954. *The story of dentistry.* 3d ed. Brooklyn, N. Y.: Dental Items of Interest Publishing Co.

Frykholm, K. O. 1957. Mercury from dental amalgam: Its toxic and allergic effects and some comments on occupational hygiene. *Acta Odontol. Scand.* 15 (Suppl. 22): 1-108.

Göthlin, G. F. 1909. Kvicksilfverhaltig luft och fall af Kronisk Kvicksilfverforgiftning vid en medicinsk laroanstalt. *Hyg. Tidskr.* 2:138-183.

Grossman, L. I., and Dannenberg, J. R. 1949. Amount of mercury vapor in air of dental offices and laboratories. *J. Dental Res.* 28:435-38.

Guerini, V. 1969. *A history of dentistry.* Pound Ridge, N. Y.: Milford House.

Hoover, A. W., and Goldwater, L. J. 1966. Absorption and excretion of mercury in man: X. Dental amalgams as a source of urinary mercury. *Arch. Environ. Health* 12: 506-8.

Isaacs, A. 1932. Mercury for dental amalgams. *J. Am. Dental Assoc.* 19:54-57.

Joselow, M. M.; Goldwater, L. J.; Alvarez, A.; and Herndon, J. 1968. Absorption and excretion of mercury in man: XV. Occupational exposure among dentists. *Arch. Environ. Health* 17:39-43.

Knapp, D. E. 1963. Hazards of handling mercury. *J. Am. Dental Assoc.* 67:79-81.

Leschke, E. 1934. *Clinical toxicology.* Trans. C. P. Stewart and O. Dorrer. Baltimore: William Wood and Co.

Lindsay, L. 1933. *A short history of dentistry.* London: John Bale, Sons and Danielsson.

Lintz, W. 1935. Prevention and cure of occupational diseases of the dentist. *J. Am. Dental Assoc.* 22:2071-2081.

Lufkin, A. W. 1948. *A history of dentistry.* 2d ed., 2 vols. Philadelphia: Lea and Febiger.

McCord, C. P. 1961. Mercury poisoning in dentists. *Ind. Med. Surg.* 30:554.

McGeorge, J. R. 1935. Mercurial stomatitis. *J. Am. Dental Assoc.* 22:60-64.

Pinkus, G. 1926. Entgegnung auf die Veröffentlichen Herrn Professor Alfred Stock über die Gefährlichkeit des Quecksilberdämpfes und der Amalgame. *Z. Angew. Chem.* 39:1534.

Preussner, S.; Klocking, H. P.; and Bast, G. 1963. Chronisch-Schleichende Quecksilbervergiftung in der zähnärtzlichen Praxis. *Arch. Toxikol.* 20:12-20.

Singer, C., and Underwood, E. A. 1961. *A short history of medicine.* 2d ed. Oxford: Oxford University Press.

Souder, W., and Sweeney, W. T. 1931. Is mercury poisonous in dental amalgam restorations? *Dental Cosmos* 73:1145-52.

Stock, A. 1926. Die Gefährlichkeit des Quecksilbersdämpfes und der Amalgame. *Z. Angew. Chem.* 39:984-89.

――――. 1928. Die Gefährlichkeit des Quecksilbers und der Amalgam-Zahnfüllungen. *Z. Angew. Chem.* 41:663-72.

Stock, A., and Cucuel, F. 1934. Die Verbreitung des Quecksilbers. *Naturwissenschaften* 22/24:390-93.

Storlazzi, E. D. and Elkins, H. B. 1941. The significance of urinary mercury: II. Mercury absorption from mercury-bearing dental fillings and antiseptics. *J. Ind. Hyg.* 23: 464–65.

Taylor, A. S. *On poisons in relation to medical jurisprudence and medicine.* 3d American ed. Philadelphia: Henry C. Lea.

Ward, M. L. 1934. Landmarks in dental education. *Dental Cosmos* 76:3–23.

Weinberger, B. W. 1948. *An introduction to the history of dentistry.* 2 vols. St. Louis: C. V. Mosby Co.

21

Mercury in Veterinary Medicine

Standard textbooks of veterinary medicine show that there are many similarities between human and veterinary medicine in the use of mercury compounds (Kirk 1951; Miller 1956). Mercury ointments have been used in treating ringworm and other skin diseases, the yellow oxides being preferred for eczematous lesions around the eyelids of dogs. Care must be taken to prevent the animals from licking the ointments, particularly dogs and oxen, which are said to be especially susceptible to mercury poisoning. Calomel, followed by a saline purge, is an historically familiar remedy for both humans and lower animals. Metallic mercury in the form of blue mass and grey powder ". . . are both used as gentle stimulant purgatives for foals and calves and the smaller animals" (Miller 1956). Mercurial diuretics and organomercurial antiseptics have been accepted into the veterinary materia medica.

The Eighth Edition of *The Merck Index*, published in 1968, contains considerable information on current uses of mercurials in veterinary medicine, and additional details can be found in *The Merck Veterinary Manual.* Mercuric chloride is listed as an antiseptic and general disinfectant and also as an irritant for the treatment of bony growth and spavin. Users are cautioned on the possibility of poisoning through licking or absorption through the skin. Mercuric cyanide and ammoniated mercury are mentioned as antiseptics, as are the organic mercurials merbromin and mercurophen. Thiomersal is listed as an antiseptic and for use in fungus infections in animals. Except for its application to fissured heel, the recommended uses for the red oxide of mercury are the same as for humans. Red mercuric iodide has a number of uses: as a vesicant and local counterirritant in horse practice, mixed with lard or petrolatum and cantharides for treating exostoses in large animals, for swellings about tendons, joints, or bursae, and for swollen glands. Multiple uses are found for mercurous chloride: as a local antiseptic, and as a desiccant in moist

289

eczema, canker, thrush, and foot rot. It is recommended for corneal ulcers and phlyctenular conjunctivitis as a local remedy, and internally as a cathartic for horses, dogs, cats, and swine. Another laxative for dogs is mercury mass (blue pill). The organic mercurial compounds, mersalyl and meralluride, are listed as diuretics for use in veterinary practice.

Concern over the presence of toxic agents in foods as a result of veterinary or agricultural applications, so prominent in the middle of the 20th century, had its counterpart a century earlier, when attention was focused on mercury. In March 1863 a mysterious disease attacked a large flock of sheep in Lincolnshire, England, killing more than 40 animals. On investigation it was found that blue (mercury) ointment had been applied to the sheep "as a dressing for the fly" and that this was common practice. ". . . the bodies of sheep thus poisoned with mercury had been sent for sale to the dead-meat market in London, and they realized more money than sound mutton sold in the county of Lincoln." The toxicologist who investigated the situation concluded that ". . . this practice of inunction with mercury should be suppressed: it is not only injurious to cattle, but is often an unsuspected source of noxious food to human beings" (Taylor 1875, p. 353). External application of cinnabar to animals (presumably as a paste) is also said to have produced fatal results (Taylor 1875, p. 384).

Strikingly similar to the above report is one which received wide circulation in the American press in January 1970. A typical headline read: "Meat Sought May be Fatal to Humans," and the story goes on to relate that government authorities were searching for meat from 11 hogs suspected of having mercury poisoning before they were slaughtered. An additional 258 hogs were also under suspicion. In a related news article it was stated that two persons had become comatose and a third was paralyzed and blinded after eating meat from a hog that had been fed seed grain treated with a methyl mercury fungicide. Investigation of these poisonings revealed the fact that 14 hogs in the area had died following the ingestion of treated grain. The U.S. Department of Agriculture moved promptly to ban any further use of alkyl mercurials in seed treatment.

Unique in the annals of veterinary medical history is the case of "canine syphilis" treated with mercury, described by Gouverneur Morris (1752–1816) in his *Diary of the French Revolution* (1939), under the date of May 19, 1789. After casually referring to his dealings with the finance minister, Necker, and the minister of marine, de Castries, Morris tells the story of "Pauvre Toutou."

> Call on Madame de La Suse. She is in great Distress, her
> Lap Dog being very ill. The pauvre Bête has suffered now
> for a long Time. At first it had the Maladie néapolitaine;
> for this it was sent to the Doctor of Dogs who by a Course
> of Mercurials eradicated the Disease and returned him as
> compleat a Skeleton as ever came out of the powdering
> Tub. The Kind Mistress by her Care and Assiduity how-

ever soon brought him up to a tolerable Embonpoint when lo! another indisposition. This is très grave and voila (sic) Madame la Fille de Chambre et un des Valets qui ne s'occupent que de cela. At three different Times in my short Visit: 'Je vous demande bien Pardon, Mr. Ms., mais c'est une Chose si désolante que de voir souffrir comme ça un (e) pauvre bête." "Ah! Madame, ne me faites point de vos excuses, je vous en prie, pour des soins si aimables; aussi merite-il toutes vos attentions." At length by peeping into his Backside she discovers a little Maggot: "Ah! Bon Dieu! mais voyez donc!" The Fundament is now wiped and extended to see a little farther into his Case. I prescribe a Clyster of sweet Oil but Madame and her fille de chambre are of Opinion that this Remedy must be deferred till ToMorrow Morning as he has already had two Clysters toDay without making any Returns, by which Means his Belly is considerably swoln. I insist however on the third and leave them to deliberate.

REFERENCES

Eaton, L. G. 1968. In *The Merck index*, 8th ed. Rahway, N. J.: Merck and Co.

Kirk, H. 1951. *Index of treatment in small-animal practice*. 2d ed. Baltimore: Williams and Wilkins Co.

Miller, W. C., and West, G. P. 1956. *Encyclopedia of animal care*. 4th ed. Baltimore: Williams and Wilkins Co.

Morris, G. 1939. *A diary of the French Revolution*. Ed. B. C. Davenport. 2 vols. Boston: Houghton Mifflin Co.

Taylor, A. S. 1875. *On poisons*. 3d American ed. Philadelphia: Henry C. Lea.

Appendix:
Notes on Terminology

Egypt

In his comprehensive discussion of Egyptian metals, Partington says of mercury that "... the Coptic and hence perhaps Egyptian name was thrim. ..." This supposition does not appear to have received wide support among Egyptologists. For Egyptian times earlier than the Coptic period, the terms "prš," "mnšt" and, to some degree "didi," are of greater importance than "thrim." All three of these common Egyptian terms have been associated with red pigments, but for none of them is the meaning clear.

A scholarly, critical analysis of the various possibilities for the meaning of prš, mnšt and also of didi is presented by Harris in his *Lexicographical Studies in Ancient Egyptian Minerals.* The relevant evidence is reviewed with an objectivity not always observed by all commentators and the result of this is proper emphasis on caution in drawing conclusions from what is presently known about these red pigments. Among the Egyptologists, Harris stands alone in introducing the possibility that any of the pigments could have been cinnabar (mercuric sulfide). This he does in his discussion of miltos and minium.

Assyria

The Assyrian term IM.KAL.GUG has been interpreted by R. C. Thompson as meaning " 'sublimate of the red (= cinnabar),' mercury" (1936). In support of this interpretation, Thompson notes that in modern times there have been active mercury mines at Kerkuk Baba and in the Afshar district of Kurdistan and that there is evidence of ancient mercury mining at the Sarshuran mines in the latter area. He lists a number of medicinal uses for this material.

292

Thompson's conclusions have been brought in question in the Chicago Assyrian Dictionary (Oppenheim 1956). This work purports to have assembled the entire material for IM.KAL.GUG and for the related kalgukku without finding substantiation for the meaning mercury or cinnabar. This pigment, whatever it is, was used for coloring glass, as a colored slip applied to clay vessels and to clay figurines for magic purposes, and also as a medicine. According to Oppenheim, the nature of kalgukku and IM.KAL.GUG remains unknown, but the uses to which they were put are not incompatible with cinnabar. There have been no reports on chemical analyses, except possibly on a few specimens of Mesopotamian glass.

China

In a brief, but well documented article, Schramm has summarized some of the Chinese methods of signifying mercury and its compounds, pointing out that "tan" and "tan-sha" are used to designate cinnabar (mercuric sulfide) but not mentioning that chu-sha is similarly used (Li Ch'iao-p'ing 1948). He gives five ways of writing quicksilver: "lien-tan-shu," "lien-chin-shu," "tan-chia," "tan-lu-chia" and "tan-ts'ao-chia." Pictographs for these and for other related materials are reproduced in Schramm's article.

India

Terms for cinnabar (mercury sulfide) given by Ray are: hingul or hingula and harada; for Aethiops Mineral (black sulfide of mercury) the name is Kajjali, and for quicksilver itself—rasa. Alchemy is known as rasayana.

Greece and Rome

Because of the interrelationships between classical Greek and Latin terminology, the two will be discussed together, particularly using the works of Theophrastus, Dioscorides, Vitruvius, Strabo, and Pliny the Elder. The concepts of these authors are important not only in themselves, but also for the clues they provide to the interpretation of terms used by Homer and to an understanding of some of the medical prescriptions of Hippocrates and Celsus.

Quicksilver (mercury). (German–*Quecksilber;* French–*Argent vif;* Spanish–*azogue*). The earliest use of a term which can with certainty be indentified with the modern quicksilver is found in the writings of Aristotle. The expression he employs is *argyros chytos* (liquid silver). Theophrastus of Eresus,

Aristotle's pupil, also uses the term *argyros chytos* in his *De Lapidibus,* the oldest known book on mineralogy (1965 translation, Chap. 105).

Two varieties of quicksilver are described by the elder Pliny (1938-63 translation, Bk. 33, line 99). He designates the naturally occurring type as *argentum vivum* and that which is derived from heated ore as *hydrargyrum.* Greek and Latin terminology for quicksilver is clear, but this is not true of several other substances which may or may not contain mercury.

Cinnabar. (German-*Zinnober;* French-*cinabre;* Spanish-*cinabrio*). In modern usage, cinnabar refers to the naturally occurring red sulfide of mercury. The Greek and Latin cognates, *kinnabari* and *cinnabris,* respectively, may or may not signify mercury sulfide.

Aristotle mentions *kinnabari* in Book III, Chapter VI of his *Meteorologica* in a context which strongly suggests that he was referring to mercury sulfide.

Two varieties of kinnabari are described by Theophrastus: natural, which comes from Spain, and "manufactured," which comes from the territory of the Cilbians near Colchis. Preparation of the latter involves the removal of impurities from the raw ore (Theophrastus 1965 translation, Chap. 103). Spanish kinnabari is almost certainly mercury sulfide, but the nature of the "manufactured" Cilbian variety is not entirely clear.

According to Dioscorides (Book 5, Chap. 109) it is incorrect to believe that *kinnabari* is the same as the substance he calls *ammium.* The latter, he says, comes from Spain while the former comes from Africa and is very expensive and scarce. In stating that *hydrargyrum* is made from *ammium,* Dioscorides establishes beyond any reasonable doubt that his *ammium* is mercury sulfide. If, as he says, *ammium* and *kinnabari* are not the same thing, then his *kinnabari* obviously must be something other than mercury sulfide. Goodyer's translation of *kinnabari* to *cinnabaris* is confusing (Dioscorides 1934).

Vitruvius and Strabo do not speak of *kinnabari* or of *cinnabaris,* although they do discuss red pigments and other related materials. Pliny, however, makes it clear that *cinnabaris* is not the *kinnabari* of Aristotle, of Theophrastus, or of Dioscorides. On the contrary, says Pliny, *cinnabaris* is the name given to the mixture of the blood of a dying elephant mixed with that of a dragon, the so-called dragon's blood. He notes that ignorant physicians confuse *cinnabari Indicae* or dragon's blood, which is beneficial, with *minium* which is poisonous (1938-63 translation, Book 33, line 119). Pliny's *minium* is the same as the Greek *kinnabari* or mercury sulfide. Obviously some of the confusion in terminology which persists into modern times was present as early as the first century A. D.

Minium (German-Mennig; French-minium; Spanish-minio). Evidence that the *minium* of Pliny is mercury sulfide and consequently the same as modern cinnabar is found in Pliny's description of *hydrargyrum* and *argentum vivum.*

Hydrargyrum extracted from *minium* is called artificial and is to be distinguished from the naturally occurring *argentum vivum*. To complicate things, however, Pliny describes a *minium secundarium* which is found in lead mines. This must have been the red oxide of lead which is the *minium* of today. An important point is that in the days of Pliny and for some time thereafter, *minium* meant mercury sulfide. It is the same as the *ammium* of Dioscorides.

There is little doubt that Vitruvius used *minium* in the same sense as did Pliny (Book 8, Chap. 1). He states that mines of *minium* had been discovered in Ephesus and that in the mining and refining of the ore large amounts of *argentum vivum* are produced. Apparently Vitruvius did not accept Pliny's distinction between the natural *argentum vivum* and the processed *hydrargyrum*.

The term *minium Sinopicum* is used by Celsus in a number of his prescriptions apparently with the same meaning as just plain *minium*. Perhaps this was the contemporary counterpart of a proprietary drug, specifying a particular brand or quality.

In addition to *kinnabari, cinnabaris, ammium* and *minium* there are several other terms used in classical Greek and Latin literature to designate reddish pigments derived from the earth. Of these *rubrica* and *ochra* certainly refer to oxides of iron. *Sinopis* is another compound which was probably iron oxide or ruddle, but the term may at times have been used for mercury sulfide as well. By far the most difficult to interpret is the Greek word *miltos*. The situation is not simplified by the existence of *miltos Sinopike* and *miltos Lemnia* as well as simple *miltos*. The fact that Homer and Hippocrates used this term invests it with great significance in the early history of mercury (and of medicine) and therefore justifies a detailed discussion of its meaning or meanings.

The context in which Homer used the term *miltos* does not give any clear indication of the nature of the material he had in mind other than that it was used to impart a red color (Iliad, Book 2; Odyssey, Book 9). Among the ships of the Achaians there were scores of black ships, but only Odysseus had red ships. The meaning of Homer's *miltos* can be nothing more than a matter of speculation, but it is not unreasonable to assume that in his mind there was something very special about the color used on the ships of him who was the peer of Zeus in counsel. Iron oxide or common ruddle was cheap and plentiful; mercury sulfide (cinnabar) was available only by importation and consequently was scarcer and more expensive than ruddle. Which would have been more appropriate for someone who occupied Odysseus's exalted position? The possibility that it was mercury sulfide cannot be dismissed out of hand.

Pliny's comments on Homer's miltos are of prime importance in the interpretation of that word. The key passage (XXXIII, 115) is: "Auctoritatem colori fuisse non miror. Iam enim Troianis temporibus rubrica in honore erat Homero teste, qui naves ea commendat, alias circa pigmenta picturasque rarus. *Milton vocant Graeci miniumque cinnabarim.* Unde natus error Indicae cinnabaris nomine."

The italicized words, above, usually have been translated to convey
the idea that the Greeks call it (rubrica) *miltos* and they call *minium cinnabar.*
Other possible and perhaps preferable translations have been proposed as, for
example, the Greeks call *minium* and *cinnabar miltos,* or the Greeks call
miltos and *minium cinnabar.* Of these two alternatives the former would seem
to be more reasonable since it confirms the idea that *miltos* was used by the
Greeks as a generic term. It also provides a good explanation for the confusion between the Latin *cinnabaris* or *cinnabaris Indicae,* which was dragon's
blood, and the Greek *kinnabari,* which was the Latin *minium* or mercuric sulfide. This interpretation is also in consonance with the evidence that Pliny had
a better understanding of the nature of red pigments than did his predecessors
and contemporaries. Grammatically, there is some question as to the most acceptable translation.

Some of the confusion surrounding the meaning of *miltos* can be
attributed to Theophrastus who uses the word several times in his *De
Lapidibus* (1965 translation, Chap. 103, 105). His statement that there are
three kinds of *miltos* strongly suggests that he employs the term in a sort of
generic sense for red, earthy pigments. The variety which he says is sometimes
found in iron mines, must have been iron oxide; another, which is made by
burning *ochra,* is also probably ferric oxide. The latter was considered by
Theophrastus to be inferior to the former. On the other hand, he speaks of
miltos and *ochra* as though they were different substances, and this implies
that one form of *miltos* might have been mercury sulfide, this representing
the third variety that he had in mind.

The complexity of Greek and Latin terminology for red earthy pigments
is reflected in Pauly's *Encyclopedia,* the article on *minium* occupying nearly
six columns with dozens of references (Kroll 1894 ff., Vol. 15-2, cols. 1848-
54).

One of the most comprehensive works dealing with pigments and colors is
J. F. John's *Die Malerei der Alten* (1836). Based primarily, but by no means
exclusively, on the writings of Pliny and Vitruvius, this book lists in its unusually complete index 25 page references to *Zinnober* alone. An even larger
number is formed by the combined total for *miltos, mennig, cinnabaris,
rothel, ocher, sinopis, sinopisroth* and *rubrica.*
The following is worthy of quotation:

> Minium of the ancients must not be considered to be the same as our red
> lead . . . only this red compound of sulfur and quicksilver, which we call
> cinnabar, always was given the name of minium by Pliny . . . our cinnabar
> and our red lead were both known in antiquity. Miltos seems to have been
> synonymous with minium, as cinnabar was known by the Romans.

In a book which is a classic of 19th century German scholarship, Blümner
has given a detailed discussion of red pigments used by the Greeks and
Romans. He covers writings not only of scientists such as Theophrastus,

Dioscorides and Pliny but also the use of the various terms for red colors by poets, historians and other writers. Among the many significant statements found in Blümner's work, the following are particularly noteworthy: "In any case, a variety of minerals were included under the name of *miltos, rubrica;* [and] a definite designation is here as in most cases difficult, indeed impossible."

Blümner correctly says that "To the Romans, *cinnabari* did not mean cinnabar, but dragon's blood." Other German scholars who have contributed their views are Lenz and Donner.

Among British and American scholars who have discussed the problem of *miltos,* mention may be made of Partington, Bailey (see Plinius Secundus, 1929 and 1932 translations), and Stillman, and in France, Berthelot.

In his classic work, *Introduction à l'Étude de la Chimie des Anciens et du Moyen Âge,* Berthelot devotes considerable space to the meaning of cinnabar and of minium, the essence of which may be summarized as follows: *cinabre* and *minium* may have been used by the ancients to designate almost any red pigment, including derivatives of iron, mercury, arsenic, antimony, lead and copper!

Controversy over the terminology of red pigments might well have ended with the wise words of Blümner that it was not possible to know just what the Greek and Latin writers meant when they used the various terms, but the debate still goes on.

A leading protagonist of the theory that the Greek *miltos* is the modern cinnabar (mercuric sulfide) is Walter Leaf who advances the following arguments:

(1) He quotes Strabo to the effect that "In Cappadocia is found what is called Sinopic miltos, the best of all, though the Iberian competes with it. It is called Sinopic because the traders used to take it to that port before the Ephesian market got through to the people in these parts." Leaf goes on to state that

> ... miltos was a trade-name covering all sorts of red pigments derived from the earth. In many cases it included clays coloured with oxide of iron or similar matters, and may be translated by our 'ruddle,' the Latin rubrica. But it also included the finest, most brilliant and most durable of all red pigments, namely, vermilion, which is given directly by the mineral cinnabar, the native ore of mercury; and I feel no doubt that the Sinopic miltos was in fact cinnabar. As a matter of trade it was called miltos; but the men of science knew it by the foreign name of kinnabari. Hence there arose a confusion which misled Theophrastus in his treatise *On Stones,* and those who followed him, into distinguishing between the miltos of Sinope and the cinnabar of other regions.

(2) Leaf also points out that Theophrastus mentions the suffocating nature of the air in the Cappadocian mines which are the source of the Sinopic miltos

and interprets this as evidence that these were mercury mines. "Thus," says Leaf, "Theophrastus himself, though ignorant of the identity of the Sinopic miltos with cinnabar, unconsciously supplies evidence in favor of it."

(3) Quoting Leaf again: ". . . Strabo, who does not mention kinnabari, equates the Sinopic earth with the miltos of Spain. And what the Spanish miltos was there can be little doubt. . . . To suppose that Strabo, when discussing the exports of Spain, should pass this over in silence while recording mere ruddle as one of the chief products of Turdetania, is clearly absurd."

(4) Another of Leaf's points is based on geography, in that "any mineral which would bear the cost of the long inland carriage from the highlands of Cappadocia to the emporium at Sinope must have been no common red clay, such as was found in abundance at many points close to the sea, notably at Lemnos and Keos, but some quite rare and unusually valuable ore; and the only ore which suits the conditions is cinnabar." To these principal points, Leaf adds additional arguments based on the writings of Strabo, Vitruvius, Pliny and others.

An opposing point of view is advanced by Caley and Richards (Theophrastus 1956 translation). These authors say "Though there can be no doubt that the word miltos usually designated what we now call red ochre, a mixture of red ferric oxide with clay, sand and other impurities, this was probably a general term, like so many of the other Greek names for minerals, which included all the pigments that owe their color to the presence of red ferric oxide. That it was applied to an artificial red iron oxide pigment is clear from what Theophrastus says in sections 53 and 54."

Caley and Richards mention that some scholars have suggested that the miltos of Sinope is not red ochre but the more costly cinnabar. "This opinion is maintained, for example, by Leaf, mainly because it would have been so expensive to transport such a common product as red ochre through the difficult country lying between Cappadocia and Sinope that it could not have been sold at a profit in Greece, where it would have to compete with the red ochre found abundantly in much nearer localities such as Ceos and Lemnos." They then continue in this cavalier manner: "Though some of his (Leaf's) other arguments are also ingenious, it is not at all likely that his identification is correct. . . . A very serious objection is that cinnabar does not occur within the confines of ancient Cappadocia, though various iron minerals such as brown iron ore and the ochres are found in many places."

Caley and Richards cite many of the same references as are given by Leaf but come to a completely opposite conclusion: "Most of the scholars who would identify Sinopic miltos or sinopis with cinnabar either ignore the descriptions of these minerals left to us by ancient writers or misinterpret them. The descriptions of Theophrastus, Vitruvius, Pliny and Dioscorides indicate clearly enough that the pigment exported from Sinope was not cinnabar" (Theophrastus 1956).

Although Caley and Richards were aware of the existence of the mercury mine at Iconium (Koniah), they apparently overlook the fact that this region was at one time embraced by Cappadocia. Thus their "very serious objection" to Leaf's arguments loses weight.

Perhaps the foregoing discussion of Greek and Latin terminology has done nothing more than to compound confusion. The summary which follows purports to give a reasonable interpretation, in modern English, of the terms which have been discussed.

Greek	English
Hydrargyros	Quicksilver or Mercury
Argyros Chytos	Quicksilver or Mercury
Kinnabari	Cinnabar or Sulfide of Mercury
Ammium	Cinnabar or Sulfide of Mercury
Ochra	Iron Oxide or Ruddle
Rubrica	Iron Oxide or Ruddle
Miltos Lemnia	Iron Oxide or Ruddle
Sinopis	Probably Iron Oxide but possibly Sulfide of Mercury
Miltos	Probably Iron Oxide and Sulfide of Mercury
Miltos Sinopike	Probably Iron Oxide but possibly Sulfide of Mercury

Latin	English
Argentum Vivum	Quicksilver or Mercury
Hydrargyrum	Quicksilver or Mercury
Minium	Cinnabar or Sulfide of Mercury
Minium Secundarium	Red Oxide of Lead
Cinnabaris	Dragon's Blood or the Resin of the tree Dracaena Draco and Pterocarpus Draco
Cinnabaris Indicae	Dragon's Blood or the Resin of the tree Dracaena Draco and Pterocarpus Draco
Minium Sinopicum	Probably the same as Minium, i. e., Sulfide of Mercury

America

Problems in terminology are not confined to ancient Assyria and Egypt and to classical Greece and Rome. Striking parallels with the latter are found in descriptions of the red pigments used in South America in pre-Columbian times.

When Greek and Roman writers spoke of a red mineral color coming from Spain, particularly from Sisapo, there could be little doubt that they were referring to cinnabar. A similar conclusion can be drawn about red pigments originating at Huancavelica in Peru. No such assurance attends the use of the terms *carayuru, bixa, ychma (ichma)* and *llimpi.*

Most writers have assumed that *llimpi* is the Inca word which refers to cinnabar (Montell 1929, pp. 81 et passim; Jiminez de la Espada 1965, p. 304), but Garcilasso de la Vega claimed that *ichma* was the word for cinnabar and that *llimpi* was a different kind of red color which was derived from some other mineral (cited in Montell 1929, p. 221). Holguin's Lexicon defines *ichma* as "Un color de fruto de arbol . . . ," this vegetable origin placing it in the same category as the Latin *cinnabaris* or dragon's blood (ibid., p. 219).

The Metamorphosis of Minium

The why, the how, and the when of the metamorphosis of *minium* from mercury sulfide to lead oxide are matters of interest and of some importance in the history of quicksilver. Failure to realize the occurrence and significance of this "transmutation" has been a pitfall to translators and to some medical historians.

As pointed out above (p. 295), Pliny and Vitruvius used the term *minium* to designate mercury sulfide or the modern cinnabar. Both describe the use of minium to produce *argentum vivum.* An early clue to the change in the meaning of *minium* is provided by Vitruvius when he describes a method for detecting adulteration of true minium with cheaper materials, of which Pliny's *minium secundarium* or lead oxide was one. Over the years, this type of adulteration became increasingly common and quantitatively more pronounced. The final result was that the commodity called minium contained more lead oxide than mercury sulfide and eventually the term *minium* was used to designate red lead.

Because of the clandestine and gradual substitution of lead oxide for the more costly mercury sulfide it is difficult or impossible to set a date for the final change in terminology. It appears that this step was taken at different times by chemists and apothecaries in different places, principally during the 17th century. Chemistry textbooks of that period generally identify minium with lead but the same is not so of the pharmacopoeias. The Pharmacopoeia Londinensis of 1618, for example, lists minium nativum and minium factitium, suggesting Pliny's influence.

The Pharmacopoeia Amstelredamensis (Amsterdam) of 1636 contains a list of 49 minerals, including argentum vivum, cinnabaris, minium and sublimatum. The juxtaposition of these substances suggests that they all belong to the mercury family, but there is no proof that the compiler meant to imply such an association. The Brussels Pharmacopoeia of 1671 contains a section which is headed: "Argenti vivi, vena Minii, sive cinabri vocatur." It also recognizes argentum vivum nativum and argentum vivum factivum.

Available evidence points to the conclusion that the chemists adopted the new designation for minium earlier than did the pharmacists and physicians. One explanation for this might be that the chemists were more progressive than their medical colleagues and accepted the reality of a situation in which lead oxide, in practice, had actually replaced mercury sulfide. The pharmacists and physicians, on the other hand, seem to have remained under the influence of Pliny and other ancient writers much longer than did the chemists.

Theophilus, in his *De Diversis Artibus,* describes the preparation of minium and of cenobrium (cinnabar). The former is made from "plumbeas tabulas" (lead) and the latter from sulfur and argentum vivum. The date of Theophilus' work is not definitely known, but it probably was written during the latter part of the 11th or early part of the 12th century. This versatile artisan was one of the first to recognize and record the metamorphosis of minium, but several centuries passed before the change was universally appreciated. This suggests that his writings on metallurgy were not widely read. Agricola adheres to Pliny's terminology of minium (nativum) and minium secundarium for cinnabar and red lead, respectively, when he states, incorrectly, that "Quicksilver refineries produce minium and an artificial minium, each of which is called *cinnabaris* today. Chemists first discovered these substances by accident" (1955 translation, pp. 199, 209).

Agricola's discussion of minium is interesting because of its inaccuracies. His statement that "Chemists first discovered these substances by accident" ignores the natural occurrence of cinnabar. He goes on to say: "When they threw sulphur on quicksilver in an attempt to produce silver or gold they produced this pigment instead." Next he says that "When the quicksilver mines of Betica ceased to produce this pigment minium took its place." This is misleading in that the mines of Betica (Almadén) have produced cinnabar continuously from the earliest times up to the present.

Agricola describes a method for producing cinnabar from sulfur and quicksilver but also tells how to prepare minium from cinnabar. In the discussion which follows these directions, Agricola quotes extensively from Theophrastus and Pliny. Apparently he uses the term *minium* to apply to pulverised or purified cinnabar.

Stillman calls attention to changes in the names of metals other than that of cinnabar and minium. He states that "The word 'stannum' (modern Latin for tin) is used by Latin writers of later ancient periods not to designate tin, but an alloy of lead and tin. . . ." "So late as the first century of our era, tin was called by the Latins white lead (plumbum candidum or album), as distinguished from our lead (plumbum nigrum)." Stillman also points out that lead was called "molybdos" by the Greeks.

Kroll, in his discussion of minium, states that:"Das Wort M. entspricht zwar etymologisch unserem MENNIG (Bleioxyd), wird aber im allgemeinen zur Bezeichnung des Zinnobers gebraucht." His article contains no definite information on the history of the change in meaning from mercury sulfide to lead oxide. It is interesting to note that a reader interested in MILTOS is referred by Kroll to Minium (Vol. 15-2, cols. 1848-54).

The Oxford English Dictionary contains a number of interesting references under MINIUM; not the least interesting is the first: "Obs. exc. Hist." A quotation from Trevisa (1398) says that: "Minium is a red colour and the Grekis founde the matere therof in Ephysym. In Spayne is more suche pigment than in other londes." The Chemical Dictionary of John French (1650) is quoted as follows: "MINIUM is the Mercury or rather the Crocus of Lead precipitated."

REFERENCES

Agricola, G. 1955. *De natura fossilium.* Trans. from the 1st Latin ed. of 1546 by Mark Chance Bandy and Jean A. Bandy. New York: Geological Society of America.

Aristotle. *Meteorologica* 4.8.

Aristotle. *De anima* 1.3.

Berthelot, M. 1889. *Introduction à l'étude de la chimie des anciens et du moyen âge.* Paris.

Blümner, H. 1884. *Technologie und Terminologie der Gewerbe und Kunste bei Griechen und Römern.* Leipzig.

Dioscorides, P. 1934. *The Greek herbal.* Englished by John Goodyer, A. D. 1655. Edited and first printed, A. D. 1933, by Robert T. Gunther. Oxford: At the University Press.

Donner, O. 1869. *Die erhalten antiken Wandmalereien in technischer Beziehung.* Leipzig.

Harris, J. R. 1961. *Lexicographical studies in ancient Egyptian minerals.* Berlin: Akademie-Verlag.

Homer *Iliad 2.*

Homer *Odyssey 9.*

Jiminéz de la Espada, M. 1965. *Relaciones geográficas de Indias: Perú.* Vol. 2. Madrid: Ediciones Atlas.

John, J. F. 1836. *Die Malerei der Alten.* Berlin.

Kroll, W. 1894. ff. In *Pauly's Real-encyclopädie der classischen Alterthumswissenschaft,* by A. F. von Pauly, ed. G. Wissowa, vol. 15-2, Stuttgart: J. B. Metzler.

Leaf, W. 1916. The commerce of Sinope. *J. Hellenic Studies* 36:10-16.

Lenz, H. O. 1861. *Mineralogie der alten Griechen und Römer.* Gotha.

Li Ch'iao-p'ing. 1948. *The chemical arts of old China.* Easton, Pa.: Journal of Chemical Education.

Montell, G. 1929. *Dress and ornaments in ancient Peru.* Goteborg: Elanders.

Oppenheim, A. L. 1956. *The Assyrian dictionary.* Chicago: The Oriental Institute.

Partington, J. R. 1935. *Origins and development of applied chemistry.* London: Longmans, Green and Co.

Plinius Secundus, C. 1938-63. *Natural history.* Trans. H. Rackham, W. H. S. Jones and D. E. Eichholz. 10 vols. London: W. Heinemann.

Plinius Secundus, C. 1929, 1932. *The Elder Pliny's chapters on chemical subjects.* Ed. and trans. K. C. Bailey. 2 vols. London: Arnold and Co.

Ray, P. 1956. *History of chemistry in ancient and medieval India,* incorporating *The history of Hindu chemistry* by A. P. C. Ray. Calcutta: Indian Chemical Society.

Schramm, G. 1961. Zur Alchemie und Technologie des Quecksilbers im alten China. *Deut. Apotheker. Z.* 101:1483-85.

Stillman, J. M. 1960. *The story of alchemy and early chemistry.* New York: Dover Publications.

Theophilus. 1961. *De diversis artibus.* Trans. with notes and introd. by C. R. Dodwell. London: Thomas Nelson and Sons.

Theophrastus. 1956. *Theophrastus on stones.* Introd., and commentary by Earle R. Caley and John F. C. Richards. Columbus: Ohio State University Press.

Theophrastus. 1965. *De lapidibus.* Ed., with introd., trans., and commentary by D. E. Eichholz. Oxford: Oxford University Press.

Thompson, R. C. 1936. *Dictionary of Assyrian chemistry and geology.* Oxford: Clarendon Press.

Vitruvius Pollio. 1914. *Vitruvius: The ten books on architecture.* Trans. Morris Hicky Morgan. Cambridge: Harvard University Press.

Index of Subjects
and Places

Index of Persons

This index does not list authors of books and articles cited in the References, except those having historical significance.

Composed in 10 pt. Press Roman by
Jones Composition Company

Printed offset by
Collins Lithography and Printing Co.
on 60 pound Gladfelter Special Book LLSS, White B32

Bound in Joanna Oxford 43601 by
Maple Press

DATE DUE